D0163792

PROBLEM SOLVING FOR CHEMISTRY

SECOND EDITION

EDWARD I. PETERS

Department of Chemistry
West Valley College
Saratoga, California

SAUNDERS GOLDEN SUNBURST SERIES

Saunders College Publishing
Harcourt Brace Jovanovich College Publishers
Fort Worth Philadelphia San Diego
New York Orlando Austin San Antonio
Toronto Montreal London Sydney Tokyo

Library of Congress Cataloging in Publication Data

Peters, Edward I.

Problem solving for chemistry.

(Saunders golden series)

Includes index.

1. Chemistry — Programmed instruction.
 I. Title.

QD42.P47 1976 540'.77 75–12493

ISBN 0–7216–7206–X

Printed in the United States of America.

Problem Solving for Chemistry
ISBN # 0–7216–7206–X

123 026 15 14 13 12 11

Harcourt Brace Jovanovich, Inc.
The Dryden Press
Saunders College Publishing

PREFACE to the Second Edition

Problem Solving for Chemistry is for the general chemistry student who wants to solve chemistry problems and understand the method he uses. Dimensional analysis, presented as more than a mere exercise in unit cancellation, is emphasized as a method, but not to the exclusion of other methods if they better serve a specific purpose. The second edition retains the auto-tutorial style of the first in which the student actively solves each example problem through a series of guided steps.

A major addition in the second edition is a non-calculus introduction to thermodynamics located before the chapters on equilibrium and oxidation-reduction reactions. Both of these chapters now have new sections that add the perspective that only thermodynamics can provide. The chapters remain complete, however, without reference to thermodynamics; the new sections may be omitted without breaking a logical and complete sequence.

Another major change-addition combination appears in the problems at the end of each chapter. There are nearly three times as many problems as in the first edition. They are presented in two matching columns. Juxtaposed examples involve the same principles and comparable methods of calculation. Complete solutions—not just answers this time—are given for all problems in the left column, a significant aid to the student when he has difficulty with some concept in his home study. Problems in the right column are not answered except in the Instructor's Guide, so they may be used for assignments.

Keeping pace with current practice, the oxidation potentials of the first edition have been replaced by reduction potentials. Attention is also given to SI units that are in growing acceptance. This change is not complete, however; we continue to talk about the atmosphere and torr, rather than the pascal; and degrees Celsius as well as degrees Kelvin, or simply Kelvin. Energy examples are in terms of calories and kilocalories, but you will find joule and kilojoule columns in tables of thermodynamic properties if you prefer this conversion to SI units. Answers to all problems and examples are given in joules and kilojoules as well as the traditional units.

Many people have contributed significantly to the development of the second edition. Comments and suggestions from students and instructors using the first edition have been invaluable. Mr. Howard Garnel of West Valley College, Dr. William L. Masterton of the University of Connecticut, and Dr. Robert J. Munn of the University of Maryland made major contributions to the development of the thermodynamics chapter. Mrs. Susan Weiner and Mr. Jack Pease, both of West Valley College, reviewed portions of the text and made valuable suggestions. The chores of manuscript preparation, checking calculations and proofreading were performed by former students Jean Reagan, Claudia Goodman, Marsha Brown, Dan Maxwell and Don Papuzynski, and at home by my daughter, Judy, and my wife, Geb. To all of these helpful people I am most thankful.

EDWARD I. PETERS

PREFACE to the First Edition

This is an auto-tutorial text. It is unlike any other book now available. Its purpose is to give the student a book from which he can learn how to solve routine chemistry problems without the assistance of a teacher, in the privacy of his own study.

The book does "tutor" the student. As a chemistry teacher listens to the questions of a student who is having difficulty, how does he formulate his response? Does he simply return a direct answer? Rarely. More often, and probably unconsciously, he tries to analyze the student's understanding of the concept involved and locate its weakness. He then returns to the student a question—a question designed to lead the student to find for himself the answer to his original question without spoon-feeding him that answer. If the tutorial question is effective, the student finds more than just an answer: he finds comprehension.

This book asks the questions the teacher tutoring an individual student asks. As a problem book, it is a hybrid, combining the text-type presentation of quantitative principles typical of chemistry problem books with a program approach to example problems. It is in the example problems that the student is tutored. *He* solves the problems, rather than the author. He is given the problem and then asked a series of tutor-type questions that lead him to think through the example himself. Solving problems is not a vicarious experience for the student of this text. It is an active one: *he* does the work; *he* does the thinking. And in the process, he learns and understands.

There are several methods by which most problems in chemistry may be solved. It is my personal conviction that no single method is "right" and the others "wrong." Any method that is sound chemically and sound mathematically is a "right" method. Selection then becomes a matter of identifying the better method. This selection is subjective. I personally favor a dimensional analysis approach, known also by various names, such as factor-label, unit conversion, and so forth. Naturally, then, it is the method of this book.

There is considerable advantage to setting up a chemistry problem as a whole, rather than as a series of parts. Actually it is the parts, each logically developed according to a well established principle, that build the whole. By deferring calculations to the end the student sees the whole problem. It is frequently possible to eliminate multiplication or division steps by suitable "cancellation." Calculations are significantly simplified and shortened, and calculation errors consequently reduced. This is one of the important advantages of the dimensional analysis approach.

I wish to acknowledge and thank Dr. William L. Masterton, who reviewed the manuscript for this book. His critical analysis was thoughtful and constructive, and his suggestions were most helpful in preparing the final manuscript. My appreciation extends also to the hundreds of students who worked with portions of the book as it was being developed. Their frank criticism and suggestions, as well as their enthusiastic encouragement to complete the work, are all manifest in the final product. To one such student named Greg, who had the unhappy fate of having his own dad as a chemistry teacher, go particular thanks. His reactions furnished much insight as to how a student receives and uses learning tools such as this. To him I gratefully dedicate this book.

EDWARD I. PETERS

TO THE STUDENT

This book is designed for you. Its purpose is to make it possible for you to teach yourself how to solve chemistry problems. But if you and the book are to succeed in your common goal, you must use the book as it is supposed to be used.

There are no "worked-out" problems in this book. Instead *you* are to work example problems. You will be guided by a series of questions designed to build in your thinking an understanding of the problems and how to solve them. You must answer these questions *yourself,* without looking ahead a few inches on the page to where you know the answer will appear. It is only if you do the thinking yourself that you will *learn* what is supposed to be learned with every question. There are no short cuts. Any deviation from the recommended procedure will probably result in either a lower comprehension level or a longer study time for a comparable level of understanding.

The question-and-answer or "programmed instruction" example problems carry you so far, and then ask a question. This question is separated from the answer by a broken line across the page:

- -

You should not look below that line until you have *answered* the question to the best of your ability. To keep yourself honest you should use an opaque shield which you move down the page from one dashed line to the next—but do not move it until the question then exposed has been answered. Many of these questions involve calculations or setups for problems, or sometimes both. This work should be done on a separate sheet of paper, so you may compare your answer with the answer that will be exposed as you move your shield down to the next line. If the answer does not agree with yours, stop right there and find out why. Most answers include an explanation of the theory behind them. If this explanation does not clear up any difficulty, back up. Re-read the material preceding the problem and the question, and stay with it until you have mastered everything up to that point. Then proceed to the next question or problem or principle.

Obviously you are on your own in the use of this book. If you play by the rules, you will learn more in less time than by any other way. Go to it!

CONTENTS

CHAPTER 1

DIMENSIONAL ANALYSIS

How many inches are in two feet?

This is hardly a problem in chemistry, you say—and rightly so. In this opening chapter we shall concern ourselves with few, if any, problems in chemistry. Our goal will be to introduce the problem solving method that will appear throughout this book. It is based upon dimensional analysis, a method of analyzing the units of physical measurements and calculations. The method is also known as the factor-label method, unit conversion, unit cancellation, and others.

In this chapter we shall be concerned not so much with answers to problems as with *how* the answers may be found. Accordingly, just *how* did you leap to the answer, "24 inches," after reading the opening sentence? You probably reasoned—in a flash—along these lines: "There are 12 inches in 1 foot. Therefore in 2 feet there are 2 × 12, or 24 inches." In this reasoning you employed an equality, 12 inches = 1 foot. Dimensional analysis uses such equalities. They are most readily seen as conversion factors, as in this example. Division of both sides of the equality by 1 foot yields

$$\frac{12 \text{ inches}}{1 \text{ foot}} = \frac{1 \text{ foot}}{1 \text{ foot}} = 1.$$

The fraction 12 inches/1 foot equals 1 because the numerator and denominator are equal.

Mathematics books identify the number 1 as the multiplicative identity. It may be symbolized by the statement

$$a \times 1 = a.$$

In other words, multiplication of a value by 1 does not change the value. And this is what you did, unconsciously, in answering the opening question. You multiplied the *given quantity*, 2 feet, by 1 in the form of 12 inches/1 foot.

In algebra, the multiplication of 2a by 12b/a would yield 24b as a result. The "a" appearing in the numerator of the first factor cancels the "a" in the denominator of the second factor. (Mathematically impure as the term "cancel" may be, it seems to persist in use, and will be employed here.) Now if 12b/a happens to be equal to 1, then 2a must be equal to 24b, by the multiplicative identity principle.

In a precisely parallel situation,

$$2\ \cancel{\text{feet}} \times \frac{12\ \text{inches}}{1\ \cancel{\text{foot}}} = 24\ \text{inches}.$$

The fraction 12 inches/1 foot does equal 1, so we may conclude that 2 feet = 24 inches—which is obviously true. In dimensional analysis, units are treated as algebraic symbols. They are "canceled" in the identical fashion that symbols are "canceled" in mathematics.

In nearly all chemistry problems there will be a *given quantity*—mass, volume, time, or something else—on which the problem is based. Your ability to identify it will enhance greatly your ability to solve the problem. Furthermore, most problems in chemistry may be expressed in an ''equivalent interpretation.'' In this interpretation the problem is essentially rewritten into the form, ''How many ______ are equivalent to ______?" The first blank is the identification of the thing sought in the problem, best expressed in its units of measurement. The second blank is the given quantity.

As an example, take the opening question: How many inches are in two feet? This question is almost in the equivalent interpretation form. It asks, how many inches are equivalent to two feet? The given quantity, two feet, is the starting point in a dimensional analysis set up, and that is followed by one or more conversions until the desired answer is reached.

The "given quantity" and "equivalent interpretation" concepts of a problem are presented here as a suggestion to the student who needs to develop an approach to interpreting problems. It is a temporary device, to be discarded as soon as the chemical or mathematical "maturity" of the student enables him to analyze and set up problems without this artificial process. In the meantime the method will be used frequently throughout this book as new concepts are introduced.

Let us turn to example problems to illustrate these ideas*:

Example 1.1. What is the distance in feet between two points three miles apart? There are 5280 feet in a mile.

First, what is the given quantity in this problem—the quantity on which the problem is to be based?

- -

1.1a. 3 miles.
5280 feet in a mile is a conversion factor, not a quantity peculiar to the problem.

Can you now rewrite the question in the equivalent interpretation form, "How many ______ are equivalent to ______?"

- -

*If you did not read "To the Student" on page vii, describing the method of working example problems, please do so now.

1.1b. How many *feet* are equivalent to *three miles*?
As indicated above, the second blank is the given quantity, and the first identifies the quantity sought, including its units.

The problem includes the conversion statement that is required: there are 5280 feet in one mile. Expressed as an equation, this becomes

$$5280 \text{ feet} = 1 \text{ mile.}$$

Two fractions equal to 1 may be written from this equation. What are they?

1.1c. $\frac{5280 \text{ feet}}{1 \text{ mile}}$ or $\frac{1 \text{ mile}}{5280 \text{ feet}}$.

Now to solve the problem you must multiply the given quantity, three miles, by one of these conversion factors so that the answer appears in the required units, feet. Write the given quantity, select the proper factor, multiply the two, and show the answer. In other words, complete the dimensional analysis setup and the problem.

1.1d. 15,840 feet.

$$3 \text{ \sout{miles}} \times \frac{5280 \text{ feet}}{1 \text{ \sout{mile}}} = 15{,}840 \text{ feet}$$

If you had chosen the wrong conversion factor your answer would have been something like this:

$$3 \text{ miles} \times \frac{1 \text{ mile}}{5280 \text{ feet}} = 0.000568 \text{ miles}^2/\text{foot!}$$

The ridiculousness of the numerical answer is obvious, but only because you have an intuitive feeling that there should be more feet than miles in a given distance. In some areas of chemistry you will not enjoy the advantage of intuition. Then the units of the answer, which are *not* what was sought, should alert you to the fact that an error has been made. So the answer for every problem should be twice reviewed: First, is the numerical answer reasonable? Second, are the units reasonable? If the answer to either question is no, the entire problem should be checked to find the error.

Distance-Rate-Time problems provide excellent examples of the use of dimensional analysis in a familiar context. First, however, a comment about rate: What is the meaning of "25 miles per hour," both physically and in terms of units? Speed is the ratio of distance traveled to time required. If you traveled 75 miles in 3 hours, what would be your average speed? 25 miles per hour, of course. But how did you get it? Did you not divide 75 by 3? In other words,

$$75 \text{ miles} \div 3 \text{ hours} = 25 \text{ miles per hour.}$$

From this it follows that "per" means "divided by." This is also the significance of the fractional form,

$$\frac{75 \text{ miles}}{3 \text{ hours}}, \text{ or } \frac{25 \text{ miles}}{1 \text{ hour}}$$

which is the form suitable for use in dimensional analysis.

Finally, speed indicates an "equivalence" of sorts between distance and time, in which a given quantity of time can be translated into distance. For example, how far will one travel in five hours at 25 miles per hour? The "equivalent interpretation" of this problem is, "How many miles are equivalent to five hours?" In this case the given quantity, five hours, must be multiplied by the conversion factor, 25 miles/1 hour:

$$5 \cancel{\text{hours}} \times \frac{25 \text{ miles}}{1 \cancel{\text{hour}}} = 125 \text{ miles.}$$

As before, the conversion factor 25 miles/1 hour equals 1. It follows that 25 miles = 1 hour. This is the equivalence interpretation of speed.

Example 1.2. How long will it take to travel the 406 miles between Los Angeles and San Francisco at an average of 58 miles per hour?

Rewrite this question in the "equivalent interpretation" form.

1.2a. How many hours are equivalent to 406 miles?

The speed simply furnishes the conversion factor by which the calculation may be made. Write this relationship as an equation.

1.2b. 58 miles = 1 hour, or 1 hour = 58 miles.

There are now two fractions, both equal to 1, that may be written from this equivalence statement. . . .

1.2c. $\frac{58 \text{ miles}}{1 \text{ hour}}$ and $\frac{1 \text{ hour}}{58 \text{ miles}}$.

If 406 miles is multiplied by one of these, the time for the Los Angeles–San Francisco trip results. Think about it. The distance-time equivalence has 58 miles "packaged" into 1 hour units. How many 58-mile packages are in 406 miles? How would you find out—by multiplying 406 by 58, or by dividing 406 by 58? Does your answer give you proper unit cancellation? Complete the problem.

1.2d. 7 hours.

$$406 \cancel{\text{miles}} \times \frac{1 \text{ hour}}{58 \cancel{\text{miles}}} = 7 \text{ hours.}$$

Obviously the arithmetical operation this time is division rather than multiplication, as in the previous examples. The miles²/hour units resulting from multiplication would have signaled the error, had you followed that wrong path. Also, the 23,500 "hour" numerical answer would have given anyone second thoughts about embarking on that journey!

The multiplicative identity may be used any number of times without changing the value of the given quantity. Sometimes problems are not completed in a single step, but require several. For example. . . .

Example 1.3. If light travels 186,000 miles per second, how many minutes are required for light to cover the 93,000,000 miles between the sun and Earth?

First, rewrite the question in its equivalent interpretation form.

1.3a. How many minutes are equivalent to 93,000,000 miles?

As in Example 1.2, the speed is simply a conversion factor. This time the given conversion factor does not connect the given quantity unit with the answer unit. It permits changing distance to time, but to seconds rather than minutes. Set up the problem that far, but do not solve to an answer.

1.3b. $$93{,}000{,}000\ \cancel{\text{miles}} \times \frac{1\ \text{second}}{186{,}000\ \cancel{\text{miles}}}$$

Now seconds must be converted to minutes. The conversion factor is obviously 60 seconds = 1 minute. Now *extend* the above expression to include a second conversion factor equal to 1 such that the final result will be in minutes. Calculate the answer.

1.3c. 8.33 minutes.

$$93{,}000{,}000\ \cancel{\text{miles}} \times \frac{1\ \cancel{\text{second}}}{186{,}000\ \cancel{\text{miles}}} \times \frac{1\ \text{minute}}{60\ \cancel{\text{seconds}}} = 8.33\ \text{minutes}$$

Two important points emerge in this problem. First, when a series of conversions is required, set up the entire problem without doing any calculations. This considerably shortens, and frequently simplifies, the calculations required, particularly if a slide rule is used. Errors are thereby reduced. Second, *think* logically through each step, and then set up the sequence of operations in such a manner that each step has physical significance. For example,

$$93{,}000{,}000\ \cancel{\text{miles}} \times \frac{1\ \text{minute}}{60\ \cancel{\text{seconds}}} \times \frac{1\ \cancel{\text{second}}}{186{,}000\ \cancel{\text{miles}}} = 8.33\ \text{minutes}$$

gives the correct answer, but multiplication of the first two factors without the third gives miles-minutes/second, a meaningless unit that indicates an illogical thought process for solving the problem. If you think through the problem, each step will have meaning and the problem will be more easily understood.

Volume problems warrant special mention:

Example 1.4. What is the volume in cubic yards of a storage bin measuring 40 inches long by 80 inches wide by 50 inches deep?

The volume of a rectangular box is found by multiplying length by width by height. Set up for volume in cubic inches, showing units, but do not calculate the numerical result.

1.4a. 40 inches × 80 inches × 50 inches.

While you might shorten this problem by recognizing there are 36 inches in one yard, let's take it through the steps from inches to feet to yards in order to gain additional practice. Our conversion factors are, therefore, 1 foot = 12 inches and 1 yard = 3 feet. Notice that, in changing *cubic* inches to *cubic* feet, there are *three* inch units to be changed to feet. The 1 foot/12 inches conversion factor must be used *three* times:

$$40\ \cancel{\text{in}} \times \frac{1\ \text{ft}}{12\ \cancel{\text{in}}} \times 80\ \cancel{\text{in}} \times \frac{1\ \text{ft}}{12\ \cancel{\text{in}}} \times 50\ \cancel{\text{in}} \times \frac{1\ \text{ft}}{12\ \cancel{\text{in}}}\ .$$

When a single factor is used three times, that factor is cubed:

$$40\ \cancel{\text{in}} \times 80\ \cancel{\text{in}} \times 50\ \cancel{\text{in}} \times \frac{1^3\ \text{ft}^3}{12^3\ \cancel{\text{in}^3}}.$$

Notice that both the number and the unit must be cubed in both numerator and denominator.

Now extend the above setup by multiplying by a conversion factor that will change the cubic feet to cubic yards. Complete the solution.

1.4b. 3.43 cubic yards.

$$40\ \cancel{\text{in}} \times 80\ \cancel{\text{in}} \times 50\ \cancel{\text{in}} \times \frac{1^3\ \cancel{\text{ft}^3}}{12^3\ \cancel{\text{in}^3}} \times \frac{1^3\ \text{yd}^3}{3^3\ \cancel{\text{ft}^3}} = 3.42\ \text{yd}^3$$

You did remember to cube the 3, did you not?

Two final examples will complete our introduction to dimensional analysis.

Example 1.5. What is the speed in feet per second of a car traveling 50 miles per hour?

Begin this problem by writing its equivalent interpretation.

1.5a. How many feet per second are equivalent to 50 miles per hour?

The given quantity this time is 50 miles/1 hour, which is to be changed to feet per second. In other words, miles must be changed to feet, and hours to seconds. These are familiar conversions. You should be able to set up and solve the problem completely.

1.5b. 73.3 feet per second.

$$\frac{50\ \cancel{\text{miles}}}{1\ \cancel{\text{hour}}} \times \frac{5280\ \text{feet}}{1\ \cancel{\text{mile}}} \times \frac{1\ \cancel{\text{hour}}}{60\ \cancel{\text{minutes}}} \times \frac{1\ \cancel{\text{minute}}}{60\ \text{seconds}} = 73.3\ \text{feet/second}.$$

In the time conversion in Example 1.5, note that you are converting a unit in the denominator. This is handled in precisely the same manner as numerator conversions. Proper application of units guides you to the correct setup for the problems.

If you've mastered the concepts of dimensional analysis, you should be able to apply them in all sorts of unexpected ways—like this, for example. . . .

Example 1.6. What would be the cost of nails for a fence 88 feet long if you required 9 nails per foot of fence, there were 36 nails in a pound, and they sold for 25 cents per pound? Answer in dollars.

A complete dimensional analysis setup, if you please. . . .

1.6a. $5.50.

$$88\ \text{\sout{feet}} \times \frac{9\ \text{\sout{nails}}}{1\ \text{\sout{foot}}} \times \frac{1\ \text{\sout{pound}}}{36\ \text{\sout{nails}}} \times \frac{25\ \text{\sout{cents}}}{1\ \text{\sout{pound}}} \times \frac{1\ \text{dollar}}{100\ \text{\sout{cents}}} = 5.50\ \text{dollars}$$

CHAPTER 2

MEASUREMENT: DENSITY

2.1 MEASUREMENT: THE STATE OF THE ART

Quantitative problems in chemistry are based on measurements. Large handbooks of chemical and physical data have been compiled from measurements made in the laboratory, and research data are analyzed from recorded measurements. Raw material and production measurements are the basis of industrial chemistry. Each problem in this book has measurement built into it in one form or another.

The world in which we live is a dynamic one. One of the active areas of change at the time of this writing is in the field of measurement. It is, in fact, two-pronged. In the United States, Canada and England, three of the more prominent holdouts still clinging to the English system of units, the change to metric measurements is quickly gaining momentum. In the sciences this change is academic, as scientific measurements have always been made in metric units, but even in the sciences we find ourselves in the midst of change. The International System of Units, abbreviated SI from its French name, *Système Internationale*, is widely accepted and used throughout the world, again with some of the English speaking countries lagging behind. Fortunately, SI units are largely a modification of metric units. Many metric units are used as they always have been, and in other cases the SI system selects as "official" one of several units currently used to identify the same measurement.

In the measurements used in this book there are few areas of conflict between SI units and the metric units in general use. The more common measurements made in chemistry are mass, length, volume, temperature, energy and pressure. Both metric and SI systems measure mass in grams, length in meters, and volume in cubic meters or, in the case of liquids and gases, liters. The SI unit of temperature is the Kelvin; the metric unit is the Celsius (formerly centigrade) degree. In energy measurements, the SI unit is the joule, whereas the more common unit used by the chemist is the calorie, with the other energy units appearing in special areas such as atomic and nuclear energy. Pressure is expressed in pascals in the SI system, while chemists generally speak in terms of atmospheres, millimeters of mercury, or torr—the recently introduced substitution for the millimeter of mercury.

In this book we shall side with the traditional chemical units.* The reason for this decision is that a problem book is generally a supplement to a major text. From the standpoint of the student, it is desirable for the text and problem book to correspond to each other, and to date there appears to be no tendency in textbooks published in the United States to adopt SI units in the areas of temperature, energy or pressure. Furthermore, there is in the chemical literature a vast storehouse of information tabulated in conventional units, and virtually all handbooks used by chemists express energy in calories or kilocalories, pressure in atmospheres or millimeters of mercury, and temperature in Celsius degrees. Therefore, with apologies to those readers where SI units are in general use, we shall remain compatible with the vast majority of literature that accompanies the use of this book.

2.2 FUNDAMENTAL METRIC UNITS

The fundamental units of metric measurement are the gram for mass, the meter for length, and the liter for fluid volume or capacity, as it is frequently identified. Current definitions for these units are:

Gram (g) 1/1000 of the mass of the platinum-iridium kilogram block maintained at the International Bureau of Weights and Measures, near Paris, France.

Meter (m) 1,650,763.73 times the wave length of a certain line in the emission spectrum of krypton.

Liter (*l*) Exactly 1000 cubic centimeters.

Chemists frequently use the terms *mass* and *weight* interchangeably, although they are not the same. Mass is a measure of the quantity of matter in a sample; weight is a measure of the gravitational attraction between a sample of matter and the earth—or any other stellar body on which it may be found. Thus, the mass of a sample is the same anywhere in the universe, whereas its weight depends on its location. The force of gravity is essentially constant on the surface of the earth, where most chemists pursue their careers. Inasmuch as weight in a constant gravitational field is proportional to mass, weight becomes a measure of mass, and is frequently referred to in mass units. We say, then, that we "weigh" something when we actually measure its mass; and we speak of its weight when we truly mean its mass.

---

*Tabulated energy data in the Appendix and answers to energy problems are given in both calories or kilocalories and joules or kilojoules.

TABLE 2.1 METRIC PREFIXES AND THEIR MEANINGS

Prefix	Symbol	Multiple	Decimal
giga-	G	10^9	1,000,000,000
mega-	M	10^6	1,000,000
*kilo-**	k	10^3	1,000
hecto-	h	10^2	100
deka-	da	10^1	10
The basic unit, *-gram*, *-meter*, *-liter*			
deci-	d	10^{-1}	0.1
*centi-**	c	10^{-2}	0.01
*milli-**	m	10^{-3}	0.001
micro-	μ	10^{-6}	0.000001
nano-	n	10^{-9}	0.000000001

*Prefixes most frequently used in chemistry

2.3 METRIC PREFIXES AND THEIR MEANINGS

Two of the three definitions above suggest that measurements are made in multiples and submultiples of the basic metric units. These are multiples and submultiples of ten. Therefore all conversions between these units may be accomplished by moving a decimal point. The larger and smaller units are identified by prefixes that are listed in Table 2.1. Thus the platinum-iridium block on which the definition of a gram is based contains 1000 grams and is called a *kilo*gram. In the other direction, 1/1000 of a gram is called a *milli*gram; and 1/100 of a gram is called a *centi*gram. Similarly, 1000 meters is a *kilo*meter, 1/100 meter is a *centi*meter and 1/1000 meter is a *milli*meter.

As cubic feet (ft^3) represent volume as the cube of length units in the English system, cubic meters (m^3) and cubic centimeters (cm^3) are volume measurements in the metric system. The cube of length measurements is generally used to express the volume of a solid. Liquids and gases are generally measured in liters. A milliliter is, as the prefix *milli-* would suggest, 1/1000 of a liter. One milliliter is equal to one cubic centimeter.

2.4 METRIC SYSTEM CONVERSIONS

One of the principal advantages of the metric system is the ease by which conversions may be made from one measurement unit to the next. This can be illustrated through the following examples.

Example 2.1. How many centimeters are in 1.42 meters?

First, what is the given quantity in this problem?

2.1a. 1.42 meters.

The given quantity in meters is now to be converted to its equivalent in centimeters. From Table 2.1 we see that 1 unit is equal to 100 centiunits, or 1 meter = 100 centimeters.

Will the number of centimeters in a given length be more or less than the number of meters? Is multiplication or division by 100 required? Set up by dimensional analysis and solve the problem.

2.1b. 142 cm.

$$1.42 \cancel{\text{meters}} \times \frac{100 \text{ cm}}{1 \cancel{\text{meter}}} = 142 \text{ cm.}$$

As this example illustrates, conversions within the metric system are accomplished by moving the decimal the proper number of places to the right if multiplication is indicated, or to the left for division. The direction is readily confirmed by the application of dimensional analysis.

Example 2.2. The first edition of this book had a mass of 430 grams. Determine the mass in kilograms.

Think it through. What is the given quantity? Will the number of kilograms be more or less than the number of grams? Write the dimensional analysis setup and solve.

2.2a. 0.430 kg.

$$430 \cancel{\text{g}} \times \frac{1 \text{ kg}}{1000 \cancel{\text{g}}} = 0.430 \text{ kilograms}$$

Example 2.3. A rectangular container is 75.0 cm long and 35.0 cm wide. How many liters of liquid will it hold when filled to a depth of 13.0 cm?

The given quantity is a bit more complicated this time, but you know what must be done to find the volume of a rectangular box. The conversions to liters should be apparent from the definition of a liter. Set up by dimensional analysis and solve.

2.3a. 34.1 liters.

$$75.0 \cancel{\text{cm}} \times 35.0 \cancel{\text{cm}} \times 13.0 \cancel{\text{cm}} \times \frac{1 \text{ liter}}{1000 \cancel{\text{cm}^3}} = 34.1 \text{ liters}$$

2.5 CONVERSIONS BETWEEN ENGLISH AND METRIC SYSTEMS

While we shall not be concerned with measurements in the English system in this book, it is helpful to do a few problems converting between the systems to gain some feeling for the metric units. The most useful conversion factors are listed in Table 2.2.

Example 2.4 'Tis said that Cinderella's glass slipper was 18.6 cm long. How much is this in inches?

TABLE 2.2 CONVERSION TABLE FOR METRIC AND ENGLISH UNITS

Mass	454 grams = 1 pound 28.3 grams = 1 ounce 1 kilogram = 2.2 pounds
Length	2.54 centimeters = 1 inch 30.5 centimeters = 1 foot 1 meter = 39.4 inches 1 meter = 1.09 yards 1.61 kilometers = 1 mile
Capacity	1 liter = 1.06 quarts 3.785 liters = 1 U.S. gallon

The problem should be straightforward from the conversion factors in Table 2.2.

2.4a. 7.32 inches.

$$18.6\ \cancel{\text{cm}} \times \frac{1\ \text{inch}}{2.54\ \cancel{\text{cm}}} = 7.32\ \text{inches}$$

Example 2.5 A European automobile has a 43.5 liter gasoline tank. Find its capacity in U.S. gallons.

2.5a. 11.5 gallons.

$$43.5\ \cancel{\text{liters}} \times \frac{1\ \text{gal}}{3.785\ \cancel{\text{liters}}} = 11.5\ \text{gallons}$$

Example 2.6. Find the mass in kilograms of a 3.40 pound chicken.

2.6a. 1.55 kilograms.

$$3.40\ \cancel{\text{pounds}} \times \frac{1\ \text{kg}}{2.20\ \cancel{\text{pounds}}} = 1.55\ \text{kg}$$

2.6. TEMPERATURE MEASUREMENT

Two temperature scales are in common use today: the Fahrenheit scale used by some of the English speaking countries and the Celsius (formerly centigrade) scale used throughout most of the world. The relationship between these scales is indicated in Figure 2.1. The Celsius scale is based on the freezing point and boiling point of water, to which are assigned the values 0°C and 100°C respectively. The same range of 100 Celsius degrees is covered by 180 Fahrenheit degrees, running from 32°F at the freezing point to 212°F at the boiling point. We thus see that 180 Fahrenheit degrees = 100 Celsius degrees, or 9 Fahrenheit degrees = 5 Celsius degrees. Because the scales do not have a common zero, however, there is no equivalence relationship between them that permits a simple dimensional analysis conversion from one temperature reading to another. Such conversions may be made from one of several algebraic formulas. One of these is

$$°F - 32 = \frac{9}{5}°C. \tag{2.1}$$

The use of Equation 2.1 involves no more than substitution of a known temperature and solving for the other.

Example 2.7. The boiling point of benzene is 80.1°C. Calculate the equivalent temperature on the Fahrenheit scale.

2.7a. 176°F.

$$°F - 32 = \frac{9}{5} \times 80.1$$

$$°F = \frac{9}{5} \times 80.1 + 32 = 176°F$$

Example 2.8. Elemental mercury freezes at −38.0°F. Find the freezing point of mercury in °C.

2.8a. −38.9°C.

$$-38.0 - 32 = \frac{9}{5}°C$$

$$°C = \frac{5}{9}(-38.0 - 32) = -38.9°C$$

As indicated earlier, the SI temperature unit is the Kelvin. It corresponds with the degree Kelvin of the *absolute* temperature scale in which the size of the temperature degree is the same as the degree Celsius. The origin of this temperature scale is discussed in Chapter 7. The relationship between the Kelvin and Celsius scales is given by the equation

$$°K = °C + 273. \tag{2.2}$$

Thus a comfortable day, in which the temperature is 20°C, has a temperature of 293°K, or "two hundred ninety three degrees Kelvin," as it would be spoken. Under the SI system, we would write the temperature as 293K, and say, "two hundred ninety three Kelvin," eliminating both the word "degree" and its symbol.

Example 2.9. Dry ice, or solid carbon dioxide, sublimes (changes directly from a solid to a vapor) at −78°C. Express this temperature in degrees Kelvin.

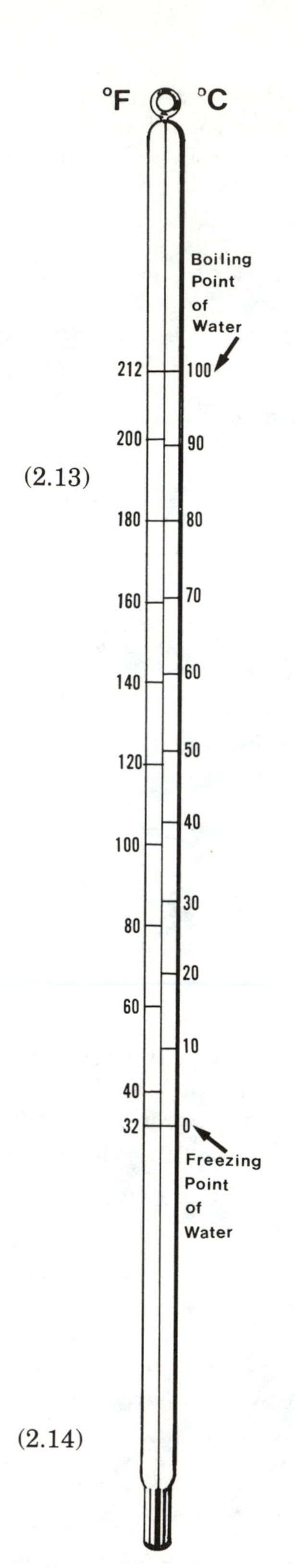

(2.13)

(2.14)

Figure 2.1. Comparison between Fahrenheit and Celsius temperature scales.

2.9a. 195°K.

$$°K = °C + 273 = -78 + 273 = 195°K$$

2.7 SIGNIFICANT FIGURES

All physical measurements have associated with them a certain number of **significant figures**, or **significant digits**, the terms *figures* and *digits* being used interchangeably. No measurement is exact. Given a more accurate measuring instrument, or an improved measuring technique, the accuracy of the measurement can be improved. Properly used, significant figures are a way of telling how certain—or how uncertain—a measurement is.

If we use a centimeter ruler to measure the length of a piece of paper, we might find that it is 12.4 cm long. In such a measurement there would be no doubt that the number of *tens* in the numerical expression of length is 1. It is certainly not zero, and it is not 2. We could also be quite sure about the number of *ones:* it is 2, not 1 or 3. When we reach the number of *tenths* of a centimeter, however, our confidence wanes. It would not take much of an error in calibrating the ruler, or in placing it on the paper properly, or in reading and counting closely spaced lines, to be off by 0.1 cm. The last digit is therefore called a **doubtful digit**. Its correct value is uncertain. We often express this uncertainty by recording a measurement as 12.4 ± 0.1 cm. In expressing a measurement to the proper number of significant figures *the last figure recorded is the figure containing uncertainty: it is the doubtful digit.*

Counting Significant Figures. *The number of significant figures in a measurement is the number of figures that are known accurately plus one that is doubtful, beginning with the first nonzero digit.* From this definition it is clear that there are three significant figures in the 12.4 cm measurement described above. The number of *tens* in the measurement, and the number of *units*, are known accurately. The number of tenths of a unit is doubtful.

It is important to realize that it is the method of measurement rather than units in which the measurement is expressed that establish the doubtful digit. For example, if the results of the 12.4 cm measurement were expressed in kilometers, it would be recorded as 0.000124 km. There are still three significant figures, with the *millionth* of a kilometer being the doubtful digit. The millionth of a kilometer is the same as a tenth of a centimeter. This illustrates the importance of beginning your counting of significant figures with the first non-zero digit, regardless of the location of the decimal point. The zeros between the decimal point and the first non-zero digit serve simply to locate the decimal point. They are not significant.

At the other extreme, the value of 12.4 cm might be expressed in nanometers. It would then be 124,000,000 nanometers, with the doubtful digit being the *millions* figure. It is thus still a three significant figure number, with the millions of nanometers being the same as the tenths of a centimeter and millionths of a kilometer.

We see in this example of 124,000,000 nanometers an ambiguity arising from the earlier statement that the last recorded figure is the doubtful digit. Interpreted in that light, 124,000,000 nanometers would have nine significant figures, with an uncertainty of plus or minus 1 nanometer. Yet the zeros that are not significant must be shown to express the value properly. This difficulty is overcome by using scientific or exponential notation (see Appendix I, page 280). In three significant figures, 124,000,000 becomes 1.24×10^8. Exponential notation, incidentally, is an excellent way of showing clearly the number of significant figures in any large or very small measurement. On the small side, it gets you away from the temptation to begin counting significant figures at the decimal point, rather than with the first non-zero digit.

Numbers having 0 as the final digit warrant a special comment. Remember the rule that the last digit shown is the doubtful digit. One time in ten times it will be zero. Show it—even if it may not seem necessary. Thus 4.80 kilograms suggests a measurement made to plus or minus 0.01 kilogram, a three significant figure number; whereas the numerically equal 4.8 kilograms suggests a two significant figure measurement made to plus or minus 0.1 kilogram. The same measurement in grams would be 4,800. Again, it is necessary to convert to exponential notation to indicate the number of significant figures. 4.8×10^3 indicates two significant figures, 4.80×10^3 indicates three, and 4.800×10^3 indicates four. Let's test your understanding of counting significant figures:

Example 2.10. For each of the following quantities, write the correct number of significant figures.

(a) 21.45 g
(b) 248 ml
(c) 0.528 in
(d) 85,300 cm
(e) 0.50 ft
(f) 0.00084 kg
(g) 5.4096×10^{12} cm
(h) 2.140×10^6 liters

2.10a. (a) 4; (b) 3; (c) 3; (d) see below; (e) 2; (f) 2; (g) 5; (h) 4.

The quantity 85,300 cm is ambiguous. Without exponential notation, there is no way to tell if the doubtful digit is in hundreds, tens or ones. The zero at the end of 2.140×10^6 liters shows it to be both the significant and the doubtful digit. The equivalent, 2,140,000 liters, would be ambiguous, as is the 85,300 cm.

Rounding Off. Sometimes when we add, subtract, multiply or divide experimentally measured quantities, the result contains figures that are not significant. When this happens, the answer must be rounded off. Various rules are used for rounding off, and there is little basis for preferring one over another, as the last digit remaining is always the doubtful digit anyway. The rules we shall follow in this book will always round off to the closest preceding digit, which some other systems do not. These rules are as follows:

1. If the first digit to be dropped is less than 5, leave the preceding digit unchanged. (Example: Round off 34.21 to the closest integer. Answer: Because 2 is less than 5, drop the two and all smaller digits, and leave the four unchanged. The result is 34.)

2. If the digit to be dropped is 5 or more, increase the preceding digit by 1. (Example: Round off 34.52 to the closest integer. Answer: Because 5 is "5 or

more" drop the 5 and all small digits, and increase the four by 1. The result is 35.)

Example 2.11. Round off each of the following to three significant figures:

(a) 1.42752 g/cc
(b) 643.349 cm^2
(c) 0.0074562 kg
(d) 2.103×10^4 cm
(e) 45,853 cm^3
(f) 0.0394498 m
(g) 3.605×10^{-7} cm
(h) 3.5000 sec.

2.11a.
(a) 1.43 g/cc
(b) 643 cm^2
(c) 0.00746 or 7.46×10^{-3} kg
(d) 2.10×10^4 cm
(e) 4.59×10^4 cm^3
(f) 0.0394 or 3.94×10^{-2} m
(g) 3.61×10^{-7} cm
(h) 3.50 sec.

Calculations with Significant Figures. The number of significant figures to which a calculated result must be rounded off is determined by applying certain rules that yield answers in which the last digit is *usually* the doubtful digit. Occasionally certain combinations of numbers produce exceptions and a more sophisticated approach is required. In this book we shall not concern ourselves with these occasional happenings, leaving them to more advanced chemistry or physics courses.

The significant figure rule for addition and subtraction is: Round off the answer to the first column that has a doubtful digit. Example 2.12 shows how this rule is applied:

Example 2.12. A student weighs four different chemicals in a pre-weighed beaker. The individual weights and their sum are as follows:

Beaker	319.542 g.
Chemical A	20.46 g.
Chemical B	0.0639 g.
Chemical C	38.2 g.
Chemical D	4.076 g.
Total	382.3419 g.

Express the sum in the proper number of significant figures.

This sum is to be rounded off to the first column that has a doubtful digit. What column is this: hundreds, tens, units, tenths, hundredths, thousandths or ten thousandths?

2.12a. Tenths. The doubtful digit in 38.2 is in the tenths column. In all other numbers the doubtful digit is in the hundredths column or smaller.

According to the rule, the answer must now be rounded off to the nearest number of tenths. What answer do you report?

2.12b. 382.3 grams

An alternate procedure is to round off before adding. In the example given rounding off first would give 382.4 as the answer:

$$\begin{array}{r} 319.5 \\ 20.5 \\ 0.1 \\ 38.2 \\ 4.1 \\ \hline 382.4 \end{array}$$

Regardless of method, uncertainty is present in the tenths digit, so both answers are acceptable.

Justification of the rule is clear in the context of Example 2.12. If the weight of Chemical C is known only to ±0.1 gram, it follows that the sum of the weights cannot be known to better than ±0.1 gram.

The same rule, procedure and rationalization holds for subtraction.

Example 2.13. In an experiment in which oxygen is produced by heating potassium chlorate in the presence of a catalyst, a student assembled and weighed a test tube, test tube holder and catalyst. He then added potassium chlorate and weighed the assembly again. The data were as follows:

Test tube, test tube holder, catalyst and potassium chlorate	26.255 grams
Test tube, test tube holder and catalyst	24.11 grams

The weight of potassium chlorate is the difference between these numbers. Express this weight in the proper number of significant figures.

2.13a. 2.15 grams. 26.255 grams − 24.11 grams = 2.145 grams. However, the hundredths digit is doubtful in the smaller weighing, and therefore is doubtful in the difference as well. Consequently, the answer must be rounded off to hundredths, or the second decimal.

The significant figure rule for multiplication and division is: Round off the answer to the same *number* of significant figures as the smallest *number* of significant figures in any factor.

Again application will be illustrated by example:

Example 2.14. The density of a certain gas is 1.436 grams per liter. Find the mass of 0.0573 liter of the gas. Express the answer in the proper number of significant figures.

The mass is found by multiplying the volume by the density. The setup of the problem is

$$0.0573\ \cancel{\text{liter}} \times \frac{1.436\ \text{grams}}{1\ \cancel{\text{liter}}}$$

According to the rule, the product may have no more significant figures than the least number of significant figures in any factor. What is the least number of significant figures in any factor?

2.14a. Three. 0.0573 is a three significant figure number: 5.73×10^{-2}. Now calculate the answer and round off to three significant figures.

2.14b. 0.0823 gram. If you showed 0.082, you forgot that counting significant figures begins at the first non-zero digit.

Justification for the rule for multiplication and division is not so apparent as for addition and subtraction. If we assume in Example 2.14 that the last digits in both quantities are higher by one than their true value, the true answer would be $0.0572 \times 1.435 = 0.0821$. If both are too low by one, the problem is $0.0574 \times 1.437 = 0.0825$. Uncertainty appears in the third significant figure just as predicted by application of the rule.

Example 2.15. Assuming the numbers are derived from experimental measurement, solve

$$\frac{(2.86 \times 10^{4})\ (3.163 \times 10^{-2})}{1.8}$$

and express the answer in the correct number of significant figures.

2.15a. 5.0×10^{2}. The answer should not be shown as 500, as the number of significant figures is ambiguous. Two significant figures are set by the 1.8.

Example 2.16. How many millimeters are in 0.6294 centimeter? Express the answer in the proper number of significant figures.

With ten millimeters per centimeter, the arithmetic is simple enough. But how many significant figures? How would you state the answer?

2.16a. 6.294 millimeters.

In this problem you are multiplying by ten—exactly 10. By definition there are *exactly* 10 mm in 1 cm. Similarly there are exactly 12 inches in 1 foot. Whenever a conversion factor is exactly an integer, the factor is *infinitely* significant, and never establishes the limit on the number of significant figures in an answer. The concept of significant figures applies specifically to measured quantities and results derived from measurements.

Significant Figures and this Book. Hopefully you are using a slide rule or calculator to perform all calculations in chemistry. Slide rules have physical limits in how they may be read, i.e., generally to three significant figures. Calculators also have limits, usually eight or more significant figures—many more than are justified by the data presented in a problem. In this text, calculations have been made on a calculator and answers rounded off to three significant figures according to the rules stated in this section. Exceptions appear when a problem is stated in a way that clearly indicates that more or fewer than three significant figures express the results more correctly. We suggest that you follow any specific instructions regarding significant figures given by the teacher in charge of your chemistry course; or, in the absence of such instructions, follow our example. With a calculator and barring "typing" errors on your part or ours, you should duplicate our answers precisely. With a slide rule, you should strive to reproduce book answers to ±1% on short calculations, or ±2% on long ones.

2.8 DENSITY AND SPECIFIC GRAVITY

We shall illustrate the use of scientific measurement in the context of the closely related physical properties, density and specific gravity.

Density

Density is defined as mass per unit volume. As with many concepts in the physical sciences, the definition establishes the units in which the concept is measured. Recalling the meaning of "per," mass per unit volume means mass/volume. Thus density is expressed in mass units divided by volume units. In the English system of units, pounds per cubic foot would be a common unit of density. In the metric system, grams per cubic centimeter is most frequently employed.

Example 2.17. A rectangular block of wood is 92.0 centimeters long, 11.3 centimeters wide, and 3.80 centimeters high. It weighs 3040 grams. Find its density.

In solving this problem you *could* compute the volume by multiplying length by width by height, and then divide this value into the mass to determine density. Alternately you can set up the entire problem at once, assembling units in such a manner that the units of density result. Set up in this manner and solve.

2.17a. 0.770 grams per cubic centimeter.

$$\frac{3040 \text{ g}}{92.0 \text{ cm} \times 11.3 \text{ cm} \times 3.80 \text{ cm}} = 0.770 \text{ g/cm}^3$$

Density can also be thought of as a form of equivalence. It combines two measurements for a given sample of a substance. In that sense, the two measured quantities must be equivalent. Therefore the statement that a certain metal has a density of 7.2 grams per cubic centimeter, or 7.2 g/cm^3, can be interpreted to mean that 7.2 grams of the substance = 1 cubic centimeter of that substance. The density relationship, then, can be used as a conversion factor between the mass of a substance and its volume.

Example 2.18. The density of aluminum is 2.70 grams per cubic centimeter. What is the weight of 84.9 cubic centimeters of aluminum?

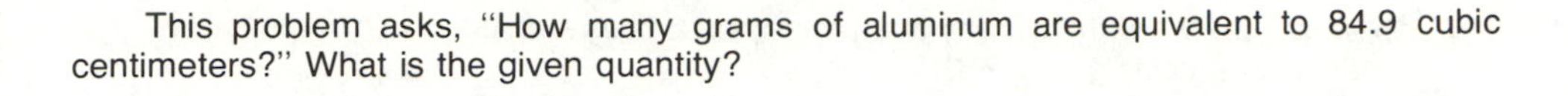

This problem asks, "How many grams of aluminum are equivalent to 84.9 cubic centimeters?" What is the given quantity?

2.18a. 84.9 cubic centimeters.

The density of aluminum is a property of the substance which holds for any quantity; it is not a "given quantity." Express the density of aluminum in the form of an equivalence.

2.18b. 2.70 grams of aluminum = 1 cubic centimeter, or, more briefly, 2.70 g Al = 1 cm^3.

From this equivalence there are two possible conversion factors that are equal to one. Multiply the given quantity by the ratio that will yield the mass of the aluminum. Set up in dimensional analysis form and solve.

2.18c. 229 grams.

$$84.9\ \cancel{\text{cm}^3\ \text{Al}} \times \frac{2.70\ \text{g Al}}{1.00\ \cancel{\text{cm}^3\ \text{Al}}} = 229\ \text{g Al}$$

Notice that approaching this problem through "logic" or common sense would yield the identical result. If *one* cubic centimeter of aluminum weighs 2.70 grams, then 84.9 cubic centimeters (84.9 times as much volume) weighs 84.9 times as many grams:

$$2.70 \times 84.9 = 229 \text{ grams of aluminum.}$$

Obviously the equivalence approach works in the reverse direction too:

Example 2.19. The density of carbon tetrachloride (CCl_4) is 1.60 grams per cubic centimeter. What volume will 26.2 grams of carbon tetrachloride occupy?

Set up completely and solve.

2.19a. 16.4 cubic centimeters.

$$26.2\ \cancel{\text{g CCl}_4} \times \frac{1\ \text{cm}^3\ \text{CCl}_4}{1.60\ \cancel{\text{g CCl}_4}} = 16.4\ \text{cm}^3\ \text{CCl}_4.$$

CCl_4 being a liquid, its volume would normally be expressed as 16.4 milliliters.

Specific Gravity

Specific gravity may be defined as the ratio of the density of a substance to the density of water at 4°C. The density of water at 4°C is 1.00 gram per cubic centimeter. Expressed as an equation,

$$\text{Specific gravity} = \frac{\text{Density of substance}}{\text{Density of water at 4°C}}. \qquad (2.3)$$

Example 2.20. If the density of aluminum is 2.70 grams per cubic centimeter, what is its specific gravity?

The solution to this problem is a simple plug-in, using Equation 2.3. Set up and solve, but be sure to include units.

2.20a. Specific gravity = 2.70.

$$\frac{D_{\text{substance}}}{D_{\text{water}}} = \frac{2.70\ \cancel{g/cm^3}}{1.00\ \cancel{g/cm^3}} = 2.70.$$

The answer has no units: they all cancel. Specific gravity is a pure, or dimensionless, number.

Observe that the specific gravity of aluminum is *numerically* equal to its density. This is so because the density of water at 4°C (the denominator in Equation 2.3) is 1.00 gram per cubic centimeter. It may be concluded that the density of any substance in grams per cubic centimeter and the specific gravity are numerically equal. As a consequence, any problem in which specific gravity is used may be interpreted as a density problem, applying units of g/cm^3 to the numerical value of the density. The solutions of Examples 2.18 and 2.19 would thus be identical if the specific gravities of aluminum and carbon tetrachloride had been given rather than their densities.

The student should be aware that the interchange of numerical values of densities and specific gravities is allowable only if the densities are expressed in grams per cubic centimeter. In the English system, for example, the density of water is 62.4 pounds per cubic foot. The density of aluminum is 168 pounds per cubic foot. The specific gravity is again found by substitution into Equation 2.3:

$$\frac{D_{\text{substance}}}{D_{\text{water}}} = \frac{168\ \cancel{lbs/ft^3}}{62.4\ \cancel{lbs/ft^3}} = 2.70.$$

The specific gravity obviously is not numerically equal to the density in English units. Note also that the specific gravity of aluminum is 2.70 regardless of the units in which density is measured. This is one of the values of the specific gravity concept: it is independent of units.

Specific gravity may also be taken to be the ratio of the mass of a given volume of a substance to the mass of the same volume of water. If V cm^3 of a given liquid weighs X grams, its density is X/V. If V cm^3 of water weighs Y grams, its density is Y/V. By definition, specific gravity equals

$$\frac{D_1}{D_w} = \frac{X/V}{Y/V} = \frac{X}{Y}.$$

A pycnometer bottle is a device by means of which the specific gravity of liquids may be found. It can be filled repeatedly to precisely the same volume. The weight of that volume of liquid is the difference between the weight filled with liquid and the weight empty.

Example 2.21. An empty pycnometer weighs 16.203 grams. Filled with water, its mass is 24.892 grams. Filled with a certain solution, it weighs 25.048 grams. Find the specific gravity of the solution.

First find the masses of the equal volumes of liquid.

2.21a. 8.689 grams H_2O and 8.845 grams liquid.

$$\begin{array}{r} 24.892 \\ -16.203 \\ \hline 8.689 \text{ g } H_2O \end{array} \qquad \begin{array}{r} 25.048 \\ -16.203 \\ \hline 8.845 \text{ g liquid} \end{array}$$

The ratio of the weight of the liquid to the weight of the same volume of water is the specific gravity. Complete the problem.

2.21b. Specific gravity = 8.845/8.689 = 1.018.

PROBLEMS

Section 2.4. Problems 2.1–2.7 and 2.36–2.42: For each measurement, insert the number in the blank space that converts the given unit to the wanted unit. Use exponential notation in the case of very large or very small numbers.

2.1. 9.46 cm = ______ mm

2.2. 734 ml = ______ l

2.3. 0.603 g = ______ cg

2.4. 81.9 dm = ______ m

2.5. 9.30×10^4 cm = ______ km

2.6. 8.99 Mg = ______ mg

2.7. 56.7 mm^3 = ______ cm^3

2.36. 6.97 mg = ______ g

2.37. 82.9 km = ______ m

2.38. 0.00526 mm = ______ cm

2.39. 61.4 m = ______ cm

2.40. 64.9 kg = ______ mg

2.41. 80.1 nm = ______ cm

2.42. 0.0723 km^2 = ______ m^2

Section 2.5.

2.8. Laboratory counter tops are customarily 36 inches high. Express this height in centimeters.

2.9. Calculate the distance in feet covered by participants in a 100 meter swimming race.

2.10. The driving distance between San Francisco, California, and Seattle, Washington, is about 850 miles. How many kilometers is this?

2.43. The main battery of World War II battleships fired 16 inch diameter projectiles. Express this diameter in millimeters.

2.44. Find the length in meters of a 100 yard football field.

2.45. The radius of the earth is 12.7 megameters. What is the equivalent distance in miles?

2.11. Express in centimeters the 9.48 foot wheel base of an automobile.

2.12. A shoe box measures 12 inches × 5.5 inches × 3.5 inches. Calculate its volume in cubic centimeters.

2.13. A European dairy farm has a land area of 0.425 square kilometers. Determine this in acres if there are 43,560 square feet in an acre.

2.14. Find the mass in pounds of a 1120 kilogram automobile.

2.15. How many grams of peas are in a container having a net weight of 12 ounces? (16 ounces = 1 pound.)

2.46. How many miles are between the earth and the moon if that distance is 3.85×10^{10} centimeters?

2.47. The floor area of a home is 1820 square feet. Find this area in square meters.

2.48. The surface area of Lake Michigan is 22,400 square miles. Calculate this area in square kilometers.

2.49. A football program lists the weight of an all-pro linebacker at 248 pounds. Find his mass in kilograms.

2.50. A human hair, weighed on an analytical balance, is found to have a mass of 0.3 milligrams. How many pounds is this?

Section 2.6.

2.16. The highest temperature recorded in the United States yesterday was a balmy 88°F in central Florida. What would this be in Celsius degrees?

2.17. Nick's Pizza Parlor, a favorite among nearby San Jose State University students, operates pizza ovens at 525°F. If Nick were to open a branch restaurant in Italy, at what Celsius temperature would he operate his European-made oven?

2.18. Temperatures in the hotter regions of a blast furnace used for extracting iron from its ore are about 2000°C. What would be the more familiar Fahrenheit temperature known to the steelworkers in Pittsburgh, Pennsylvania, and Gary, Indiana?

2.19. If SI units are ever adopted for common use, we are apt to hear weather forecasters predict a high of 296K, or 296°K as Kelvin temperatures are generally written today. Find the Celsius and Fahrenheit equivalents of this absolute temperature.

2.51. It was cold in northern Montana last night. The thermometer dropped all the way down to −28°F. How would this be read on a thermometer calibrated in Celsius degrees?

2.52. Modern steam driven ocean liners are propelled by superheated steam that leaves the boilers at temperatures around 1000°F. To what number would the needle point on the Celsius thermometer gauge in a German vessel if operated at the same temperature?

2.53. Liquid nitrogen looks much like water, but because of its extremely low temperature it behaves very differently. One method of preparation is to liquify air, and then to permit the temperature to rise to about −188°C, which is half way between the boiling points of nitrogen (−196°C) and oxygen (−180°C); the pure nitrogen then distills off. Find the Fahrenheit equivalent of this half-way temperature.

2.54. In Chapter 7 of this book you will learn that the temperature, pressure and volume of a gas are interrelated, but the temperature must be expressed in absolute units, such as degrees Kelvin. How would we state in degrees Kelvin the temperature of the exhaust gases of an automobile that leave the tailpipe at 230°F?

Section 2.7. 2.20–2.22 and 2.55–2.57: Indicate the number of significant figures in which each of the following quantities is stated:

2.20. 3.09 liters

2.21. 1.280×10^{-2} kilograms

2.22. 275°C

2.55. 21.90 kilograms

2.56. 0.0056 liters

2.57. 34,296.0 milligrams

2.23–2.25 and 2.58–2.60: Round off each of the following quantities to three significant figures:

2.23. 4.22107 kilograms

2.24. 278.5 cubic centimeters

2.25. 7.9795×10^3 grams

2.58. 175816 grams

2.59. 10.95 milliliters

2.60. 6400 milligrams

2.26. A solution is prepared by dissolving 2.86 grams of sodium chloride, 3.9 grams of ammonium sulfate and 0.896 grams of potassium iodide in 246 grams of water. Calculate and express in the correct number of significant figures the total mass of the solution.

2.27. If you were to calculate the percentage by weight of ammonium sulfate in the solution described in Problem 2.26, in how many significant figures would you be entitled to express the answer? Why?

2.61. A moving van crew picks up the following items: a couch that weighs 126 pounds; a chair that weighs 45.8 pounds; a piano at 2.4×10^2 pounds; and several boxes having a total weight of 374.82 pounds. Calculate and express in the correct number of significant figures the total weight of the load.

2.62. What would be the correct number of significant figures in which you could express the percentage of the total load represented by the chair in Problem 2.61? Explain.

Section 2.8.

2.28. Magnesium is among the least dense of our common metals. Calculate its density if an 865 cubic centimeter block has a mass of 1510 grams.

2.29. A rectangular block of iron 4.60 cm × 10.3 cm × 13.2 cm weighs 4.92 kilograms. Calculate its density.

2.30. Calculate the mass in grams of 35.3 milliliters of pure alcohol if its density is 0.790 grams/cubic centimeter.

2.31. An empty pycnometer bottle weighs 12.047 grams. When filled with water it weighs 17.003 grams; and with a certain oil, 16.723 grams. Find the specific gravity of the oil.

2.32. What is the weight in pounds of 50.0 liters of coconut oil if its specific gravity is 0.925?

2.63. 182 grams of benzene, one of the most important of organic compounds, fills a graduated cylinder to the 207 milliliter mark. Find the density and specific gravity of benzene.

2.64 Densities of gases are usually measured in grams per liter. Calculate the density of air if 24.5 liters has a mass of 29,0 grams.

2.65. Ether, the well-known anesthetic, has a density of 0.736 grams/cubic centimeter. What is the volume of 225 grams of ether?

2.66. When the pycnometer bottle described in Problem 2.31 is filled with carbon tetrachloride, its mass is 19.927 grams. Calculate the specific gravity of carbon tetrachloride.

2.67. The density of balsa wood is 7.8 pounds per cubic foot. What is the weight in kilograms of a piece of balsa wood 4.0 inches by 6.0 inches by 20.0 inches?

2.33. How many pounds does a quart of mercury weigh? Its specific gravity is 13.6. There are four quarts in a gallon, 3.785 liters in a gallon, and 454 grams in a pound.

2.34. The density of copper is 557 pounds per cubic foot. Find (a) its specific gravity and (b) the weight in grams of a piece 12.0 cm $\times$ 4.00 cm $\times$ 0.50 cm.

2.35. On December 31, 1974, citizens of the United States were legally allowed to own gold again. On January 17, 1975, gold was quoted at $174 per troy ounce (1 troy ounce = 31.1 grams). Calculate the price of 250 grams of gold in U.S. dollars at the quoted figure.

2.68. A 1.00 inch by 4.00 inch piece of platinum 0.0200 inches thick weighs 1.00 ounce. Find the density of platinum in grams per cubic centimeter and pounds per cubic foot.

2.69. What is the weight in pounds of the contents of a 55-gallon drum of turpentine? Its specific gravity is 0.870. Water weighs 8.34 pounds per gallon.

2.70. How many liters of turpentine (See Problem 2.69) weigh 200 pounds?

CHAPTER 3

PROBLEMS INVOLVING CHEMICAL FORMULAS

3.1 ATOMIC WEIGHT; MOLECULAR WEIGHT

Atoms have mass. This mass is so small, however, that it cannot be determined by the most sensitive balances known today, to say nothing of the early days of atomic theory. This notwithstanding, chemists of the first half of the 19th century were able to isolate bulk samples of matter that contained equal numbers of atoms from the various elements known. From mass measurements of these samples, the *relative* weights of atoms of the elements were determined.* By selecting the mass of an atom of one element as a standard and assigning to it an atomic weight, the atomic weights of all other elements were—and are—organized into scales of *relative* atomic weights. The standard of comparison used today is a certain isotope (see below) of carbon identified as carbon-12 or ^{12}C. Accordingly, **the atomic weight of an element is the number that expresses the average weight of its atoms compared to an atom of carbon-12 at exactly 12.**

Atomic weight, as here defined, is dimensionless; it has no unit. Atomic weight is given a unit, however, in a different approach to the concept. It is said, arbitrarily, that the mass of a carbon-12 atom is exactly 12 atomic mass units. It follows, therefore, that **one atomic mass unit (amu) is exactly 1/12 of the mass of a carbon-12 atom.** By this definition, the average mass of the atoms of an element in atomic mass units is numerically equal to the atomic weight by the earlier definition.

*The ratio of masses of equal numbers of atoms of two elements is the same as the ratio of masses of the individual atoms:

$$\frac{\text{n atoms of A} \times \text{grams per atom of A}}{\text{n atoms of B} \times \text{grams per atom of B}} = \frac{\text{grams per atom of A}}{\text{grams per atom of B}}.$$

The atomic weight of carbon is 12.01115. This indicates that all carbon atoms are not carbon-12 atoms, and points to the importance of the word *average* in the definition of atomic weight. Atoms of the same element that differ in mass are called isotopes. Nearly 99% of the naturally occurring atoms of carbon are the isotope identified as carbon-12, or ^{12}C, having a mass of 12 amu. The remainder have a mass of 13 amu, and are identified as carbon-13 or ^{13}C. The presence of the heavier isotope in any random sampling of carbon raises the average mass of atoms in that sampling to 12.01115 amu.

Presently accepted values of the atomic weights of all elements are tabulated inside the back cover of this book. The better source for these values, however, is the periodic table inside the front cover. Using the periodic table will increase your familiarity with the table and enhance your recognition of its other applications.

The molecular weight of molecular substances is conceptually similar: **the molecular weight of a compound is the number that expresses the average weight of its molecules compared to an atom of carbon-12 at exactly 12.** As defined, molecular weight is unitless; but this molecular weight number is equal to the average mass of molecules in atomic mass units.

3.2 THE MOLE

Neither of the definitions of atomic weight are very useful for "bulk" or weighable-quantity chemistry until we include the concept of the **mole.** One mole of any substance is **that quantity of a species that contains the same number of units as the number of atoms in exactly 12 grams of carbon-12.** This rather strange definition immediately suggests the question, "What is the number of atoms in exactly twelve grams of carbon-12?" It is a large number, determined experimentally as 6.02×10^{23}. It is frequently called **Avogadro's Number,** after the man whose experiments with gases contributed to the formulation of the mole concept. It is commonly identified by the symbol N.

Many chemists think of, and many textbooks present, the mole simply as the number 6.02×10^{23}. In this sense the word *mole* is precisely analogous to the term *dozen*. If you are ever in doubt as to how to interpret "mole" in a sentence, substitute the word "dozen" to give it more familiar meaning, and then transfer that meaning back to "mole."

The principal utility of the mole concept is that it enables us to "count" atoms, molecules or formula units by weighing. For example, suppose we wanted to prepare 6.02×10^{23} carbon dioxide molecules—one mole of CO_2. Each molecule contains 1 atom of carbon and 2 atoms of oxygen. We need, therefore, one mole of carbon atoms and two moles of oxygen atoms. By weighing out 12.0 grams of carbon—its atomic weight—we would have a sample containing the required number of carbon atoms. The atomic weight of oxygen is 16.0. This means that the mass of one mole of oxygen atoms is 16.0 grams. To obtain two moles of oxygen atoms, we would weigh out 2×16.0 or 32.0 grams of oxygen. It follows that the mass of one mole of carbon dioxide is $12.0 + 2(16.0) = 44.0$ grams.

3.3 MOLAR WEIGHT

From the foregoing discussion we see that the weight in grams of any element that corresponds to its atomic weight contains one mole of atoms of that element. This weight in grams is equal to the atomic weight and is called the **gram atomic weight.** The quantity of an element that weighs the gram atomic weight is called a **gram atom** of the element. Similarly, the **gram molecular weight** of a compound is the weight in grams of a compound that corresponds to the molecular weight of a compound, and a **gram molecule** is that quantity of a compound that weighs the gram molecular weight. All of these terms are related to the mole, and all represent one mole of the element or compound. We shall, in this book, use only the term mole for gram atom and gram molecule. We shall further identify **molar weight** as **the mass in grams of one mole of any species,** and use this all-inclusive term in most of our discussions. It follows from the definition that *the units of molar weight are grams per mole.*

Chemical compounds may be classified broadly into molecular compounds, such as carbon monoxide, carbon dioxide and water, and ionic compounds, such as sodium chloride, which is ordinary table salt. Molecular compounds exist as distinct individual units. Ionic compounds do not. Ionic compounds are assembled in a crystal lattice structure of a large and indefinite number of ions, but are always present in the ratio indicated by the chemical formula. Such a substance cannot appropriately be said to have a "molecular weight," because it does not form molecules. Instead, we speak of its **formula weight**, or **gram formula weight**, depending on the context, and refer to one mole of "formula units" of the substance. The term *molar weight* recognizes no distinction between ionic and molecular compounds, and is used for both.

3.4 THE CHEMICAL FORMULA

Having referred to the "formula unit," let us establish the significance of the chemical formula. In its simplest sense, the chemical formula indicates the ratio of atoms of different elements making up a chemical species. In the case of molecular substances, the formula tells us the number of atoms of each element present in the molecule. We have indicated that the carbon dioxide molecule has one carbon atom and two oxygen atoms. Its chemical formula, therefore, is CO_2. Subscript numbers are used behind the elemental symbol to indicate the number of atoms of that element in the molecule. Absence of a subscript, as for carbon, indicates that the number of atoms in the molecule is 1. We interpret the formula H_2O as meaning the molecule contains two atoms of hydrogen and one atom of oxygen. Carbon monoxide, CO, has one carbon and one oxygen atom in each molecule. In the case of ionic compounds, the subscripts indicate the *ratio* of ions* of the elements in the crystal structure. NaCl for sodium chloride, for example, indicates the presence of sodium and chlorine ions in a 1:1 ratio; and $CaCl_2$ identifies calcium and chlorine ions in a 1:2 ratio.

---

*Ions are electrically charged atoms or groups of atoms.

3.5 CALCULATION OF MOLAR WEIGHT

The example used to illustrate that we can count atoms by weighing concluded by showing how the molar weight of a chemical species is determined. Let's use the same procedure with a different compound:

Example 3.1. What is the molar weight of ammonia, NH_3?

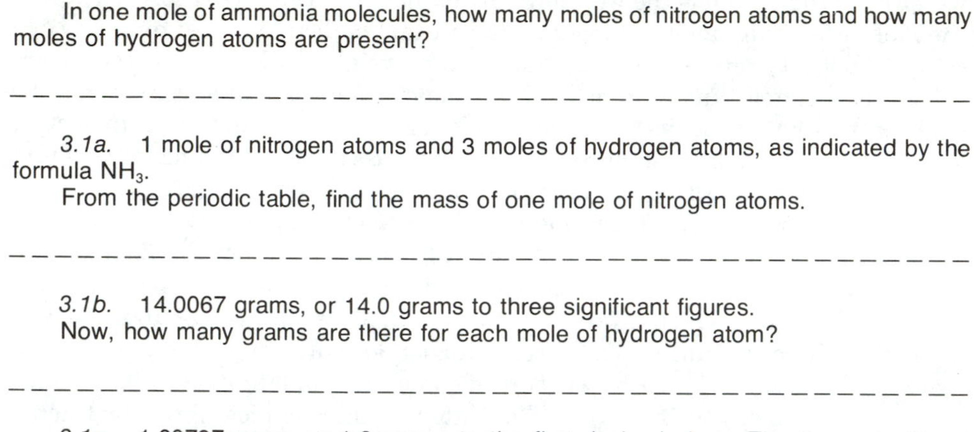

In one mole of ammonia molecules, how many moles of nitrogen atoms and how many moles of hydrogen atoms are present?

3.1a. 1 mole of nitrogen atoms and 3 moles of hydrogen atoms, as indicated by the formula NH_3.

From the periodic table, find the mass of one mole of nitrogen atoms.

3.1b. 14.0067 grams, or 14.0 grams to three significant figures.

Now, how many grams are there for each mole of hydrogen atom?

3.1c. 1.00797grams, or 1.0 grams to the first decimal place. The three significant figures in the atomic weight of nitrogen establish the tenths column as the smallest digit that can be present in any addition in which it is involved, so listing the second decimal for hydrogen is of no value. This is almost always true with hydrogen.

Now let's combine this information to find the molar weight of NH_3. Be careful. . . .

3.1d. 17.0 grams per mole.

There are, of course, three moles of hydrogen atoms. Therefore, 14.0 grams nitrogen + 3(1.0 grams hydrogen) = 17.0 grams NH_3/mole.

Rather than give additional examples of increasingly complex compounds, these will be incorporated, with suitable comment, into other applications of molar weight.

3.6 MOLAR WEIGHTS OF CERTAIN ELEMENTS

The question, "What is the molar weight of oxygen?" is ambiguous. Does it mean the gram *molecular* weight (the molar weight of oxygen molecules), or the gram *atomic* weight (the molar weight of oxygen atoms)? This ambiguity arises in the case of certain gaseous and easily vaporized elements that form molecules containing two atoms—diatomic molecules. The elements are nitrogen, oxygen, hydrogen, fluorine, chlorine, bromine, and iodine. Their molecular formulas are, in order, N_2, O_2, H_2, F_2, Cl_2, Br_2, and I_2.

More often than not, the student will have to decide which—gram atomic weight or gram molecular weight—is required in a given problem. If *atoms* are specifically involved, gram atomic weight is required; if molecules, gram molecular weight is used.

A single illustration is sufficient to show the difference: The molar weight of nitrogen atoms, formula N, is the gram atomic weight, 14.0 grams per mole of atoms; the molar weight of nitrogen molecules, formula N_2, is the gram molecular weight, 28.0 grams per mole of molecules. There are two moles of nitrogen atoms in one mole of nitrogen molecules. Molar weight should always be based on one mole of units described by the chemical formula.

3.7 PERCENTAGE COMPOSITION

The objective in percentage composition problems is to find the weight percentage of an element in a chemical compound. Weights required are derived from atomic weights. As is always the case, percentage is equal to the part quantity divided by the total quantity, multiplied by 100:

$$\frac{\text{Part}}{\text{Total}} \times 100 = \text{per cent.}$$

The "quantities" are measured in grams, unless specified otherwise.

The method can best be illustrated by an example:

Example 3.2. Find the percentage by weight of each element in $CaSO_4$.

Begin by determining the molar weight of $CaSO_4$. In doing so, keep track of the number of grams of each element entering the calculation, as these will be the "part" quantities in the percentage calculations.

3.2a. 136 grams per mole.

$$\begin{aligned} \text{Ca: } 1 \times 40.1 &= 40.1 \\ \text{S: } 1 \times 32.1 &= 32.1 \\ \text{O: } 4 \times 16.0 &= \underline{64.0} \\ & 136.2 = 136 \text{ g/mole, to three significant figures} \end{aligned}$$

If you have the correct answer you realize that a total of 136 grams of $CaSO_4$ contains 40.1 grams of calcium as a part of that total. From these figures, what is the percentage of calcium in $CaSO_4$?

3.2b. 29.5% calcium.

$$\frac{40.1 \text{ g Ca}}{136 \text{ g CaSO}_4} \times 100 = 29.5\% \text{ Ca}$$

Now compute the percentages of the other two elements.

3.2c. 23.6% sulfur and 47.1% oxygen.

$$\frac{32.1 \text{ g S}}{136 \text{ g } CaSO_4} \times 100 = 23.6\% \text{ S}$$

$$\frac{64.0 \text{ g O}}{136 \text{ g } CaSO_4} \times 100 = 47.1\% \text{ O}$$

Can you think of a simple way of checking your results, assuming you did not have the above answers to compare against your own? If so, do it, using your numbers.

3.2d. The sum of your percentages should be 100.
For the figures above,

29.5%
23.6½
47.1%
100.2%

Minor deviations from 100% are caused by round-off steps within the problem.

3.8 GRAMS OF AN ELEMENT IN GIVEN WEIGHT OF COMPOUND

Suppose, building on Example 3.2, it is necessary to find the number of grams of calcium in 235 grams of calcium sulfate. Using the results of the previous problem, this could be done simply by multiplying 235 grams by 29.5% converted to its decimal equivalent, 0.295:

$$235 \text{ g Ca} \times 0.295 = 69.3 \text{ g Ca} .$$

Assume, however, that you had not computed the percentage of calcium. From the calculation of the molar weight of calcium sulfate it appears that 136 grams of $CaSO_4$ contains 40.1 grams of calcium. In other words, for $CaSO_4$,

$$136 \text{ g } CaSO_4 = 40.1 \text{ g Ca} .$$

This leads to a direct dimensional analysis setup:

$$235 \cancel{\text{g } CaSO_4} \times \frac{40.1 \text{ g Ca}}{136 \cancel{\text{g } CaSO_4}} = 69.3 \text{ g Ca} .$$

The pattern is established. Try it:

Example 3.3. How many grams of iron could be extracted from 207 grams of Fe_2O_3?

First, determine the molar weight of Fe_2O_3.

3.3a. 160 grams per mole.

$$\begin{aligned} \text{Fe: } 2 \times 55.8 &= 111.6 \\ \text{O: } 3 \times 16.0 &= \underline{\ \ 48.0} \\ &\ 159.6 = 160 \text{ g/mole, to three significant figures} \end{aligned}$$

What equivalence can be established between weights of iron and Fe_2O_3 from the above information? In other words, how many grams of iron are equivalent to how many grams of Fe_2O_3?

3.3b. 112 grams of Fe = 160 grams of Fe_2O_3.
Again, numbers have been rounded off to three significant figures.
Using this equality, set up and solve the entire problem.

3.3c. 145 g Fe.

$$207 \cancel{\text{g Fe}_2\text{O}_3} \times \frac{112 \text{ g Fe}}{160 \cancel{\text{g Fe}_2\text{O}_3}} = 145 \text{ g Fe}.$$

Example 3.4. How many grams of nitrogen are in 86.2 grams of calcium nitrate tetrahydrate, $Ca(NO_3)_2 \cdot 4\ H_2O$?

The formula here is that of a "hydrate." The "dot" in the formula indicates that each formula unit of $Ca(NO_3)_2$ has locked into its crystal structure four molecules of water. The molar weight of the compound is, as always, based on the chemical formula, and therefore includes the four molecules of water. We must find that molar weight. One mole of $Ca(NO_3)_2 \cdot 4\ H_2O$ contains how many *moles* of calcium atoms?

3.4a. One mole of calcium atoms.
Now how about the nitrogen atoms: how many moles are there in one mole of formula units of the compound?

3.4b. Two moles of nitrogen atoms.
The subscript 2 indicates a ratio of two NO_3^- ions to one Ca^{2+} ion in the formula unit. Each NO_3^- ion contains one nitrogen atom; hence there are two moles of nitrogen atoms per mole of $Ca(NO_3)_2 \cdot 4\ H_2O$.
Now how about the oxygen: How many moles of oxygen atoms are in one mole of $Ca(NO_3)_2 \cdot 4\ H_2O$? Careful!

3.4c. Ten moles of oxygen atoms per mole of formula units.
If you said six, you probably overlooked the oxygen in the water molecules. Each nitrate ion contains three oxygen atoms, and there are two such ions per formula unit. This makes six oxygen atoms per formula unit from the nitrate alone. Add to that one from each of four molecules of water, and we reach ten. In terms of moles, there are ten moles of oxygen atoms per mole of $Ca(NO_3)_2 \cdot 4\ H_2O$.
Finally, the hydrogen: How many moles of hydrogen atoms are in one mole of $Ca(NO_3)_2 \cdot 4\ H_2O$?

3.4d. Eight moles of hydrogen atoms per mole of $Ca(NO_3)_2 \cdot 4\ H_2O$. There are four water molecules for each formula unit, and two hydrogen atoms for each water molecule. This makes eight atoms per unit, or eight moles of atoms per mole of units.

Now add up the weights of each element and find the molar weight of $Ca(NO_3)_2 \cdot 4\ H_2O$.

3.4e. 236 grams per mole.

$$\begin{array}{lrcr} \text{Ca:} & 1 \times 40.1 & = & 40.1 \\ \text{N:} & 2 \times 14.0 & = & 28.0 \\ \text{O:} & 10 \times 16.0 & = & 160.0 \\ \text{H:} & 8 \times 1.01 & = & \underline{8.1} \\ & & & 236.2 \end{array}$$

Now, the equivalence between nitrogen and $Ca(NO_3)_2 \cdot 4\ H_2O$: How many grams of nitrogen are equivalent to 236 grams of $Ca(NO_3)_2 \cdot 4\ H_2O$?

3.4f. 28.0 grams N = 236 grams $Ca(NO_3)_2 \cdot 4\ H_2O$, as shown in the weight summary above.

Now complete the problem. "Convert" 86.2 grams of $Ca(NO_3)_2 \cdot 4\ H_2O$ to grams of nitrogen.

3.4g. 10.2 grams of nitrogen.

$$86.2\ \cancel{\text{g}\ Ca(NO_3)_2 \cdot 4\ H_2O} \times \frac{28.0\ \text{g}\ N}{236\ \cancel{\text{g}\ Ca(NO_3)_2 \cdot 4\ H_2O}} = 10.2\ \text{g nitrogen}$$

3.9 CONVERSION BETWEEN GRAMS AND MOLES

A calculation that is made in all areas of chemistry is finding how many moles are in a given weight of a reagent, or its converse, finding how many grams are in a given number of moles of a particular species. Either way, it is a simple and straightforward dimensional analysis setup.

The units of molar weight are grams per mole. This forms the basic equality from which problems of this type can be solved. In the case of the calcium nitrate tetrahydrate of the previous example, 1 mole $Ca(NO_3)_2 \cdot 4\ H_2O$ = 236 grams $Ca(NO_3)_2 \cdot 4\ H_2O$.

Example 3.5. How many moles of aluminum sulfate, $Al_2(SO_4)_3$, are in 218 grams?

Begin by determining the molar weight of aluminum sulfate.

3.5a. 342 grams per mole.

$$\begin{aligned} \text{Al: } 2 \times 27.0 &= 54.0 \\ \text{S: } 3 \times 32.1 &= 96.3 \\ \text{O: } 12 \times 16.0 &= 192 \\ \hline & \quad 342 \text{ g/mole} \end{aligned}$$

Now write the equivalence between grams and moles of $Al_2(SO_4)_3$.

3.5b. 1 mole $Al_2(SO_4)_3$ = 342 grams $Al_2(SO_4)_3$.
Complete the problem.

3.5c. 0.637 mole $Al_2(SO_4)_3$.

$$218 \cancel{\text{ g } Al_2(SO_4)_3} \times \frac{1 \text{ mole } Al_2(SO_4)_3}{342 \cancel{\text{ g } Al_2(SO_4)_3}} = 0.637 \text{ mole } Al_2(SO_4)_3$$

Example 3.6. You have an experiment in which you wish to use 1.50 moles of ammonium chloride, NH_4Cl. How many grams should you weigh out?

Set up and solve.

3.6a. 80.3 grams NH_4Cl.
The molar weight of NH_4Cl is 53.5 g/mole.

$$1.50 \cancel{\text{ mole } NH_4Cl} \times \frac{53.5 \text{ g } NH_4Cl}{1 \cancel{\text{ mole } NH_4Cl}} = 80.3 \text{ g } NH_4Cl$$

The procedure may also be used for conversion between any combination of the mass of a compound, moles of a compound or any of its parts. To illustrate. . . .

Example 3.7. How many moles of water are in 90.6 grams of $Ca(NO_3)_2 \cdot 4\,H_2O$?

The molar weight of $Ca(NO_3)_2 \cdot 4H_2O$ is 236 grams/mole, from Example 3.4. Set up, but do not solve, the conversion from grams to moles of formula units.

3.7a. $$90.6 \cancel{\text{ g } Ca(NO_3)_2 \cdot 4H_2O} \times \frac{1 \text{ mole } Ca(NO_3)_2 \cdot 4H_2O}{236 \cancel{\text{ g } Ca(NO_3)_2 \cdot 4H_2O}}$$

Now, by examining the chemical formula, how many moles of water molecules are present in 1 mole of $Ca(NO_3) \cdot 4H_2O$? In other words, what equivalency exists between water molecules and formula units of $Ca(NO_3)_2 \cdot 4H_2O$?

3.7b. 1 mole of $Ca(NO_3)_2 \cdot 4H_2O$ = 4 moles of H_2O.
You are now ready to complete the setup and the problem.

3.7c. 1.54 moles H_2O.

$$90.6\ \cancel{g\ Ca(NO_3)_2 \cdot 4\ H_2O} \times \frac{1\ \cancel{mole\ Ca(NO_3)_2 \cdot 4H_2O}}{236\ \cancel{g\ Ca(NO_3)_2 \cdot 4H_2O}} \times \frac{4\ moles\ H_2O}{1\ \cancel{mole\ Ca(NO_3)_2 \cdot 4H_2O}}$$

$$= 1.54\ moles\ H_2O$$

3.10 NUMBER OF ATOMS, MOLECULES OR FORMULA UNITS IN A SAMPLE

The "number" extension of the definition of a mole makes possible the determination of the number of atoms, molecules or formula units in a known number of grams. We have just seen how mass can be converted to moles. From moles to atoms is as direct as finding the number of eggs in three dozen. To illustrate. . . .

Example 3.8. How many atoms of iron are in a sample having a mass of 25.0 grams?

The unit path is from grams to moles to atoms. Set up, but do not solve, the first step—mass to moles.

3.8a. $25.0\ \cancel{g\ Fe} \times \frac{1\ mole\ Fe}{55.8\ \cancel{g\ Fe}}$

The foregoing question about eggs should guide you to completing the setup and solving the problem. Recall that there are 6.02×10^{23} atoms in one mole.

3.8b. 2.70×10^{23} atoms Fe.

$$25.0\ \cancel{g\ Fe} \times \frac{1\ \cancel{mole\ Fe}}{55.8\ \cancel{g\ Fe}} \times \frac{6.02 \times 10^{23}\ atoms\ Fe}{1\ \cancel{mole\ Fe}} = 2.70 \times 10^{23}\ atoms\ Fe$$

In the reverse problem, finding the mass of a given number of units, the procedure is logically reversed.

Example 3.9. Calculate the mass of a million sulfur dioxide molecules—1.00×10^6 molecules of SO_2.

Before embarking on the unit path, let's determine the equivalency between moles and grams of SO_2. The molar weight of sulfur dioxide, please. . . .

3.9.a. 64.1 grams SO_2/mole.

$$1 \text{ mole } SO_2 \times 32.1 \text{ grams S} + 2(16.0 \text{ grams O}) = 64.1 \text{ grams } SO_2$$

Now, starting with the given quantity, convert molecules to moles, and then change moles to mass in grams.

3.9b. 1.06×10^{-16} grams SO_2.

$$1.00 \times 10^{6} \cancel{\text{molecules } SO_2} \times \frac{1 \cancel{\text{mole } SO_2}}{6.02 \times 10^{23} \cancel{\text{molecules } SO_2}} \times \frac{64.1 \text{ g } SO_2}{1 \cancel{\text{mole } SO_2}}$$
$$= 1.06 \times 10^{-16} \text{ grams } SO_2$$

To summarize our work with moles to this point, we have seen how to make conversions between mass in grams, moles of formula units, moles of parts of formula units, and the number of formula units. All are based on quantities that are chemically equivalent to each other, as derived from the chemical formula of the substance. Returning to $Ca(NO_3)_2 \cdot 4H_2O$, all of the equivalences listed below may be derived from the formula. A given quantity in any of the units may be converted to any other unit.

1 moles $Ca(NO_3)_2 \cdot 4H_2O$ units
1 moles $Ca(NO_3)_2$
1 moles Ca atoms
1 moles Ca^{2+} ions
2 moles N atoms
10 moles O atoms
2 moles NO_3^- ions
8 moles H atoms
4 moles H_2O molecules
6.02×10^{23} Ca atoms
12.0×10^{23} N atoms
48.2×10^{23} H atoms
60.2×10^{23} O atoms

236 grams $Ca(NO_3)_2 \cdot 4H_2O$
164 grams $Ca(NO_3)_2$
40.1 grams Ca atoms
40.1 grams Ca^{2+} ions
28.0 grams N atoms
160 grams O atoms
124 grams NO_3^- ions
8.06 grams H atoms
72.0 grams H_2O molecules
6.02×10^{23} Ca^{2+} ions
12.0×10^{23} NO_3^- ions
24.1×10^{23} H_2O molecules
6.02×10^{23} formula units of $Ca(NO_3)_2 \cdot 4H_2O$

3.11 CALCULATION OF EMPIRICAL FORMULAS

The empirical formula of a compound is the formula that is determined by chemical analysis; it gives the simplest atom ratio of the elements making up the compound. Data appear in a variety of ways: the percentage of each element may be given, or the number of grams of the different elements in a sample of the compound may be stated. In either case, the essence of the problem is to determine the ratio of the number of *moles of atoms* of each element. It is the numbers that express this ratio that appear as subscripts in the formula. The method of finding these numbers is illustrated in the following examples:

Example 3.10. A 1.338-gram sample of a compound contains 0.366 gram of sodium and 0.220 gram of nitrogen. The balance is oxygen. Find the empirical formula of the compound.

In this problem the quantity of oxygen is not specified, but it may be found readily from the given data. The total weight is known, as are the weights of the two elements other than oxygen. The weight of oxygen may be found by difference. Make this calculation now.

3.10a. 0.752 gram of oxygen.

Na: 0.366 g	Total weight: 1.338 g
N: 0.220 g	Na + N: 0.586 g
Na + N: 0.586 g	O: 0.752 g

From this point a tabulation organizes the remaining steps of the problem. This tabulation begins as follows:

Element	Grams	Moles	Mole Ratio	Formula Ratio
Na	0.366			
N	0.220			
O	0.752			

Your next step is to convert the grams of each element into moles of atoms. Do so now, filling in the second column of the table with your results. (Note: remember you are looking for moles of *atoms*, not moles of *molecules.*)

3.10b.

Element	Grams	Moles	Mole Ratio	Formula Ratio
Na	0.366	0.0159		
N	0.220	0.0157		
O	0.752	0.0470		

$$0.366 \cancel{\text{g Na}} \times \frac{1 \text{ mole Na atoms}}{23.0 \cancel{\text{g Na}}} = 0.0159 \text{ mole Na atoms}$$

$$0.220 \cancel{\text{g N}} \times \frac{1 \text{ mole N atoms}}{14.0 \cancel{\text{g N}}} = 0.0157 \text{ mole N atoms}$$

$$0.752 \cancel{\text{g O}} \times \frac{1 \text{ mole O atoms}}{16.0 \cancel{\text{g O}}} = 0.0470 \text{ mole O atoms}$$

To convert moles of *atoms* of nitrogen you use *atomic* weight, not *molecular* weight.

The numbers in the "Moles" column express the ratio of moles of atoms of each element—just what is wanted in the chemical formula. In the formula, however, the ratio should be expressed in integers. If each number in the "Moles" column is divided by the smallest number in that column, the quotients will still be in the same ratio, but they will be integers or fractions easily converted to integers. Do this, entering the result in the "Mole Ratio" column, which means the number of moles divided by the minimum number of moles.

3.10c.

Element	Grams	Moles	Mole Ratio	Formula Ratio
Na	0.336	0.0159	1	1
N	0.220	0.0157	1	1
O	0.752	0.0470	3	3

$$\frac{\text{Moles Na}}{\text{Moles N}} = \frac{0.0159}{0.0157} = 1.01 \qquad \frac{\text{Moles O}}{\text{Moles N}} = \frac{0.0470}{0.0157} = 3.00$$

These quotients, along with the 1.57/1.57 = 1.00 for nitrogen, are recorded as integers in the table. The difference between 1.01 and 1.00 may be attributed to experimental error.

The mole ratio is here expressed entirely in integers—exactly what is required for the formula. These numbers, therefore, are entered into the "Formula Ratio" column. The significance of these quotients is that there is a 1:1 ratio between the number of sodium and nitrogen atoms in the formula of the compound and a 3:1 ratio between the number of oxygen atoms and nitrogen atoms in the compound. From these, write the empirical formula of the compound.

3.10d. $NaNO_3$.

$Na_1N_1O_3$ is written as $NaNO_3$, it being understood that *no* subscript means that the subscript is 1.

Example 3.11. A compound is found to be 53% aluminum and 47% oxygen. Find its empirical formula.

In order to begin this example, the elemental quantities must be known in grams rather than in percentages. The conversion is simple. Per cent means "parts per hundred," so assume your sample weight is 100 grams. On that basis, begin a tabulation similar to that of Example 3.10. Enter the number of grams of each element in 100 grams of sample.

3.11a.

Element	Grams	Moles	Mole Ratio	Formula Ratio
Al	53			
O	47			

53% of 100 grams is 53 grams of aluminum; 47% of 100 grams is 47 grams of oxygen.

Complete the Mole and Mole Ratio columns as in the previous example.

3.11b.

Element	Grams	Moles	Mole Ratio	Formula Ratio
Al	53	2.0	1.0	
O	47	2.9	1.5	

$$53 \, \cancel{\text{g Al}} \times \frac{1 \text{ mole Al atoms}}{27.0 \, \cancel{\text{g Al}}} = 2.0 \text{ moles Al atoms}$$

$$47 \, \cancel{\text{g O}} \times \frac{1 \text{ mole O atoms}}{16.0 \, \cancel{\text{g O}}} = 2.9 \text{ moles O atoms}$$

$$\frac{\text{Moles Al}}{\text{Moles Al}} = \frac{2.0}{2.0} = 1.0 \qquad \frac{\text{Moles O}}{\text{Moles Al}} = \frac{2.9}{2.0} = 1.5$$

This time the numbers in the Mole Ratio column are neither integers nor very close to integers. But they can be changed to integers and kept in the same ratio by multiplying both of them by the same small integer. Find the smallest whole number that will yield integers when used as a multiplier for 1.0 and 1.5; use it to obtain the formula ratio figures; complete the table; and write the empirical formula of the compound.

3.11c. Al_2O_3.

Element	Grams	Moles	Mole Ratio	Formula Ratio
Al	53	2.0	1.0	2
O	47	2.9	1.5	3

Multiplication of *both* mole ratio values by two yields two integers that are in the same ratio. These integers become the subscript numbers in the formula of the compound.

The identification of multipliers to convert "Mole Ratio" numbers to integers is shown in Table 3.1.

The procedure for determining empirical formulas may be summarized as follows:

1. Determine the relative weights of different elements that are combined with each other in the compound.
2. Express these combining quantities in terms of relative numbers of moles of atoms of the elements.
3. Express the moles of atoms as a ratio of integers. These integers become the subscripts in the empirical formula.

TABLE 3.1 CONVERSION OF MOLE RATIO NUMBERS TO INTEGERS

Decimal Fraction	Equivalent Rational Fraction	Multiplier
0.5	1/2	2
0.33 0.67	1/3 2/3	3
0.25 0.75	1/4 3/4	4
0.2 0.4 0.6 0.8	1/5 2/5 3/5 4/5	5

PROBLEMS

Sections 3.5 and 3.6. Problems 3.1–3.6 and 3.22–3.27: For each substance identified by name or formula, calculate the molar weight.

3.1. Chlorine atoms, Cl

3.2. Chlorine gas, Cl_2

3.3. LiF

3.4. K_2SO_4

3.5. $Ba(NO_3)_2$

3.6. $MgCl_2 \cdot 6H_2O$

3.22. Helium gas, He

3.23. Phosphorus, P_4

3.24. NH_4Br

3.25. $C_{12}H_{22}O_{11}$

3.26. $(NH_4)_2CO_3$

3.27. $CuSO_4 \cdot 5H_2O$

Section 3.7. Problems 3.7–3.9 and 3.28–3.30: Calculate the percentage composition of each compound, selected from those above.

3.7. K_2SO_4

3.8. $Ba(NO_3)_2$

3.9. $MgCl_2 \cdot 6H_2O$

3.28. $C_{12}H_{22}O_{11}$

3.29. $(NH_4)_2CO_3$

3.30. $CuSO_4 \cdot 5H_2O$

Section 3.8.

3.10. How many grams of lithium are in 65.4 grams of LiF?

3.11. How many grams of potassium are in 16.3 grams of K_2SO_4?

3.31. How many grams of bromine are in 7.50 grams of NH_4Br?

3.32. What is the mass of oxygen in 445 grams of $C_{12}H_{22}O_{11}$?

Section 3.9.

3.12. How many moles of chlorine gas are in 65.2 grams? (Caution: what is the formula for chlorine gas?)

3.13. 52.0 grams of $MgCl_2 \cdot 6H_2O$ contains how many moles of the compound?

3.14. Calculate the number of grams of LiF that must be weighed out to obtain 0.353 mole.

3.33. What is the number of moles of $(NH_4)_2CO_3$ in a sample that weighs 18.0 grams?

3.34. How many moles of sugar, $C_{12}H_{22}O_{11}$, are in one pound (454 grams)?

3.35. 0.522 mole of $CuSO_4 \cdot 5H_2O$ weighs how many grams?

Section 3.10.

3.15. Calculate the mass of four trillion (4.00×10^{12}) atoms of aluminum.

3.16. How many atoms of carbon has a young man given his bride-to-be if the engagement

3.36. Would you require a truck to transport 10^{25} atoms of copper? Knowing the mass in grams would help you plan the job.

3.37. At $174 per troy ounce (1 troy ounce = 31.1 grams), the price quoted for gold on the

ring has a 0.500 carat diamond? There are 200 milligrams in a carat. (The price of diamonds doesn't seem so high when figured at dollars per atom.)

3.17. A typical glass of water weighs 500 grams. Calculate the number of water molecules you drink when quenching your thirst with this quantity of our most common liquid.

3.18. The quantitative significance of "take a deep breath" varies, of course, with the individual. When one person did so he found that he inhaled 2.95×10^{22} molecules of the mixture of nitrogen and oxygen we call air. Assuming this mixture has an average molar weight of 29 grams of air per mole of molecules, what is his apparent lung capacity in grams of air?

financial pages on January 18, 1975, calculate the cost of a single atom of gold.

3.38. One who sweetens his coffee with two teaspoons of sugar, $C_{12}H_{22}O_{11}$, uses about 0.600 grams. To three significant figures, how many sugar molecules is this?

3.39. Assuming gasoline to be pure octane, C_8H_{18}—it is actually a mixture of octane and other hydrocarbons—an automobile getting 15.0 miles per gallon would consume 9.37×10^{23} molecules per mile. Calculate the mass of this amount of fuel.

Section 3.11.

3.19. A certain compound is 52.2% carbon, 13.0% hydrogen, and 34.8% oxygen. Find the empirical formula of the compound.

3.20. 11.89 grams of iron are exposed to a stream of oxygen until they react to produce 16.99 grams of a pure oxide of iron. What is the empirical formula of the product?

3.21. A 2.500 gram sample of a compound of lithium, chlorine and oxygen is heated for an extended period to drive off the oxygen. The residue weighs 1.170 grams. It is then dissolved in water and treated with silver nitrate. All of the chloride precipitates as AgCl, which weighs 3.963 grams. Calculate the empirical formula of the lithium-chlorine-oxygen compound.

3.40. A compound analyzes 29.1% sodium, 40.5% sulfur and 30.4% oxygen. Calculate the empirical formula of the compound.

3.41. 27.65 grams of a certain compound containing only carbon and hydrogen are burned completely in oxygen. 35.5 grams of water and 86.9 grams of carbon dioxide are formed as the only products. What is the empirical formula of the original compound? (Hint: calculate the number of grams of carbon and hydrogen in the original compound.)

3.42. A certain hydrate has the general formula $Co_aS_bO_c \cdot xH_2O$. 43.0 grams of the compound are heated to drive off the water, leaving 26.1 grams of anhydrous compound. Further analysis shows that the percentage composition of the anhydride is 42.4% Co, 23.0% S and 34.6% O. Find the empirical formula of the (a) anhydrous compound and (b) hydrate.

CHAPTER 4

STOICHIOMETRY

Stoichiometry refers to the quantities of reactants and products involved in chemical reactions. For a hypothetical reaction, $A + B \rightarrow C + D$, we will be concerned with questions such as these: How much A is required to react with X grams of B? How much C will be produced in the reaction of A with X grams of B? How much D will be produced along with Y grams of C? Chemical quantities, the "how much" part of the above questions, can be measured in several ways. Solids are usually measured in grams, liquids in milliliters, and gases in liters. All of these quantity units may also be expressed in one other unit, the mole. In fact, the mole is the connecting link that welds them all together in the framework of stoichiometry.

4.1 THE CHEMICAL EQUATION

Every problem in stoichiometry is based on a chemical equation and its interpretation in terms of moles. The equation $2\ C + O_2 \rightarrow 2\ CO$, for example, may be interpreted as saying that two *moles* of carbon combine with one *mole* of oxygen to form two *moles* of carbon monoxide. The coefficients in the equation express the relative reacting quantities in terms of moles. From this it follows that, if the reaction quantity of carbon is four moles, twice as much as indicated by the equation, the reaction quantities of oxygen and carbon monoxide are two moles and four moles respectively. Similarly, the formation of *one* mole of carbon monoxide requires *one* mole of carbon and *one-half* mole of oxygen. The molar quantities of species taking part in a reaction are proportional to their coefficients in the equation.

The conversion of one species in a chemical reaction to another follows from the proportionality established by the equation coefficients. The method to be presented here emphasizes this point: it makes the mole to mole conversion of one species to another the central step in the procedure.

Referring again to the reaction $2\ C + O_2 \rightarrow 2\ CO$, a series of equivalences may be derived from the equation. *For this reaction* it can be said that 2 moles C = 1 mole O_2 = 2 moles CO. This expresses the rigid relationship between quantities of carbon, oxygen, and carbon monoxide *for this reaction*. Note the emphasis on *this reaction*. The chemical equivalences of moles of different species vary with the reaction. For example, carbon may also combine with oxygen to form carbon dioxide: $C + O_2 \rightarrow CO_2$. In this reaction 1 mole C = 1 mole O_2—not the same as in the case of the formation of carbon monoxide. The molar relationship between species in the reaction is based on the coefficients of the chemical equation describing that *specific* reaction.

4.2 MOLE-MOLE PROBLEMS

The central step in the solving of stoichiometry problems should now be clear. The equation yields a series of chemical molar equalities. If the number of moles of any species is known, it is readily converted into the number of moles of any other species.

Example 4.1. When carbon reacts with oxygen to produce carbon dioxide by the equation $C + O_2 \rightarrow CO_2$, how many moles of CO_2 will be produced if 4.86 moles of carbon are used in the reaction?

The chemical equation gives the molar equality relationships between all species in a reaction, but the solution of any problem is concerned only with those species involved in the problem. Accordingly, from the equation, what molar equality may be established between carbon and carbon dioxide?

4.1a. 1 mole carbon = 1 mole carbon dioxide. The coefficients of both species are 1. Now, although the answer is no doubt obvious, set up the problem in dimensional analysis form and solve.

4.1b. 4.86 moles CO_2.

$$4.86 \cancel{\text{moles C}} \times \frac{1 \text{ mole } CO_2}{1 \cancel{\text{mole C}}} = 4.86 \text{ moles } CO_2$$

Example 4.2. Nitrogen and hydrogen react to form ammonia according to the following equation: $N_2 + 3\ H_2 \rightarrow 2\ NH_3$. How many moles of ammonia will be produced by the reaction of 7.2 moles of hydrogen?

First, what is the equivalence between hydrogen and ammonia in this reaction?

4.2a. 3 moles H_2 = 2 moles NH_3.
Set up and solve.

4.2b. 4.8 moles NH_3.

$$7.2 \cancel{\text{moles } H_2} \times \frac{2 \text{ moles } NH_3}{3 \cancel{\text{moles } H_2}} = 4.8 \text{ moles } NH_3$$

More often than not, in the reaction between two or more chemicals, the reactants are not present in stoichiometric quantities, i.e., those precise quantities that are required to consume each other completely. In both cases above, it was assumed that the reactant for which a quantity was *not* given was present in sufficient quantity to permit complete consumption of the reacting quantity identified. But what if this assumption cannot be made? The following example will illustrate both the problem and the solution.

Example 4.3. Under certain conditions carbon reacts with steam to produce hydrogen and carbon monoxide thus:

$$C + H_2O \rightarrow H_2 + CO.$$

Suppose 2.4 moles of carbon and 3.1 moles of steam react until one is completely consumed. (a) How many moles of hydrogen will be produced, and (b) how many moles of which reactant will remain unreacted?

Here we are dealing with "excess" stoichiometry: one reactant is present in excess of that quantity required to consume all of the other. In this case the equation tells us that carbon and steam react in equimolar quantities, or one mole of C = one mole of H_2O. With the initial quantities as given, which of the two substances will be exhausted first?

4.3a. Carbon.

It takes 2.4 moles of steam to react with 2.4 moles of carbon, and 3.1 moles of steam, more than enough, are available. Conversely, it would take 3.1 moles of carbon to react with 3.1 moles of steam, but only 2.4 are available. Either reasoning approach leads to the identification of carbon as the first material to be exhausted. The balance of the problem can be worked, therefore, on the basis of 2.4 moles of carbon as the reacting quantity. Proceed to set up and solve Part (*a*) of the problem.

4.3b. 2.4 moles H_2.

$$2.4 \cancel{\text{moles C}} \times \frac{1 \text{ mole } H_2}{1 \cancel{\text{mole C}}} = 2.4 \text{ moles } H_2$$

Part (*b*) asks how many moles of steam remain unreacted after the carbon is gone. By the reasoning above, 2.4 moles of steam were required to react with all of the carbon. If you started with 3.1 moles of steam, how many will be left at the end of the reaction?

4.3c. 0.7 moles of steam.

3.1 moles steam at start
2.4 moles steam used
0.7 mole steam left

Problems of this type are not always so simple, but the principle is always the same. Try this one:

Example 4.4. Nitric acid reacts with benzene by the following equation: $2\ HNO_3 + C_6H_6 \rightarrow C_6H_4(NO_2)_2 + 2\ H_2O$. If 0.64 mole of HNO_3 and 0.37 mole of C_6H_6 are brought together so that they react to the complete consumption of one, (a) how many moles of which compound will remain unreacted, and (b) how many moles of $C_6H_4(NO_2)_2$ will be produced?

The numbers are still sufficiently simple that you may be able to "see" the answer, but let's develop a thought process for these problems. Knowing how many moles of each reactant there are, select either of them and determine how many moles of the other would be required to consume it completely. In this case, take the C_6H_6. How many moles of HNO_3 would be required to react with 0.37 mole of C_6H_6 in the above reaction? Set up and solve.

4.4a. 0.74 moles of HNO_3.

$$0.37\ \text{mole}\ C_6H_6 \times \frac{2\ \text{moles}\ HNO_3}{1\ \text{mole}\ C_6H_6} = 0.74\ \text{moles}\ HNO_3$$

The problem states that 0.64 mole of HNO_3 is available, but 0.74 mole would be required to react with all of the C_6H_6; obviously there is not enough HNO_3. The HNO_3 is the limiting species, and the balance of the problem may be based on 0.64 mole of HNO_3.

Before completing the problem, observe what would have happened if you had made your first estimate on the basis of 0.64 mole of HNO_3 instead of the C_6H_6. How many moles of benzene would be required to react with 0.64 mole of HNO_3? Set up and solve.

4.4b. 0.32 mole C_6H_6.

$$0.64\ \text{mole}\ HNO_3 \times \frac{1\ \text{mole}\ C_6H_6}{2\ \text{moles}\ HNO_3} = 0.32\ \text{mole}\ C_6H_6$$

0.32 mole of benzene is required; 0.37 mole, more than enough, is available. Again HNO_3 is identified as the limiting species, and should be used in solving Part (*b*). But first, on the basis of the quantities stated above, how much benzene will remain at the end of the reaction?

4.4c. 0.05 mole C_6H_6 in excess.

0.37 moles present − 0.32 moles used = 0.05 moles left.

Now, using the limiting quantity of 0.64 mole of HNO_3, set up and solve for the quantity of $C_6H_4(NO_2)_2$ produced in the reaction.

4.4d. 0.32 mole $C_6H_4(NO_2)_2$.

$$0.64\ \text{mole}\ HNO_3 \times \frac{1\ \text{mole}\ C_6H_4(NO_2)_2}{2\ \text{moles}\ HNO_3} = 0.32\ \text{mole}\ C_6H_4(NO_2)_2$$

Additional practice in excess stoichiometry problems will appear later in this chapter.

4.3 PATTERN FOR STOICHIOMETRY PROBLEMS

Once the chemical equation for a process is established, there is a single pattern for the solution of all stoichiometry problems. It consists of three steps, as follows:

1. **Convert the quantity of the given species to moles.**
2. **Convert the moles of given species to moles of wanted species.**
3. **Convert the moles of wanted species to the quantity units required.**

All three of these steps have been illustrated already, but as individual problems. Step 2 is illustrated by all the problems covered thus far in this chapter. Steps 1 and 3, assuming quantities are expressed in grams, are illustrated by Examples 3.5 and 3.6 (page 34). It remains only to tie the three steps together into a single solution.

4.4 WEIGHT-WEIGHT PROBLEMS

Example 4.5. How many grams of oxygen are required to burn 12.9 grams of propane, C_3H_8, in the reaction $C_3H_8 + 5\ O_2 \rightarrow 3\ CO_2 + 4\ H_2O$?

To fit this problem into the three-step procedure outlined, the grams of propane must first be converted to moles; moles of propane must be converted to moles of oxygen; and moles of oxygen must then be converted to grams. Set up, but do not solve, the first step. (See Example 3.5, page 34, if necessary.)

4.5a.

$$12.9\ \cancel{g\ C_3H_8} \times \frac{1\ mole\ C_3H_8}{44.0\ \cancel{g\ C_3H_8}}$$

If you performed the indicated calculation you would have the number of moles of propane. This intermediate answer is not the objective, however, so it is not calculated as such. Now you should repeat the setup and extend it to complete Step 2 of the pattern, the conversion of moles of propane to moles of the wanted species, oxygen. Again, do not solve to an answer, just extend the setup.

4.5b.

$$12.9\ \cancel{g\ C_3H_8} \times \frac{1\ \cancel{mole\ C_3H_8}}{44.0\ \cancel{g\ C_3H_8}} \times \frac{5\ moles\ O_2}{1\ \cancel{mole\ C_3H_8}}.$$

The indicated calculation would give the moles of oxygen required. Again this is an intermediate answer. Now add the final step in the stoichiometric pattern, extending the setup to convert moles of oxygen to grams. (See Example 3.6, page 35, if necessary.) This time solve the problem to an answer.

4.5c. 46.9 grams of oyxgen.

$$12.9\ \cancel{g\ C_3H_8} \times \frac{1\ \cancel{mole\ C_2H_8}}{44.0\ \cancel{g\ C_3H_8}} \times \frac{5\ \cancel{moles\ O_2}}{1\ \cancel{mole\ C_3H_8}} \times \frac{32.0\ g\ O_2}{1\ \cancel{mole\ O_2}} = 46.9\ g\ O_2$$

Study carefully the manner in which the three steps in the stoichiometric pattern are set up. This will be required in many problems throughout the course.

Example 4.6. 86.3 grams of CO_2 are produced by the reaction

$$Fe_3O_4 + 4\ CO \rightarrow 3\ Fe + 4\ CO_2.$$

How many grams of iron are also produced in the reaction?

Set up all three steps and solve.

4.6a. 82.0 grams of iron.

$$86.3\ \cancel{g\ CO_2} \times \frac{1\ \cancel{mole\ CO_2}}{44.0\ \cancel{g\ CO_2}} \times \frac{3\ \cancel{moles\ Fe}}{4\ \cancel{moles\ CO_2}} \times \frac{55.8\ g\ Fe}{1\ \cancel{mole\ Fe}} = 82.1\ g\ Fe$$

The next example involves excess stoichiometry:

Example 4.7. A solution containing 14.0 grams of silver nitrate is added to a solution containing 4.83 grams of calcium chloride. AgCl precipitates according to the equation

$$2\ AgNO_3 + CaCl_2 \rightarrow 2\ AgCl + Ca(NO_3)_2.$$

a. Determine the weight of the silver chloride that precipitates.
b. How many grams of which compound, $AgNO_3$ or $CaCl_2$, are in excess?

In previous excess stoichiometry problems you learned it was necessary to determine which reactant is the limiting species by comparing the number of moles of different species available. This is true here. Therefore, your first step is to convert *both* reacting quantities to moles.

4.7a. 0.0824 mole $AgNO_3$ and 0.0435 mole $CaCl_2$.

$$14.0\ \cancel{g\ AgNO_3} \times \frac{1\ mole\ AgNO_3}{170\ \cancel{g\ AgNO_3}} = 0.0824\ mole\ AgNO_3$$

$$4.83\ \cancel{g\ CaCl_2} \times \frac{1\ mole\ CaCl_2}{111\ \cancel{g\ CaCl_2}} = 0.0435\ mole\ CaCl_2$$

You are now ready to determine the limiting species, as in Example 4.4. Which one is it?

4.7b. $AgNO_3$.

$$0.0824\ \cancel{mole\ AgNO_3} \times \frac{1\ mole\ CaCl_2}{2\ \cancel{moles\ AgNO_3}} = 0.0412\ mole\ CaCl_2$$

0.0412 mole $CaCl_2$ is required. 0.0435 mole $CaCl_2$ is available, so it is in excess, and $AgNO_3$ is the limiting species. Alternately,

$$0.0435\ \cancel{\text{mole } CaCl_2} \times \frac{2 \text{ moles } AgNO_3}{1\ \cancel{\text{mole } CaCl_2}} = 0.0870 \text{ mole } AgNO_3 \text{ required.}$$

0.0824 mole $AgNO_3$ is available, and $AgNO_3$ is the limiting species.

You are now ready to slip into the stoichiometric pattern at the appropriate place and determine the number of grams of AgCl that precipitate. Caution: What is that appropriate place? Notice that you have already determined the number of moles of $AgNO_3$. Step 1 of the pattern is already completed. You can start there. Complete and solve.

4.7c. 11.8 grams of AgCl precipitate.

$$0.0824\ \cancel{\text{mole } AgNO_3} \times \frac{2\ \cancel{\text{moles } AgCl}}{2\ \cancel{\text{moles } AgNO_3}} \times \frac{143 \text{ g } AgCl}{1\ \cancel{\text{mole } AgCl}} = 11.8 \text{ g } AgCl$$

Part (b) of the problem asks for the number of grams of calcium chloride in excess of that required to react with the silver nitrate. The easiest way to find this is first to determine the number of *moles* in excess. The necessary figures are all in answer 4.7b. What is this number of moles?

4.7d. 0.0023 mole $CaCl_2$.

$$\begin{array}{ll} 0.0435 & \text{mole } CaCl_2 \text{ available} \\ \underline{0.0412} & \text{mole } CaCl_2 \text{ used} \\ 0.0023 & \text{mole } CaCl_2 \text{ left} \end{array}$$

Now convert the excess calcium chloride to grams:

4.7e. 0.26 gram $CaCl_2$ excess.

$$0.0023\ \cancel{\text{mole } CaCl_2} \times \frac{111 \text{ g } CaCl_2}{1\ \cancel{\text{mole } CaCl_2}} = 0.26 \text{ g } CaCl_2$$

4.5 PERCENTAGE YIELD

In actual practice chemical reactions rarely yield the amount of product that stoichiometric calculations would lead us to expect. Impure reactants, incomplete reactions, side reactions and unavoidable procedural losses account for some of these deviations. It is customary, therefore, to consider the **percentage yield** in a chemical reaction. Percentage yield is calculated by determining the *theoretical yield* according to the principles of stoichiometry; measuring the *actual yield*, by weighing the substance actually produced in the reaction; and then finding what percentage of the theoretical yield is represented by the actual yield. In an equation,

$$\text{percentage yield} = \frac{\text{actual yield}}{\text{theoretical yield}} \times 100.$$

Looking back to *4.7c* in the foregoing example, if the actual AgCl weighed was 11.4 grams, the percentage yield would be

$$\frac{11.4}{11.8} \times 100 = 96.6\%.$$

Example 4.8. A solution containing excess* barium nitrate is added to a second solution containing 6.24 grams of sodium sulfate. Barium sulfate precipitates. The precipitate is filtered, dried and weighed, yielding 9.98 grams. Find the (a) theoretical yield and (b) percentage yield.

First the equation. . . . .

$$Ba(NO_3)_2 + Na_2SO_4 \rightarrow BaSO_4 + 2\ NaNO_3.$$

Theoretical yield is found in the usual way. . . .

4.8a. 10.2 grams $BaSO_4$.

$$6.24\ \cancel{g\ Na_2SO_4} \times \frac{1\ \cancel{mole\ Na_2SO_4}}{142\ \cancel{g\ Na_2SO_3}} \times \frac{1\ \cancel{mole\ BaSO_4}}{1\ \cancel{mole\ Na_2SO_4}} \times \frac{233\ g\ BaSO_4}{1\ \cancel{mole\ BaSO_4}} = 10.2\ g\ BaSO_4$$

Now calculate the percentage yield when the actual precipitate weighed 9.98 grams.

4.8b. 97.8%.

$$\frac{9.98}{10.2} \times 100 = 97.8\%$$

Example 4.9. A process for producing $Mg(OH)_2$ by the reaction between magnesium oxide and water, $MgO + H_2O \rightarrow Mg(OH)_2$, is known to operate at 81.3% efficiency. How many grams of MgO must be used to produce 785 grams of the hydroxide?

If 785 grams of $Mg(OH)_2$ is the *actual* yield, and the efficiency is 81.3%, what is the theoretical yield? In other words, 81.3% of what number equals 785?

4.9a. 966 g $Mg(OH)_2$.

$$\frac{785\ g\ Mg(OH)_2}{0.813} = 966\ g\ Mg(OH)_2$$

Dividing by a partial quantity—785 grams—by percent expressed as a decimal fraction—0.813—gives the total quantity. This step might also be approached from the dimensional analysis standpoint that 81.3 grams actual = 100 grams theoretical:

*The word "excess" as used here means "more than enough;" more than enough barium nitrate than is required to precipitate all the sulfate ion in 6.24 grams of sodium sulfate.

$$785\ \cancel{g}\ Mg(OH)_2\ \cancel{actual} \times \frac{100\ g\ theoretical}{81.3\ \cancel{g\ actual}} = 966\ g\ Mg(OH)_2\ theoretical.$$

Now, by the usual stoichiometric procedure, find the grams of MgO that must be used to produce a theoretical yield of 966 grams.

4.9b. 668 g MgO.

$$966\ \cancel{g\ Mg(OH)_2} \times \frac{1\ \cancel{mole\ Mg(OH)_2}}{58.3\ \cancel{g\ Mg(OH)_2}} \times \frac{1\ \cancel{mole\ MgO}}{1\ \cancel{mole\ Mg(OH)_2}} \times \frac{40.3\ g\ MgO}{1\ \cancel{mole\ MgO}} = 668\ g\ MgO$$

4.6 GAS VOLUME STOICHIOMETRY

The volume of a gas is a measure of its quantity, just as liquid volume (a liter of water) or solid volume (a cubic centimeter of iron) expresses the quantity of these substances. But gas volume as a quantity measurement is relatively undependable because the volume of a fixed quantity of gas depends upon its temperature and pressure. It is an experimental fact, however, that equal volumes of all gases at the *same* temperature and pressure contain the same number of molecules. This statement is known as Avogadro's Hypothesis (see page 101). For convenience, a "standard" temperature and pressure (abbreviated STP) are used for reference. These are respectively, 1 atmosphere—the normal atmospheric pressure at sea level—and 0°C (see p. 94).

The volume occupied by one mole of a gas is called its **molar volume. At STP the molar volume of any gas is 22.4 liters.** In an equivalence form,

$$1\ \text{mole of gas at STP} = 22.4\ \text{liters.}$$

Molar volume provides a second method by which the mass or volume of a substance can be converted to moles and vice versa. It is used in precisely the same manner as molar weight:

	Units	Quantity to Moles	Moles to Quantity
Molar weight	g/mole	grams ÷ MW = moles	moles × MW = grams
Molar volume	l/mole	liters ÷ MV = moles	moles × MV = liters

There are two important differences between the conversions from quantity to moles by molar volume and by molar weight:

(a) The molar weight of a chemical is a unique property of each individual chemical, whereas all gases at the same temperature and pressure have the same molar volume.

(b) The molar weight of a chemical is independent of temperature and pressure, whereas molar volume is a function of temperature and pressure. Only at STP is the molar volume 22.4 l/mole.

The pattern for solving stoichiometry problems is the same: Convert quantity of given species to moles; convert moles of given species to moles of wanted species; convert moles of wanted species to required quantity units.

Example 4.10. How many liters of hydrogen at STP will be released by the complete reaction of 0.982 gram of magnesium with hydrochloric acid by the reaction $Mg + 2\ HCl \rightarrow H_2 + MgCl_2$?

The first step is to convert the 0.982 gram of magnesium to moles. Begin the setup.

4.10a. $0.982\ \cancel{g\ Mg} \times \frac{1\ mole\ Mg}{24.3\ \cancel{g\ Mg}}$.

From the equation, extend the setup to convert from moles of magnesium to moles of hydrogen.

4.10b. $0.982\ \cancel{g\ Mg} \times \frac{1\ \cancel{mole\ Mg}}{24.3\ \cancel{g\ Mg}} \times \frac{1\ mole\ H_2}{1\ \cancel{mole\ Mg}}$.

Finally, convert moles of hydrogen to liters at STP. Complete the setup and solve.

4.10c. 0.905 liter of H_2 at STP.

$$0.982\ \cancel{g\ Mg} \times \frac{1\ \cancel{mole\ Mg}}{24.3\ \cancel{g\ Mg}} \times \frac{1\ \cancel{mole\ H_2}}{1\ \cancel{mole\ Mg}} \times \frac{22.4\ l\ H_2}{1\ \cancel{mole\ H_2}} = 0.905\ l\ H_2$$

Example 4.11. 344 ml oxygen, measured at STP, reacts with hydrogen to form water via the reaction: $2\ H_2 + O_2 \rightarrow 2\ H_2O$. How many grams of water will be formed?

This problem is similar to Example 4.8, except that the given quantity is volume of a gas and the required quantity is grams. You should be able to set up and solve the entire problem. Careful, though, on that volume unit for oxygen. . . .

4.11a. 0.553 gram of H_2O will be formed.

$$344\ \cancel{ml\ O_2} \times \frac{1\ \cancel{l\ O_2}}{1000\ \cancel{ml\ O_2}} \times \frac{1\ \cancel{mole\ O_2}}{22.4\ \cancel{l\ O_2}} \times \frac{2\ \cancel{moles\ H_2O}}{1\ \cancel{mole\ O_2}} \times \frac{18.0\ g\ H_2O}{1\ \cancel{mole\ H_2O}} = 0.553\ g\ H_2O$$

According to Avogadro's Hypothesis, equal volumes of all gases at the same temperature and pressure contain the same number of molecules—or equal numbers of moles. It follows, therefore, that if one mole of gas A reacts with two moles of gas B, the volume ratio at a given temperature and pressure will likewise be one to two. At STP, for example, one mole of A would occupy 22.4 liters; two moles of B would occupy 44.8 liters. The mole ratio, 1:2, is the same as the volume ratio, 22.4:44.8. The coefficients of the equation, therefore, represent a volume equivalence as well as the customary mole equivalence. Applied to the reaction $2\ H_2(g) + O_2(g) \rightarrow 2\ H_2O(g)$, 2 moles H_2 = 1 mole O_2 and

2 liters H_2 = 1 liter O_2, provided both gases are measured at the same temperature and pressure.

Volume to volume conversions at a given temperature and pressure are among the easiest of stoichiometry problems:

Example 4.12. How many liters of oxygen at STP are required to react with 1.39 liters of ethane, C_2H_6, also at STP? The equation is $2\ C_2H_6(g) + 7\ O_2(g) \rightarrow 4\ CO_2(g) + 6\ H_2O(l)$.

To begin, what equivalence is there between reacting volumes of C_2H_6 and O_2, both at STP?

4.12a. 2 liters C_2H_6 = 7 liters O_2.

The volume equivalence is the same as the mole equivalence.

From here the setup and solution are straightforward. Complete the problem.

4.12b. 4.87 liters O_2.

$$1.39\ \cancel{l\ C_2H_6} \times \frac{7\ l\ O_2}{2\ \cancel{l\ C_2H_6}} = 4.87 \text{ liters } O_2$$

PROBLEMS

Sections 4.1–4.4

Problems 4.1–4.7: Butane, C_4H_{10}, is a common fuel for heating rural homes in areas not serviced by natural gas sources. The equation for its combustion is $2\ C_4H_{10} + 13\ O_2 \rightarrow 8\ CO_2 + 10\ H_2O$. Questions 4.1–4.7 relate to this reaction.

Problems 4.19–4.25: The first step in the Ostwald process for manufacturing nitric acid involves the reaction between ammonia and oxygen described by the equation $4\ NH_3 + 5\ O_2 \rightarrow 4\ NO + 6\ H_2O$. Questions 4.19–4.25 relate to this reaction.

4.1. How many moles of oxygen are required to burn 3.40 moles of butane?

4.2. How many moles of carbon dioxide will be produced in the burning of 4.68 moles of butane?

4.3. How many moles of water will be produced along with 0.568 mole of carbon dioxide?

4.4. How many grams of butane can be burned by 1.42 moles of oxygen?

4.5. 9.43 grams of oxygen are used in burning butane. How many moles of water result?

4.19. How many moles of ammonia will react with 49.6 moles of oxygen?

4.20. How many moles of NO will result from the reaction of 9.45 moles of ammonia?

4.21. If 10.3 moles of water are produced, how many moles of NO will also be produced?

4.22. How many moles of ammonia can be oxidized by 303 grams of oxygen?

4.23. If the reaction consumes 37.8 moles of ammonia, how many grams of water will be produced?

4.6. Calculate the number of grams of carbon dioxide that will be produced by the burning of 78.4 grams of butane.

4.7. How many grams of oxygen are used in a reaction that produces 43.8 grams of water?

4.8. A reaction that produces great heat for welding and incendiary bombs is the "thermit" reaction: $Fe_2O_3 + 2\ Al \rightarrow Al_2O_3 + 2\ Fe$. How many grams of Fe_2O_3 can be converted to aluminum oxide by the reaction of 47.1 grams of aluminum?

4.9. A biological process whereby large starch molecules are converted to a sugar may be represented by the equation $C_{600}H_{1000}O_{500} + 50\ H_2O \rightarrow 50\ C_{12}H_{22}O_{11}$. Calculate the number of grams of sugar that will result from the reaction of 100 grams of starch by this reaction.

4.10. A common type of fire extinguisher depends upon the reaction of sodium hydrogen carbonate, $NaHCO_3$, with sulfuric acid to produce carbon dioxide that develops a pressure to squirt water or foam onto a fire. The equation is $2\ NaHCO_3 + H_2SO_4 \rightarrow Na_2SO_4 + 2\ H_2O + 2\ CO_2$. If a fire extinguisher were designed to hold 600 grams of sodium hydrogen carbonate, how many grams of sulfuric would be required to react with all of it?

4.11. One of the methods for manufacturing sodium sulfate, widely used in making the kraft paper for grocery bags, involves the reaction $4\ NaCl + 2\ SO_2 + 2\ H_2O + O_2 \rightarrow 2\ Na_2SO_4 + 4\ HCl$. Calculate the number of grams of NaCl required to produce 5.00 kilograms of sodium sulfate.

4.12. 45.0 grams of zinc and 28.0 grams of sulfur are intimately mixed and caused to react until one is completely consumed. How many grams of zinc sulfide will be formed? How many grams of which element will remain unreacted?

4.13. A solution containing 1.46 grams of barium chloride is added to a solution containing 2.14 grams of sodium chromate, Na_2CrO_4. Find the number of grams of barium chromate that precipitate. Determine also which reactant was in excess, as well as the number of grams over the amount required by the limiting species.

4.24. How many grams of ammonia are required to produce 307 grams of NO?

4.25. If 7.05 grams of water result from the reaction, what will be the yield of NO?

4.26. The reaction of a dry cell may be represented by $Zn + 2\ NH_4Cl \rightarrow ZnCl_2 + 2\ NH_3 + H_2$. Calculate the number of grams of zinc consumed during the release of 9.32 grams of ammonia in such a cell.

4.27. The explosion of nitroglycerine is described by the equation $4\ C_3H_5(NO_3)_3 \rightarrow 12\ CO_2 + 10\ H_2O + 6\ N_2 + O_2$. How many grams of carbon dioxide are produced by the explosion of 69.7 grams of nitroglycerine?

4.28. Soaps are produced by the reaction of sodium hydroxide with naturally-occurring fats. The equation for one such reaction is $C_3H_5(C_{17}H_{35}COO)_3 + 3\ NaOH \rightarrow C_3H_5(OH)_3 + 3\ C_{17}H_{35}COONa$, the last compound being the soap. Calculate the number of grams of NaOH required to produce 165 grams of soap by this method.

4.29. One way of making sodium thiosulfate, the "hypo" in photographic developing, is described by the equation $Na_2CO_3 + 2\ Na_2S + 4\ SO_2 \rightarrow 3\ Na_2S_2O_3 + CO_2$. How many grams of sodium carbonate are required to produce 494 grams of sodium thiosulfate?

4.30. 4.62 grams of oxygen and 2.98 grams of carbon monoxide are placed in a closed reaction vessel and the mixture is ignited. Combustion occurs until one of the gases is totally consumed. Calculate the grams of carbon dioxide produced. Identify the reactant that is in excess, and determine the number of unreacted grams that remain.

4.31. 2.71 grams of sodium iodide are dissolved in water, and the solution is added to a second solution containing 4.47 grams of lead(II) nitrate. Find the number of grams of lead(II) iodide that will precipitate, which compound was in excess, and by how many grams.

Section 4.5

4.14. Copper is extracted from chalcocite, a copper(I) sulfide ore, by a reaction with oxygen that may be represented by the equation $Cu_2S + O_2 \rightarrow 2\ Cu + SO_2$. If treatment of 41.9 grams of pure Cu_2S by the process yields 29.2 grams of copper, calculate the percentage yield.

4.15. In an industrial process for making phosphorus, an intimate mixture of sand, charcoal and calcium phosphate is heated in an electric furnace. The reaction is essentially $Ca_3(PO_4)_2 + 3\ SiO_2 + 5\ C \rightarrow 5\ CO + 3\ CaSiO_3 + 2\ P$. The percentage yield of phosphorus is only 63%, however, some of the phosphorus being exhausted as P_2O_5. How many tons of calcium phosphate must be processed in order to yield one ton of phosphorus? (Hint: If you convert the ton figures to grams, use exponential notation. But then, is this really necessary. . . ?)

4.32. In the analysis of chloride ion in a commercial nickel plating solution, chloride is precipitated in a reaction that may be represented by the equation $2\ AgNO_3 + NiCl_2 \rightarrow 2\ AgCl + Ni(NO_3)_2$. In analyzing a solution known to contain 0.239 grams of $NiCl_2$, 0.525 grams of AgCl precipitate. Calculate the percentage yield. (Note: Use 130 g/mole as the MW of $NiCl_2$, and 144 g/mole as the MW of AgCl in solving this problem.)

4.33. In the Deacon process for manufacturing chlorine, a dry mixture of hydrogen chloride and air is passed over a heated catalyst. Oxidation occurs by the following reaction: $4\ HCl + O_2 \rightarrow 2\ Cl_2 + 2\ H_2O$. If the conversion is 58% complete, how many tons of chlorine can be recovered from 1.4 tons of HCl?

Section 4.6.

4.16. The discovery of oxygen occurred from the decomposition of mercury(II) oxide: $2\ HgO \rightarrow 2\ Hg + O_2$. What volume of oxygen would be produced by the reaction of 28.9 grams of the oxide, the gas being measured at STP?

4.17. How many liters of oxygen, measured at STP, are required for the complete combustion of 72.0 grams of heptane, C_7H_{16}, a component of gasoline? If air is 21% oxygen by volume, how many liters of air are necessary for this amount of fuel? (Note: CO_2 and H_2O are the only products.)

4.18. When nitric oxide, NO, is released in air it immediately combines with oxygen to produce nitrogen dioxide, NO_2. If 0.345 liter of NO, measured at STP, react, how many liters of NO_2, also measured at STP, will be formed?

4.34. Calculate the number of grams of HCl that must react with limestone, $CaCO_3$, in order to liberate 650 ml of CO_2, measured at STP. The equation is $2\ HCl + CaCO_3 \rightarrow CaCl_2 + CO_2 + H_2O$.

4.35. The conventional laboratory method for preparing oxygen is to heat potassium chlorate, $KClO_3$, in the presence of a catalyst. The oxygen is driven off, leaving a residue of potassium chloride. How many milliliters of the gas, measured at STP, will result from the decomposition of 2.06 grams of $KClO_3$?

4.36. Most of the destructive effect of the explosion of nitroglycerine is the result of the rapid production of large volumes of gaseous products (see Problem 4.27). Assuming the volumes of all gases are measured at the same temperature and pressure, how many liters of CO_2, $H_2O(g)$ and O_2 will accompany the formation of 2.40 liters of N_2?

CHAPTER 5

ENERGY IN CHEMICAL AND PHYSICAL CHANGES

The chemical equation, as it has been used to this point, does not tell the whole story of a chemical reaction. Each reaction also involves changes in thermodynamic properties. The most apparent of these properties to the student working with chemicals in the laboratory is energy in the form of heat. In this chapter we shall introduce some fundamentals of heat measurement; in the next chapter we shall look at the thermal characteristics of chemical reactions; and in Chapter 12 we shall identify the more sophisticated concepts of thermodynamics.

Chemical energy is usually measured in kilocalories, each of which is equal to 1000 calories. The calorie is variously defined, depending on the purpose and precision required of the calculated results. The currently accepted formal definition equates one calorie to 4.1840 joules. The joule is the energy unit used in the SI system. A functional definition of a calorie is the amount of heat required to raise the temperature of one gram of water one degree Celsius. This amount of heat is somewhat variable, depending on the particular temperature range considered, but within the limits of accuracy in general chemistry problems the variations are negligible.

Temperature is not a unit of heat or heat energy, a misconception frequently held by beginning students in chemistry. Temperature, as will be seen later, is a measure of the average kinetic energy of the particles composing a substance. It has been called a measure of the "degree of heat." There are two major scales of temperature, the Fahrenheit scale and the Celsius scale. (The Celsius scale was formerly known as the centigrade scale.) Though the Fahrenheit scale is commonly used in the United States, it is rarely used in scientific work and will not be employed in this text.*

---

*The Fahrenheit and Celsius temperature scales are related by the equation

$$°F - 32 = \frac{9}{5}\ °C.$$

To convert between the scales, substitute the given value and solve for the other.

5.1 SPECIFIC HEAT: MOLAR HEAT CAPACITY

In order to raise the temperature of a substance, you must heat it: heat must flow into the substance. If the substance is to be cooled, there must be heat flow from the substance to its surroundings. For a fixed temperature change, the heat flow required is proportional to the mass of the substance; for a fixed mass, the heat flow is proportional to the change in temperature. These proportionalities are expressed mathematically as

$$Q \propto m \qquad \text{and} \qquad Q \propto \Delta T,$$

where Q is heat flow in calories, m is mass in grams, and ΔT is the temperature change—final temperature minus initial temperature*—in degrees Celsius. Combining the proportionalities and introducing a proportionality constant, c, yields the equation

$$Q = m \times c \times \Delta T \tag{5.1}$$

The proportionality constant, c, is a property of a pure substance called **specific heat. Specific heat is the heat flow required to change the temperature of one gram of a substance one degree Celsius.** The units of specific heat may be inferred from the definition or derived from Equation 5.1. Solving the equation for c gives

$$\text{Specific heat} = c = \frac{\text{calories}}{\text{gram-degree}},$$

or calories per gram degree. Recalling the definition of the calorie—the amount of heat required to raise the temperature of one gram of water one degree—the specific heat of water is one calorie per gram degree.

Specific heats of substances other than water are listed in Table IV in the Appendix (page 289). As with water, most of these specific heats are not constant over all ranges of temperature. While differences are significant over a large temperature change, it will be assumed for purposes of this book that all specific heats given are average values over the temperature range involved. With this assumption, all sensible heat problems, as they are sometimes called, may be solved with Equation 5.1.

Significant comparisons between physical and chemical properties of many substances are possible when we consider their magnitudes for equal numbers of particles, or for equal numbers of moles of particles. One such property is known as **molar heat capacity, the amount of heat flow required to change the temperature of one mole of a substance one degree Celsius.** Its units are calories per mole degree. Heat flow may be calculated from the molar heat capacity by the equation

$$Q = n \times C \times \Delta T, \tag{5.2}$$

---

*The Greek symbol delta, Δ, is used to represent the change in some value. It is always calculated by *subtracting the initial value from the final value.* The difference may be a positive number or a negative number, depending on the relative magnitudes of the initial and final values.

where n is the number of moles and C is molar heat capacity.

Specific heat and molar heat capacity problems are usually straightforward plug-ins to Equations 5.1 and 5.2.

Example 5.1. How many calories are required to raise the temperature of 45.0 milliliters of water from 14.0°C to 48.0°C? (Recall the density of water: 1 gram/milliliter.)

5.1a. 1530 calories, or 1.53 kilocalories.

$$Q = 45.0\ \not{g} \times \frac{1.00\ \text{cal}}{\not{g}\text{-}\not{°C}} \times (48.0 - 14.0)\not{°C} = 1530\ \text{cal}\,.$$

Notice that, at one gram per milliliter, immediate conversion was made from milliliters of water to grams.

Example 5.2. The specific heat of ethanol is 0.561 calorie per gram-degree, and its specific gravity is 0.789. How much energy is required to warm 250 milliliters of ethanol from 22.0°C to 37.0°C?

In this problem the mass of ethanol is not stated, but it may be calculated from the data given. Set up this step, but do not solve.

5.2a. $250\ \not{ml} \times \frac{0.789\ \text{g}}{\not{ml}}$.

Recall that specific gravity is numerically equal to density in the metric system.

From this point the problem is identical to Example 5.1. Extend the setup and complete the problem.

5.2b. 1660 calories, or 1.66 kilocalories.

$$Q = 250\ \not{ml} \times \frac{0.789\ \not{g}}{1\ \not{ml}} \times \frac{0.561\ \text{cal}}{\not{g}\text{-}\not{°C}} \times (37.0 - 22.0)\not{°C} = 1660\ \text{cal}$$

As this problem illustrates, the conversion of volume to grams via density is sometimes required.

Example 5.3. How much heat energy is released as 81.5 grams of zinc cools from 134°C to 30°C if its average molar heat capacity over the range is 6.02 calories per mole-degree?

In this example you will be working with the molar heat capacity. You must convert, therefore, the given quantity of 81.5 grams of zinc to moles, and then proceed as before. Set up and solve.

5.3a. −780 calories.

$$Q = 81.5\ \cancel{g\ Zn} \times \frac{1\ \cancel{mole\ Zn}}{65.4\ \cancel{g\ Zn}} \times \frac{6.02\ cal}{\cancel{mole-°C}} \times (30\text{-}134)\cancel{°C} = -780 \text{ calories}$$

The negative sign indicates simply that heat is released rather than absorbed—that the final temperature is lower than the initial temperature.

5.2 LATENT HEAT

Latent heat is the heat required to change one gram of a substance from one state to another. It may also be considered as the heat required to change one mole of a substance from one state to another, in which case it is called the **molar latent heat.** Either term is frequently modified to describe specifically the change taking place with the word "latent" omitted. In melting, for example, it is called **heat of fusion** or **molar heat of fusion.** The heat flow for the reverse process, freezing, is the **heat of solidification** or **molar heat of solidification.** These latent heats are equal in magnitude, but opposite in sign. Corresponding terms for changes between the liquid and gaseous states are latent **heats of vaporization and condensation.** On occasion a substance will change from a solid directly to a gas, as with "dry ice." The heat involved here is the latent **heat of sublimation,** and it equals the sum of the heats of fusion and vaporization at the given temperature.

Units of latent heat are implied by the definition: calories per gram or calories per mole. Problems involving latent heat are solved by conversions based on the definition:

$$Q = \text{mass} \times \text{latent heat} = \text{grams} \times \frac{\text{calories}}{\text{gram}} = \text{calories} \qquad (5.3)$$

or

$$Q = \text{moles} \times \text{molar latent heat.} \qquad (5.4)$$

Example 5.4. How much heat is required to boil 76 grams of water if its latent heat of vaporization is 540 calories per gram?

Select the given quantity, set up, and solve.

5.4a. 41,000 calories, or 41 kilocalories.

$$Q = 76\ \cancel{g} \times \frac{540\ cal}{1\ \cancel{g}} = 41{,}000\ cal = 41\ kcal$$

Example 5.5. How many kilocalories of heat are required to melt 63 grams of sodium at its melting point if the molar heat of fusion is 620 calories per mole?

This time the problem involves molar heat of fusion. The quantity of sodium must therefore be converted to moles. The problem is otherwise like Example 5.4. Set up and solve.

5.5a. 1.7 kilocalories.

$$Q = 63 \cancel{\text{g Na}} \times \frac{1 \cancel{\text{mole Na}}}{23 \cancel{\text{g Na}}} \times \frac{0.62 \text{ kcal}}{1 \cancel{\text{mole Na}}} = 1.7 \text{ kcal}$$

This example asked specifically for kilocalories, so the molar heat of fusion was converted from 620 calories per mole to 0.62 kilocalorie per mole.

Example 5.6. How much heat must a refrigerator remove from 2.00 liters of water at the freezing point in order to convert it to ice? The heat of solidification of water is 80 calories per gram.

The given quantity of water must be converted to grams, and the problem then solved as before. Go all the way.

5.6a. 160,000 calories, or 160 kilocalories.

$$Q = 2.00 \cancel{\text{l}} \times \frac{1000 \cancel{\text{ml}}}{1 \cancel{\text{l}}} \times \frac{1.00 \cancel{\text{g}}}{1 \cancel{\text{ml}}} \times \frac{80 \text{ cal}}{1 \cancel{\text{g}}} = 160{,}000 \text{ cal}$$

This problem does not describe all that takes place when two liters of water are placed in a refrigerator. The water does not enter at the freezing point, nor does the ice remain at that temperature after the water has frozen. There are sensible heat changes that occur before freezing begins and after it is completed.

The whole story of temperature and energy for a pure substance from a solid below its freezing point to a vapor above its boiling point is shown in Figure 5.1. This curve, quantitatively for 1 gram of water, begins at point A, $-10°C$. The specific heat of ice is 0.49 calorie/gram–°C. Adding 4.9 calories to the ice will raise its temperature to the melting point (B):

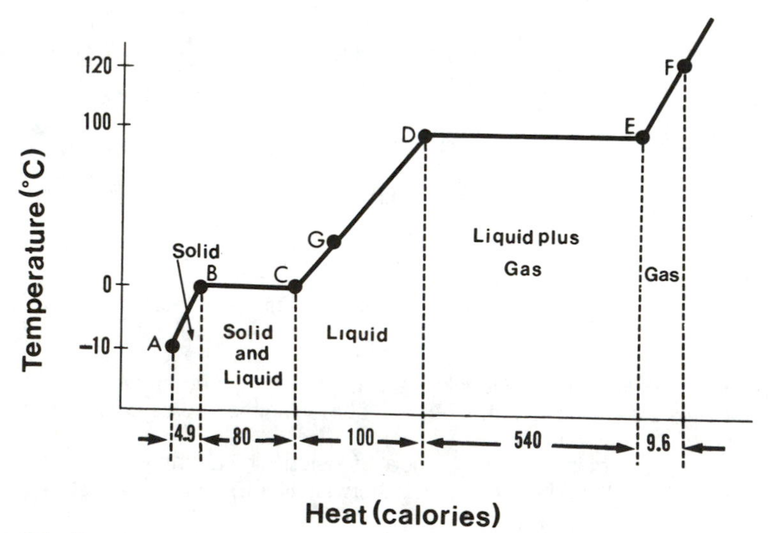

Figure 5.1 Temperature vs. energy absorbed as solid below its freezing point is heated to a gas above its boiling point.

$$1 \text{ gram} \times \frac{0.49 \text{ cal}}{\text{gram}-°\text{C}} \times [0-(-10)]°\text{C} = 4.9 \text{ calories.}$$

With additional heat the temperature remains constant, the added energy being used to melt the ice (C). 80 calories are required at a heat of fusion of 80 calories per gram. Once in the liquid state the continued addition of heat raises the temperature according to Equation 5.1, 100 calories being required to heat one gram of water from 0°C to the boiling point at 100°C(D). After that, the temperature again remains constant as the added energy boils the liquid to a gas (E), with a heat of vaporization of 540 calories per gram. Once the conversion is complete, additional energy raises the temperature according to the specific heat of steam, 0.48 calorie per gram-degree. The 9.6 calories in the illustration would boost the steam temperature to 120°C(F).

The total energy between any two points on a temperature-energy curve is the sum of the individual sensible and latent heats between those points. In the following example you may trace this process from what corresponds with point A on the graph to point F.

Example 5.7. A 120-gram piece of ice is removed from a refrigerator at −11°C, placed in a glass, allowed to melt, and finally warmed up to room temperature, 21°C. The specific heat of ice is 0.49 calorie per gram-degree. How much heat has the 120 grams of water absorbed from the atmosphere?

In solving this problem you must compute three heats: the sensible heat as the ice warms to the melting point; the latent heat required to melt the ice; and finally, the heat absorbed in warming the water to 21°C.

Calculate first the heat required to raise the temperature of ice from −11°C to 0°C.

5.7a. 650 calories.

$$Q = 120 \text{ g} \times \frac{0.49 \text{ cal}}{\text{g-°C}} \times 11°\text{C} = 647 \text{ cal}$$

Now find the heat required to melt the ice, using the latent heat given in Example 5.6. Recall that heat of fusion and heat of solidification are numerically equal.

5.7b. 9600 calories = 9.6×10^3 calories.

$$Q = 120 \text{ g} \times \frac{80 \text{ calories}}{\text{g}} = 9600 \text{ cal} = 9.6 \times 10^3 \text{ cal}$$

Finally, calculate the sensible heat as the water is warmed from 0°C to 21°C.

5.7c. 2520 calories = 2.5×10^3 calories.

$$Q = 120 \text{ g} \times \frac{1.0 \text{ cal}}{\text{g-°C}} \times 21°\text{C} = 2520 \text{ cal} = 2.5 \times 10^3 \text{ cal}$$

The total heat absorbed by the 120 grams of water is the sum of the three individual steps. Find the total.

5.7d. 12,800 calories, or 12.8 kilocalories.

$$Q = 647 + 9600 + 2520 = 12{,}800$$

Since Q in 5.7b and 5.7c are expressed only to the nearest 100 calories, the hundreds column is the smallest which should be kept in the summation. The answer is therefore rounded off to 12,800 calories, or 12.8 kilocalories.

5.3 CALORIMETRY

The heat effects of physical and chemical changes are studied in a device known as a calorimeter. A calorimeter is ideally a thermally isolated segment of the universe from which no heat can be lost and to which no heat can be added. It is further governed by the law of conservation of energy, which provides that energy may neither be created nor destroyed in any ordinary (non-nuclear) phenomenon. Therefore any energy lost by one part of the calorimeter or its contents must be gained by another. The net overall energy change within the calorimeter system must consequently be zero.

Consider a simple experiment in calorimetry. If a piece of hot metal is dropped into an insulated vessel—a calorimeter—containing water, the metal will be cooled by the water and the water will be warmed by the metal. In the ideal calorimeter, where there is no heat exchange with the surroundings, the heat lost by the hot metal will be exactly equal to the heat gained by the water, with a small amount of energy going to the calorimeter itself. The net change will be zero. Expressed as an equation,

$$Q_w + Q_c + Q_m = 0, \qquad (5.5)$$

where Q represents the heat lost or gained by the water (Q_w), the calorimeter (Q_c), and the metal (Q_m). All Q values are sensible heat changes, assuming none of the water is vaporized, and are therefore found by Equation 5.1.

Determining Q_c is somewhat complex. The calorimeter consists of the vessel itself, a thermometer, and perhaps some sort of stirring device. Together these have a constant "heat capacity" which is determined by the masses and specific heats of the components. The sum total of these is called the calorimeter constant, K_c. The calorimeter constant is the number of calories absorbed by the calorimeter per degree change in the temperature of its contents. Its units are calories per degree. Consequently

$$Q_c = K_c \times \Delta T = \frac{\text{calories}}{\cancel{\text{degrees}}} \times \cancel{\text{degrees}} = \text{calories}\,. \qquad (5.6)$$

Substituting into Equation 5.3,

$$Q_w + K_c\Delta T + Q_m = 0\,. \qquad (5.7)$$

Determination of K_c and calorimeter calculation techniques are illustrated in the following example:

Example 5.8. 220 grams of bismuth initially at 98.0°C are dropped into 100 grams of water in a calorimeter. Both water and calorimeter are initially at 20.00°C. The entire system reaches temperature equilibrium at 24.50°C. The average specific heat for bismuth over the temperature range is 0.0300 calorie per gram-degree. Calculate K_c.

As a first step, set up and calculate Q_w, using Equation 5.1.

5.8a. Q_w = 450 calories.

$$Q_w = \text{mass} \times \text{specific heat} \times \Delta T$$

$$= 100\ \text{g} \times \frac{1.00\ \text{cal}}{\text{g-°C}} \times (24.50 - 20.00)\text{°C} = 450\ \text{cal}$$

Now do the same for the bismuth.

5.8b. Q_m = −485 calories.

$$Q_m = \text{mass} \times \text{specific heat} \times \Delta T$$

$$= 220\ \text{g} \times \frac{0.0300\ \text{cal}}{\text{g-°C}} \times (24.50 - 98.0)\text{°C} = -485\ \text{cal}$$

Note the negative sign. The sign of ΔT must be accounted for in calorimeter problems.

The middle term remains. What is ΔT for the calorimeter's contents?

5.8c. 4.50°C. ΔT for the water is the "degree change in the temperature of [the calorimeter's] contents."

Assuming there is no heat exchange to or from the surroundings, we may now substitute the known values of Q_w, Q_m, and ΔT into Equation 5.7 and solve for K_c.

5.8d. K_c = 7.8 calories per degree.

$$450\ \text{cal} + 4.50\ K_c - 485\ \text{cal} = 0; \qquad K_c = 35\ \text{cal}/4.50\text{°C} = 7.8\ \text{cal/°C}$$

Once K_c is known, the calorimeter may be used for other experiments. . . .

Example 5.9. 95.0 grams of water are placed in a calorimeter having a K_c of 13.3 calories per degree. The calorimeter and water reach temperature equilibrium at 17.2°C. A piece of unknown metal at 93.2°C, weighing 72.4 grams, is placed into the calorimeter. The entire system reaches temperature equilibrium at 23.0°C. Calculate the specific heat of the metal.

The solution of this problem is essentially the same as the example just completed, using Equation 5.5 or 5.7. Calculate Q_w, as in Example 5.8.

5.9a. Q_w = 551 calories.

$$Q_w = \text{mass} \times \text{specific heat} \times \Delta T$$

$$= 95.0\ \text{g} \times \frac{1.00\ \text{cal}}{\text{g-°C}} \times (23.0 - 17.2)\text{°C} = 551\ \text{cal}$$

Now find the heat absorbed by the calorimeter—the second term in Equation 5.5 or 5.7.

5.9b. Q_{cal} = 77.1 calories.

$$Q_{cal} = K_c \Delta T$$

$$= \frac{13.3\ \text{cal}}{\text{°C}} \times (23.0 - 17.2)\text{°C} = 77.1\ \text{cal}$$

The first two terms of Equation 5.5 are known; the third term, Q_m, may be replaced by its equivalent from Equation 5.1, mass × specific heat × ΔT:

$$551 + 77 + \text{mass} \times \text{specific heat} \times \Delta T = 0$$

Substitute known values and complete the problem to the specific heat of the metal. Watch your signs.

5.9c. Specific heat = 0.124 calorie per gram-degree.

$$551\ \text{cal} + 77\ \text{cal} + (72.4\ \text{g})(\text{specific heat})(23.0 - 93.2)\text{°C} = 0$$

$$\text{specific heat} = \frac{-628\ \text{cal}}{72.4\ \text{g} \times (-70.2\text{°C})} = 0.124\ \text{cal/g-°C}$$

When chemical reactions occur in a calorimeter the heat flow of the reaction, designated ΔH, may be found. The calorimeter equation becomes

$$Q_s + Q_c + \Delta H = 0, \qquad (5.8)$$

where Q_s refers to the heat absorbed by the material, usually water or an aqueous solution, in the calorimeter. The ΔH found directly is, of course, for the *quantities of chemical used*. If you wish to find the ΔH for a specific amount of reactant, the ΔH values of the experiment must be corrected to that amount—usually the number of moles in the equation.

Example 5.10. 50.0 milliliters of 7.50% NaOH (specific gravity 1.08) and 25.0 milliliters of 14.0% HCl (specific gravity 1.07) react to form a water solution of NaCl in a calorimeter having a K_c of 7.4 calories per degree. The initial temperature is 19.4°C. As a result of the reaction the temperature rises to 35.2°C. The specific heat of the NaCl solution resulting from the reaction is 0.928 calorie per gram-degree. Find (a) the ΔH for the reacting quantities, and (b) the ΔH per mole of water produced by the reaction. The equation is

$$NaOH + HCl \rightarrow NaCl + HOH$$

In analyzing the problem, we note that two reactants are combined. They react and release heat. The heat raises the temperature of the resulting solution and the calorimeter. Our objective is to determine how much heat is absorbed by the solution and the calorimeter, Q_s and Q_c in Equation 5.8. Both are sensible heat changes, found by Equation 5.1.

First, let's look at the sodium chloride solution produced by the reaction. What is its mass? It is formed by the reaction of two solutions, 50.0 milliliters of a solution of specific gravity 1.08, and 25.0 milliliters of a solution of specific gravity 1.07. The mass of the combined solutions is obviously the sum of the individual masses. Compute this.

5.10a. Mass of solution = 80.8 grams.

$$50.0\ \cancel{ml} \times \frac{1.08\ g}{\cancel{ml}} + 25.0\ \cancel{ml} \times \frac{1.07\ g}{\cancel{ml}} = 80.8\ \text{grams}$$

The specific heat of the solution and its initial and final temperatures are given. With Equation 5.1 you can now find the heat absorbed by the solution, Q_s.

5.10b. $Q_s = 1.18 \times 10^3$ calories.

$$Q_s = 80.7\ \cancel{g} \times \frac{0.928\ cal}{\cancel{g}\text{-}\cancel{°C}} \times (35.2 - 19.4)\cancel{°C} = 1.18 \times 10^3\ cal$$

The heat absorbed by the calorimeter is found as in the previous example.

5.10c. $Q_c = 117$ calories.

$$Q_c = K_c \Delta T = \frac{7.4\ cal}{\cancel{°C}} \times (35.2 - 19.4)\cancel{°C} = 117\ cal$$

ΔH now follows from Equation 5.8.

5.10d. $\Delta H = -1.30 \times 10^3$ calories.
Solving Equation 5.8 for ΔH,

$$\Delta H = -(Q_s + Q_c) = -\ (1180 + 117) = -1.30 \times 10^3\ cal\ .$$

This is the answer to part (a). The negative sign indicates that heat is released in the reaction. It is the number of calories for the reacting quantities. Part (b) calls for the calories per mole of water produced. Therefore, it is necessary to find the number of moles of water produced in the reaction, in which *one* mole of NaOH reacts with *one* mole of HCl to form *one* mole of H_2O and *one* mole of NaCl. You can find the number of moles of HCl and NaOH present. By the principles of excess stoichiometry, the lesser of these will equal the number of moles of water. Let us, then, calculate first the number of moles of HCl and then the number of moles of NaOH.

How many grams are in 25 milliliters of HCl solution, if specific gravity = 1.07? You have already answered this question in the first step of the problem. This time, however, write the setup only—do not solve.

5.10e. $25.0\ \cancel{\text{ml solution}} \times \dfrac{1.07\ \text{g solution}}{1\ \cancel{\text{ml solution}}}$.

Now we must extend the setup to grams of HCl, knowing that it is a 14% solution. The customary way to find 14% of a quantity is to convert to the decimal equivalent, 0.14, and multiply. A thought process more compatible with a dimensional analysis method of solving problems is to realize that **percent means parts per hundred.** Therefore, a 14% HCl solution means there are 14 grams of HCl in 100 grams of solution—or 14 grams HCl = 100 grams of solution. If you develop a unit conversion factor from this equivalence and use it to extend your problem to grams of HCl, you will find that it is mathematically the same as multiplying by 0.14. Try it.

5.10f. $25.0\ \cancel{\text{ml solution}} \times \dfrac{1.07\ \cancel{\text{g solution}}}{1\ \cancel{\text{ml solution}}} \times \dfrac{14.0\ \text{g HCl}}{100\ \cancel{\text{g solution}}}$

The parts per 100 approach is a simple and effective way of solving percentage problems with dimensional analysis. It's worth adding to your kit of mathematical tools.

Conversion of grams of HCl to moles completes the setup. Carry through to the answer this time.

5.10g. 0.103 mole HCl.

$$25.0\ \cancel{\text{ml solution}} \times \frac{1.07\ \cancel{\text{g solution}}}{1\ \cancel{\text{ml solution}}} \times \frac{140\ \cancel{\text{g HCl}}}{100\ \cancel{\text{g solution}}} \times \frac{1\ \text{mole HCl}}{36.5\ \cancel{\text{g HCl}}} = 0.103\ \text{mole HCl}$$

Similarly, find the moles of NaOH, and determine how many moles of water will be formed.

5.10h. 0.101 mole NaOH = 0.101 mole H_2O.

$$50.0\ \cancel{\text{ml solution}} \times \frac{1.08\ \cancel{\text{g solution}}}{1\ \cancel{\text{ml solution}}} \times \frac{7.50\ \cancel{\text{g NaOH}}}{100\ \cancel{\text{g solution}}} \times \frac{1\ \text{mole NaOH}}{40.0\ \cancel{\text{g NaOH}}} = 0.101\ \text{mole NaOH}$$

The NaOH is the limiting species (0.101 mole NaOH $<$ 0.103 mole HCl), and therefore establishes the number of moles of water formed.

ΔH was -1.30×10^3 calories for 0.101 of water. What is ΔH in kilocalories per mole of water?

5.10i. $\Delta H = -12.9$ kilocalories per mole.

$$\frac{-1.30\ \text{kcal}}{0.101\ \text{mole}} = -12.9\ \text{kcal/mole}$$

The heat of solution of a compound, ΔH_s, measured in calories per mole or kilocalories per mole, may also be determined by calorimeter experiments.

Example 5.11. 150 milliliters of warm water are placed into a calorimeter and the two equilibrate at 33.3°C. 34.0 grams of potassium nitrate are dissolved in the water, and temperature equilibrium is reached at 15.1°C. If the calorimeter constant is 2.4 calories per degree, and the specific heat of the KNO_3 solution produced is 0.832 calorie per gram-degree, find ΔH_s for potassium nitrate.

To find Q_s, the first term in Equation 5.8, it is necessary to know the mass of the solution, its specific heat, and ΔT. The specific heat is given. What about ΔT? Careful!

5.11a. $\Delta T = -18.2°C$.

$$\Delta T = T_{final} - T_{initial} = 15.1 - 33.3 = -18.2$$

Now what is the mass of the solution? Check the problem for the necessary data.

5.11b. 184 grams of solution.

$$150 \text{ ml water} = 150 \text{ g water}$$

$$150 \text{ g water} + 34.0 \text{ g } KNO_3 = 184 \text{ g solution}$$

You now have the three factors that make up Q_s. Q_c is equal to $K_c \times \Delta T_c$, as in previous examples. Substituting all known values into Equation 5.8 leaves ΔH as the only unknown. Solve for ΔH.

5.11c. $\Delta H = +2830$ calories $= 2.83$ kilocalories.

$$184 \text{ g} \times \frac{0.832 \text{ cal}}{\text{g-°C}} \times (-18.2°C) + \frac{2.4 \text{ cal}}{°C} \times (-18.2°C) + \Delta H = 0$$

$$\Delta H = +2830 \text{ cal}$$

We see that 2830 calories are absorbed when 34.0 grams of potassium nitrate dissolve. This could be expressed as 2830 calories/34.0 grams. But the problem asks for calories per *mole* of potassium nitrate. Conversion of grams to moles is a familiar operation. Set up and solve.

5.11d. 8410 calories per mole, or 8.41 kilocalories per mole.

$$\Delta H_s = \frac{2830 \text{ cal}}{34.0 \text{ g}} \times \frac{101 \text{ g}}{1 \text{ mole}} = 8410 \text{ cal/mole}$$

PROBLEMS

Section 5.1.

5.1. How much heat is required to raise the temperature of 204 grams of lead from 22.8°C to 64.9°C?

5.2. To what temperature will one pound of nickel be raised, if, beginning at 25.0°C, it absorbs 750 calories of heat?

5.3. Calculate the number of kilocalories of heat lost to the atmosphere per kilogram of CO_2 leaving a smokestack if the stack gases are at 252°C and the surrounding temperature is 19°C. The molar heat capacity of carbon dioxide is 8.87 calories per mole–degree.

5.10. Find the number of calories released as 186 grams of sodium cool from 68°C to 31°C.

5.11. One dollar's worth of pennies weighs 318 grams. If 100 pennies at an outside temperature of 23°C lose 351 calories when they are tossed into a fountain and drop to the fountain's water temperature, what is that temperature?

5.12. If 15.0 grams of pure alcohol are cooled from 21°C to 3°C, how much heat has been removed? The molar heat capacity of alcohol is 27.3 calories per mole–degree.

Section 5.2.

5.4. Table salt, sodium chloride, must be heated to 801°C before it will begin to melt. How many calories are necessary to change one tablespoon of salt—17 grams—from a solid at the melting point to a liquid?

5.5. One of the uses of ammonia, NH_3, is as a refrigerant. When liquid ammonia is allowed to vaporize at −20°F, its molar heat of vaporization at that temperature is 5.51 kcal per mole. Calculate the kilocalories of heat absorbed by the vaporization of 250 grams of ammonia under these conditions.

5.6. If an ice tray in your refrigerator holds 265 grams of water, calculate the number of calories that must be removed from a tray full of water when reducing its temperature from 17° to 0°C, freezing it, and then reducing the temperature of the ice to −8°C. Disregard the heat removed from the tray itself.

5.13. Not only does lead melt at a relatively low temperature, but it also requires a relatively small amount of heat to melt it. To compare with salt (Problem 5.4), for example, calculate the number of calories required to melt 17 grams of lead.

5.14. The coils behind household refrigerators are usually warm. This is because the heat released by the refrigerant, Freon-12 (CCl_2F_2), as it changes from a gas to a liquid is dissipated to the atmosphere through these coils. How many kcal of heat are lost to the surroundings in this way if 725 grams of Freon-12 condense when the molar heat of condensation is 40.4 calories per gram.

5.15. Calculate the energy required to warm 75.0 grams of ice at −24°C to the melting point and melt it, then heat it to the boiling point and boil it, and then superheat the steam to 130°C.

Section 5.3.

5.7. A calorimeter contains 120 grams of water at a temperature of 20.4°C. 110 grams of water at 39.3°C are poured in. The contents are

5.16. A student determines the calorimeter constant of a Styrofoam coffee cup calorimeter by pouring 50.0 grams of water at 23.5°C

stirred, and the system reaches temperature equilibrium at 29.1°C. Calculate the calorimeter constant.

5.8. A calorimeter similar to the one in Problem 5.7 is found to have a calorimeter constant of 5.2 calories per degree. This calorimeter has 72.0 grams of water poured into it, and they reach the same temperature at 19.2°C. A piece of tin weighing 140 grams is heated to a temperature of 89.0°C and dropped into the water. The system reaches temperature equilibrium at 25.5°C. Compute the specific heat of tin in calories per gram.

5.9. The experiment described in Example 5.10 is repeated with acetic acid instead of hydrochloric acid, using the same calorimeter (K_c = 7.4 calories per degree). 100 milliliters of 6.00% acetic acid, $HC_2H_3O_2$ (specific gravity 1.01) and 100 milliliters of 3.85% NaOH (specific gravity 1.03) are combined with a resulting temperature increase of 5.7°C for the calorimeter and its contents. If the specific heat of the solution formed is 0.960 calorie per gram-degree, find ΔH per mole of water formed by the reaction, which may be written $HC_2H_3O_2 + NaOH \rightarrow NaC_2H_3O_2 + HOH$.

into it. He then drops 44.8 grams of copper at 95.2°C into the cup. The entire system eventually comes to a temperature of 28.6°C. Calculate K_c in calories/°C.

5.17. 100 grams of water are placed into a calorimeter (K_c = 8.2 calories/°C), and they reach a temperature of 21.3°C. Then, 4.34 grams of NH_4Cl are dissolved in the water, during which the temperature drops to 18.5°C. The specific heat of the solution is 0.952 calorie per gram-degree. Compute the heat of solution, ΔH_s, of ammonium chloride in kilocalories per mole.

5.18. One gram of benzoic acid is burned in a bomb-type calorimeter containing 2.30 kilograms of water. The original temperature of the container and water is 25.00°C; their final temperature is 27.50°C. The heat of combustion of benzoic acid is −6.315 kilocalories per gram. Calculate (a) the calorimeter constant in calories per degree, and (b) the heat of combustion, in kilocalories per gram, of a certain hydrocarbon, given that 1.08 grams of this compound, when burned in the same calorimeter containing 2540 grams of water, brings about a temperature increase of 4.28°C.

CHAPTER 6

ENTHALPY CHANGES IN CHEMICAL REACTIONS

6.1 THERMOCHEMICAL EQUATIONS

Almost without exception, chemical reactions are accompanied by the release or consumption of energy. Though the energy may be in one or more of several different forms, it most frequently appears as heat, in which form it may be measured in calorimeter experiments as described in the preceding chapter. A substance at a given temperature and pressure may be said to have a certain "heat content," or **enthalpy.** While enthalpy cannot be measured directly or on an absolute scale, the heat released or absorbed in a reaction, which is the difference in enthalpy between the products and the reactants, can be measured. This enthalpy change is given the symbol ΔH, which is defined as the enthalpy of the products minus the enthalpy of the reactants.

In a reaction in which heat is *absorbed*, the heat content of the products is greater than that of the reactants. Consequently the sign of ΔH is positive. This type of reaction is referred to as an *endothermic* reaction. More commonly, heat is *released* in a chemical reaction. The heat of the products being less than that of the reactants, ΔH has a negative sign. A reaction of this type is said to be *exothermic*.

The amount of heat absorbed or given off in a reaction is sometimes made a part of the chemical equation, which is then called a *thermochemical* equation. For an exothermic reaction the energy term appears on the right, as if it were a product of the reaction. With endothermic reactions the amount of heat absorbed is written on the left side of the equation.

As an example, consider the decomposition of water into its elements. Electrical energy is introduced to make this endothermic reaction occur. The thermochemical equation is

$$2\ H_2O(l) + 137\ \text{kcal} \rightarrow 2\ H_2(g) + O_2(g). \qquad (6.1)$$

Quantitative consideration of the energy change in a chemical reaction requires that the state—gas, liquid, or solid—of the reactants and products be specified. State is indicated by a symbol in parentheses after each formula in an equation. For example, in Equation 6.1, (l) indicates the liquid state and (g) indicates a gas. Other symbols are (s) for solid and (aq) for aqueous solution. In the table of heats of formation which you will use shortly, values are tabulated for the natural state of each substance at 25°C and 1 atmosphere of pressure. Water appears twice in the table, once as a liquid and once as a gas, with a different heat of formation for each state. This introduces a possible ambiguity that is avoided by using state symbols.

Different chemists handle this matter in different ways, and your procedure should conform to that required in your chemistry class. In this text the following practice has been adopted: If the state of the species in a chemical equation is *not* indicated, it is understood that the substance is in its normal state at 25°C and 1 atmosphere pressure. If, for any reason, the state of any substance in an equation is indicated, the states of all substances will be shown. State will always be indicated in an equation involving water because of its dual appearance in the heat of formation table.

There is a second, more common, form of thermochemical equation—one in which the heat term, expressed as an enthalpy change, ΔH, is written to the side of the equation. Equation 6.1, written in that form, is

$$2\ H_2O\ (l) \rightarrow 2\ H_2(g) + O_2(g) \qquad \Delta H = 137\ \text{kcal} \tag{6.2}$$

Note that ΔH is a positive quantity for an endothermic reaction; the enthalpy of 2 moles of $H_2(g)$ and 1 mole of $O_2(g)$ is 137 kcal greater than that of 2 moles of liquid water.

Examining the above reaction in reverse, hydrogen and oxygen combine explosively to form water via an exothermic reaction. The thermochemical equation may be written as

$$2\ H_2(g) + O_2(g) \rightarrow 2\ H_2O\ (l) + 137\ \text{kcal} \tag{6.3}$$

or, more commonly, as

$$2\ H_2(g) + O_2(g) \rightarrow 2\ H_2O\ (l) \qquad \Delta H = -\ 137\ \text{kcal.} \tag{6.4}$$

Note that ΔH is a negative quantity for an exothermic reaction.

A comparison of equations 6.2 and 6.4 shows the ΔH values to be numerically equal but opposite in sign. This is to be expected. The numerical difference between the heat content of the products and reactants is fixed by the chemicals. The sign of ΔH is fixed by which chemicals are reactants and which are products, Δ always representing *final* value minus *initial* value. An algebraic parallel would be: If $A - B = X$, then $B - A = -X$.

In comparing the two ways of writing thermochemical equations, notice that the sign of ΔH always corresponds with the sign of the heat term as it appears on the *left* side of a thermochemical equation. For Equation 6.1, the heat term on the left side of the equation is +137 kcal, and ΔH is +137 kcal. If 137 is subtracted from each side of Equation 6.3, the result is

$$2\ H_2(g) + O_2(g) - 137\text{ kcal} \rightarrow 2\ H_2O\ (l)\ . \qquad (6.5)$$

ΔH for Equation 6.5 is −137 kcal, as shown in Equation 6.4.

6.2 THERMOCHEMICAL STOICHIOMETRY

Equation 6.1 indicates that it requires 137 kcal of energy to decompose two moles of water molecules. How much energy is required to decompose one mole of water molecules? Logically, half as much. The quantity of energy is proportional to the quantities of reactants or products. This is shown by extending the molar equivalencies of a chemical equation to include the energy term. Thus, for Equation 6.1,

$$2\text{ moles } H_2O = 137\text{ kcal} = 2\text{ moles } H_2 = 1\text{ mole } O_2\ . \qquad (6.6)$$

The dimensional analysis setup to answer the above question is

$$1\ \cancel{\text{mole } H_2O} \times \frac{137\text{ kcal}}{2\ \cancel{\text{moles } H_2O}} = 68.5\text{ kcal}\ .$$

The solution of thermochemical stoichiometry problems is the same in principle as in all stoichiometry problems, following the pattern given on page 47. It differs only in that equivalences such as those in Equation 6.6 permit direct conversion of any given energy quantity to moles of the wanted species, or from moles of a given species to energy. This modification in the stoichiometric pattern becomes apparent in the following examples:

Example 6.1. How much heat will be released in burning 78.4 grams of C_6H_{14} by the reaction

$$2\ C_6H_{14}(l) + 19\ O_2\ (g) \rightarrow 12\ CO_2\ (g) + 14\ H_2O\ (l) + 2005\text{ kcal?}$$

Your procedure is to convert the given quantity of C_6H_{14} to moles, and then moles of C_6H_{14} to kcal. Set up, but do not solve, the conversion of 78.4 grams of C_6H_{14} to moles—the first step of the stoichiometry pattern on page 47.

6.1a. $78.4\ \cancel{g\ C_6H_{14}} \times \dfrac{1\text{ mole } C_6H_{14}}{86.0\ \cancel{g\ C_6H_{14}}}$.

Now extend the setup to convert moles of C_6H_{14} to kcal, and solve to a numerical answer.

6.1b. 914 kcal.

$$78.4\ \cancel{g\ C_6H_{14}} \times \frac{1\ \cancel{\text{mole } C_6H_{14}}}{86.0\ \cancel{g\ C_6H_{14}}} \times \frac{2005\text{ kcal}}{2\ \cancel{\text{mole } C_6H_{14}}} = 914\text{ kcal}$$

Example 6.2. How many grams of magnesium must be burned to release 265 kcal by the reaction 2 Mg (s) + O_2(g) $\rightarrow$ 2 MgO (s) + 288 kcal?

This time energy is the given quantity. Convert to grams of magnesium in two steps and solve for the answer.

6.2a. 44.7 grams Mg.

$$265 \text{ kcal} \times \frac{2 \text{ moles Mg}}{288 \text{ kcal}} \times \frac{24.3 \text{ g Mg}}{1 \text{ mole Mg}} = 44.7 \text{ g Mg}$$

6.3 HESS' LAW OF ENTHALPY SUMMATION

The enthalpy changes in many thermochemical equations have been measured experimentally. There are, however, numerous reactions for which it is very difficult, and perhaps impossible, to determine ΔH directly. Yet these enthalpy changes are known. Enthalpy is a thermodynamic property of every substance. At a given temperature and pressure, it is the same regardless of how the substance was produced. If, therefore, a series of reactions having known ΔH values can be arranged in such a way that, when added up, they produce a desired reaction, the ΔH value for the desired reaction will be equal to the sum of the ΔH values for the individual reactions.

The determination of the **heat of formation** of carbon monoxide illustrates both the problem and the process. **Heat of formation,** designated ΔH_f, **is the ΔH for a chemical reaction in which one mole of a compound is produced from its elements, both product and reactants being in their natural state at 25°C and 1 atmosphere of pressure.** Thus the heat of formation for carbon monoxide is the ΔH for the following reaction:

$$C(s) + \tfrac{1}{2}\, O_2(g) \rightarrow CO(g). \qquad (6.7)$$

Again, the letters in parentheses indicate the state, solid for carbon and gaseous for oxygen and carbon monoxide, that is natural to each species at the temperature and pressure indicated. *The elements from which the compound is formed have a heat of formation of zero in their natural state at 25°C and 1 atmosphere pressure.* Heats of formation and the other thermodynamic values to be used in Chapter 12 are listed in Appendix III, page 285.

It is difficult to burn carbon in the stoichiometrically required quantity of oxygen for carbon monoxide without leaving some carbon unburned and burning some carbon to carbon dioxide. Because of these problems, the direct determination of ΔH_f for CO is not readily accomplished in the laboratory. On the other hand, one can readily measure the heats of combustion of both carbon and carbon monoxide. **Heat of combustion, ΔH_c, is the ΔH when one mole of a substance is burned in an excess of oxygen.** If the substance contains only carbon and/or hydrogen, and possibly oxygen, the products are CO_2(g) and $H_2O(l)$. The heats of combustion of C(s) and CO(g) are, respectively, -94.05 and -67.63 kcal per mole. The thermochemical equations are

$$C(s) + O_2(g) \rightarrow CO_2(g) \qquad \Delta H = -94.05 \text{ kcal} \qquad (6.8)$$

$$CO(g) + \tfrac{1}{2}O_2(g) \rightarrow CO_2(g) \qquad \Delta H = -67.63 \text{ kcal} \qquad (6.9)$$

The desired equation can be found by reversing Equation 6.9 and adding it to Equation 6.8.

	ΔH (kcal)	
$CO_2(g) \rightarrow CO(g) + \tfrac{1}{2}O_2(g)$	+67.63	(6.10)
$C(s) + O_2(g) \rightarrow CO_2(g)$	−94.05	(6.8)
$CO_2(g) + C(s) + O_2(g) \rightarrow CO(g) + \tfrac{1}{2}O_2(g) + CO_2(g)$	−26.42	(6.11)

If one CO_2 and one-half O_2 are subtracted from both sides of Equation 6.11, the thermochemical equation for the heat of formation of CO results:

$$C(s) + \tfrac{1}{2}O_2(g) \rightarrow CO(g) \qquad \Delta H = -26.42 \text{ kcal.} \qquad (6.12)$$

Thus ΔH_f for CO has been determined by algebraic manipulation of thermochemical equations for which ΔH values are known.

There is a shorter solution to this problem. From the enthalpy summation principle, the general equation

$$\Delta H = \Sigma\, \Delta H_f(\text{products}) - \Sigma\, \Delta H_f(\text{reactants}) \qquad (6.13)$$

for any reaction may be derived. In this equation the enthalpy change is the ΔH for the reaction equation as written; $\Sigma\Delta H_f$(products) is the sum of the heats of formation of all product species, each multiplied by the number of moles in the reaction equation; and $\Sigma\Delta H_f$(reactants) is the corresponding summation for the reacting species.

Equation 6.13 may be applied to the combustion of CO to find the ΔH_f of CO:

$$CO(g) + \tfrac{1}{2}O_2(g) \rightarrow CO_2(g) \qquad \Delta H = -67.63 \text{ kcal} \qquad (6.9)$$

$$1\,(\Delta H_f) \quad \tfrac{1}{2}(0) \quad 1(-94.1)$$

Beneath O_2 and CO_2 are their ΔH_f values in kcal/mole, multiplied by their coefficients in the equation. The symbol, ΔH_f, is used for CO, also multiplied by its coefficient in the equation. Substituting into Equation 6.13 and solving for ΔH_f,

$$\Delta H = \Sigma\Delta H_f(\text{products}) - \Sigma\Delta H_f(\text{reactants})$$

$$-67.63 = 1(-94.05) - [1(\Delta H_{f\,CO}) + \tfrac{1}{2}(0)]$$

$$\Delta H_{f\,CO} = -94.05 + 67.63 = 26.42 \text{ kcal/mole}$$

Use Equation 6.13 in solving the following problems.

Example 6.3. If the heat of combustion of hexene, $C_6H_{12}(l)$, is −952.6 kcal/mole, find its heat of formation.

Begin by writing the heat of combustion equation. In addition to writing the combustion equation, write the appropriate ΔH_f values, multiplied by the number of moles in the equation, beneath each chemical formula.

6.3a. $C_6H_{12}(l) + 9\ O_2(g) \rightarrow 6\ CO_2(g) + 6\ H_2O(l)$

$1(\Delta H_f) \quad 9(0) \quad 6(-94.05) \quad 6(-68.32)$

Now substitute into Equation 6.13 and solve for ΔH_f of C_6H_{12}.

6.3b. $\Delta H_f = -21.6$ kcal/mole.

$$\Delta H = \Sigma \Delta H_f(\text{products}) - \Sigma \Delta H_f(\text{reactants})$$

$$-952.6 = 6(-94.05) + 6(-68.32) - \Delta H_f$$

$$\Delta H_f = -564.30 - 409.92 + 952.6 = -21.6 \text{ kcal/mole}$$

The use of Equation 6.13 to find ΔH of a reaction is similar, and even a bit easier algebraically:

Example 6.4. Find ΔH for the reaction $3\ PbO(s) + \frac{1}{2}\ O_2(g) \rightarrow Pb_3O_4(s)$.

Reproduce the equation, with appropriate ΔH_f values from the heat of formation table.

6.4a. $3\ PbO(s) + \frac{1}{2}\ O_2(g) \rightarrow Pb_3O_4(s)$

$3(-52.40) \quad \frac{1}{2}(0) \quad 1(-175.6)$

Now substitute into Equation 6.13 and solve.

6.4b. $\Delta H = -18.4$ kcal.

$$\Delta H = \Sigma \Delta H_f(\text{products}) - \Sigma \Delta H_f(\text{reactants})$$

$$= -175.6 - 3(-52.40) = -175.6 + 157.20 = -18.4 \text{ kcal}$$

As a final example, a bit more complicated. . . .

Example 6.5. Find ΔH for the combustion of n-hexane, C_6H_{14}, by the equation $2\ C_6H_{14}(l) + 19\ O_2(g) \rightarrow 12\ CO_2(g) + 14\ H_2O(l)$. $\Delta H_f = -40.0$ kcal/mole for n-hexane.

Note that this problem does not ask for the heat of combustion of n-hexane, which would be ΔH for *one* mole, but rather for the equation as written, which has two moles of n-hexane. Reproduce the equation with appropriate ΔH_f values, as before.

6.5a. $2\ C_6H_{14}(l) + 19\ O_2(g) \rightarrow 12\ CO_2(g) + 14\ H_2O(l)$

$2(-39.96) \quad 19(0) \quad 12(-94.05) \quad 14(-68.32)$

Substitute into Equation 6.13 and solve.

6.5b. $\Delta H = -2005.16$ kcal.

$$\Delta H = \Delta H_f(\text{products}) - \Delta H_f(\text{reactants})$$
$$= 12(-94.05) + 14(-68.32) - 2(-39.96) = -2005.16\ \text{kcal}$$

PROBLEMS

Section 6.2.

6.1. How many milliliters of water can be decomposed into its elements by 85.0 kilocalories of energy? The thermochemical equation is $2\ H_2O(l) \rightarrow 2\ H_2(g) + O_2(g)$; $\Delta H = 136.6$ kcal.

6.2. The photosynthesis reaction, whereby plants use energy from the sun to form carbohydrates from carbon dioxide and water, is described by the equation $6\ CO_2(g) + 6\ H_2O(l) \rightarrow C_6H_{12}O_6(s) + 6\ O_2(g)$; $\Delta H = 673$ kcal. How much energy is required to form one pound (454 grams) of the simple sugar, $C_6H_{12}O_6$?

6.3. One of the principal fuels sold as "bottled gas" is butane, C_4H_{10}. Calculate the amount of energy that may be obtained by the combustion of 1.50 kilograms of butane if $\Delta H = -690$ kcal for the reaction $C_4H_{10}(g) + \frac{13}{2} O_2(g) \rightarrow 4CO_2(g) + 10H_2O(l)$.

6.4. In 1866 a young chemistry student conceived the electrolytic method of obtaining aluminum from its oxide. This method is still used in the production of this valuable industrial product. $\Delta H = 235$ kcal for the reaction: $Al_2O_3(s) + \frac{3}{2}C(s) \rightarrow 2\ Al(s) + \frac{3}{2}CO_2(g)$. The large amount of energy required restricts the process to areas of cheap electric power. How many kilowatt hours of energy are needed to produce one pound (454 grams) of aluminum by this process if 1 kw-hr = 860 kcal?

6.13. Quicklime, the common name of calcium oxide, CaO, is made by heating limestone in slowly rotating kilns about eight feet in diameter and nearly 200 feet long. The reaction is $CaCO_3(s) \rightarrow CaO(s) + CO_2(g)$; $\Delta H = 42.5$ kcal. How many kilocalories of energy are required to decompose 5.00 kilograms of limestone?

6.14. The quicklime produced as described in the foregoing problem is frequently converted to calcium hydroxide, or, as it is sometimes called, slaked lime, by a highly exothermic reaction with water: $CaO(s) + H_2O(l) \rightarrow Ca(OH)_2(s) + 15.9$ kcal. How many grams of quicklime must be processed if the energy evolution is 55.0 kcal?

6.15. How many grams of octane, a component of gasoline, would you have to use in your automobile to convert it to 3.00×10^4 kcal of usable energy? $\Delta H = -1.30 \times 10^3$ kcal for the reaction $C_8H_{18}(l) + \frac{25}{2}\ O_2(g) \rightarrow 8\ CO_2(g) + 9\ H_2O(l)$.

6.16. Nitroglycerine is the explosive ingredient in many industrial dynamites. The thermochemical equation for its reaction is $2\ C_3H_5(NO_3)_3(l) \rightarrow 3\ N_2(g) + \frac{1}{2} O_2(g) + 6\ CO_2(g) + 5\ H_2O(l)$; $\Delta H = -737$ kcal. Calculate the number of pounds of nitroglycerine that must be used in a blasting operation that requires 5,000 kcal.

Section 6.3. Problems 6.5–6.7 and 6.17–6.19: Calculate ΔH for each of the following reactions, using ΔH_f° values found in Table III, page 285.

6.5. $NO(g) + \frac{1}{2} O_2(g) \rightarrow NO_2(g)$

6.6. $3\ Fe_2O_3(s) \rightarrow 6\ Fe(s) + \frac{9}{2} O_2(g)$

6.7. $2\ AgBr(s) + H_2O(g) \rightarrow Ag_2O(s) + 2\ HBr(g)$

6.8. The principal constituent of "natural gas" is methane, CH_4. From the information in Table III, page 285, determine the heat of combustion of methane.

6.9. Natural gas from the oil fields of Oklahoma and Texas is 96% methane, the remainder being inert, whereas natural gas from Pennsylvania oil fields is 68% methane and 31% ethane, C_2H_6. Calculate ΔH_c for ethane.

6.10. One step in the manufacture of sulfuric acid is the oxidation of sulfur dioxide to sulfur trioxide. Calculate the energy change when 28.6 grams of sulfur dioxide are changed to sulfur trioxide. (Hint: First obtain ΔH for the reaction, relating it to a specific equation. Then, by stoichiometric calculations, determine the heat involved with 28.6 grams.)

6.11. The heat of combustion of ethylene glycol, $C_2H_4(OH)_2$, the permanent anti-freeze widely used in automobile cooling systems, is −281.9 kcal/mole. Calculate the heat of formation of ethylene glycol.

6.17. $2\ ZnO(s) + 2\ S(s) \rightarrow 2\ ZnS(s) + O_2(g)$

6.18. $2\ SO_2(g) + 2\ H_2O(g) + O_2(g) \rightarrow 2\ H_2SO_4(l)$

6.19. $Fe_2O_3(s) + 3\ CO(g) \rightarrow 2\ Fe(s) + 3\ CO_2(g)$

6.20. In listing the heats of combustion of industrial fuels, engineering handbooks frequently list a *gross* value and a *net* value. The net value is the more practical value, acknowledging that the product water is in the vapor state rather than liquid water. Though still unrealistic because of other reasons,* ΔH_c computed from $H_2O(g)$ as a product will be closer to the practical heat of combustion of methane than $H_2O(l)$ would be. Calculate ΔH for the combustion of one mole of methane to $H_2O(g)$.

6.21. Using the results of Problems 6.8 and 6.9, and assuming Oklahoma-Texas natural gas to be 100% methane and Pennsylvania natural gas to be 68% methane and 32% ethane, determine and compare the kcal of energy derived from one pound (454 grams) of each fuel. (Percentages are based on mass.)

6.22. The best known use of acetylene, C_2H_2, is in the oxyacetylene torches found in welding shops. Find the amount of heat released in the burning of two pounds of acetylene.

6.23. Find the heat of formation of the gasoline constituent octane, C_8H_{18}, if its heat of combustion is −1300 kcal/mole.

6.12. Example 5.10 (page 65) describes a calorimetry experiment in which ΔH for the reaction between $NaOH(aq)$ and $HCl(aq)$ is found to be −12.9 kcal/mole. Problem 5.9 (page 64) is similar: $\Delta H = -11.7$ kcal/mole for the reaction between $NaOH(aq)$ and $HC_2H_3O_2(aq)$. The *precise chemical changes* to which these ΔH values apply are properly described by net ionic equations.

$$H^+(aq) + OH^-(aq) \rightarrow HOH(l); \qquad \Delta H = -12.9 \text{ kcal}$$

$$HC_2H_3O_2(aq) + OH^-(aq) \rightarrow HOH(l) + C_2H_3O_2^-(aq); \quad \Delta H = -11.7 \text{ kcal}$$

*Values in Table V are for water vapor at 25°C and 1 atmosphere.

From these experimental data and the principles of enthalpy summation, calculate the energy required to ionize one mole of acetic acid by the equation

$$HC_2H_3O_2(aq) \rightarrow H^+(aq) + C_2H_3O_2^-(aq).$$

6.24. Sodium carbonate, Na_2CO_3, is one of our most important industrial chemicals and is used as a base in chemical processes. It is manufactured by the multistep Solvay process, which illustrates how the chemical engineer recycles some of the products of late steps in the process as reactants for the early steps. The thermochemical equations for the process may be approximated as follows:

$$CaCO_3 \rightarrow CaO + CO_2 \qquad \Delta H = +\ 2.5\ \text{kcal}$$
$$CaO + H_2O \rightarrow Ca(OH)_2 \qquad \Delta H = -\ 15.6\ \text{kcal}$$
$$NH_3 + H_2O + CO_2 \rightarrow NH_4HCO_3 \qquad \Delta H = -\ 30.3\ \text{kcal}$$
$$NH_4HCO_3 + NaCl \rightarrow NH_4Cl + NaHCO_3 \qquad \Delta H = +\ 0.1\ \text{kcal}$$
$$2\ NaHCO_3 \rightarrow Na_2CO_3 + CO_2 + H_2O \qquad \Delta H = +\ 20.3\ \text{kcal}$$
$$2\ NH_4Cl + Ca(OH)_2 \rightarrow 2\ NH_3 + CaCl_2 + 2\ H_2O \qquad \Delta H = +\ 37.8\ \text{kcal}$$

Examine these equations; adjust them as may be necessary so you may add them and their enthalpies to produce the net equation for the Solvay process:

$$CaCO_3 + 2\ NaCl \rightarrow Na_2CO_3 + CaCl_2\ .$$

Calculate ΔH for the process as written. You will find that the reaction is endothermic. The energy required for the process is obtained by burning coke or coal, the carbon dioxide from which is added to the carbon dioxide from the first and fifth steps to serve as a reactant in the third step.

CHAPTER 7

GASES: PRESSURE–VOLUME–TEMPERATURE–QUANTITY RELATIONS

The "gas laws" are embodied in a group of experimentally determined relationships between the measurable features of a gas, namely its mass, volume, temperature, and pressure. This material will be covered in two chapters. In this chapter we shall consider the gas laws themselves, and in Chapter 8 we shall extend and apply them to other situations.

7.1 GAS MEASUREMENTS

Quantity. The quantity of a gas may be measured in mass units, usually grams. Quantity is also often expressed in terms of numbers of moles. Volume is not a suitable quantity measurement because it is temperature- and pressure-dependent. It may be used if temperature and pressure are specified.

Volume. Because a gas fills its container, the volume of a gas is equal to the volume of the vessel that holds it.

Temperature. Gas temperatures are measured in conventional Celsius (centigrade) or Fahrenheit degrees. For use in gas law problems, temperature is converted to an absolute scale, usually degrees Kelvin.

Pressure. By definition, pressure is force per unit area. Pounds per square inch or grams per square centimeter are typical pressure units, though not commonly used with gases. In a liquid system, pressure is proportional to the depth of liquid or the height of a liquid column. This height is therefore a measure of pressure. The pressure of a gas is commonly indicated as the height of a column of mercury the gas is capable of supporting. Customary units are cm Hg or mm Hg, now commonly known as the torr. Pressure may also be measured in *atmospheres*. One atmosphere is a pressure commonly exerted by the atmosphere at sea level. It is equal to 760 mm Hg, or 760 torr. The SI pressure unit is the pascal. There are 1.013×10^5 pascals in one atmosphere.

7.2 PRESSURE MEASURING DEVICES

While gauges may be used to measure gas pressure, open end or closed end manometers are more frequently used in the chemistry laboratory. There are two operational principles of manometers:

1. Total pressure at any point in a liquid system is the sum of the pressures of each gas or liquid phase above that point.
2. Total pressures at any two points at the *same level* in a liquid system are always the same.

A closed end manometer is most easily read: the pressure of the gas is simply the height of the column. Figure 7.1 shows a closed end manometer connected to a source of gas. A vacuum exerting zero pressure is presumed to exist above the mercury column on the right. (The very small pressure exerted by mercury vapor above the mercury must be accounted for in extremely precise work, but will be neglected in the problems in this book.) The dotted line drawn at the *lower* liquid surface establishes two points in the liquid system, A and B, that are at the same total pressure. The pressure at these points equals "the sum of the pressures of each gas or liquid phase above that point." In the left leg only the pressure exerted by the gas, P_g, is above that level. In the right leg only the pressure exerted by the mercury, P_{Hg}, is above that point, the vacuum exerting zero pressure. Consequently, $P_g = P_{Hg}$.

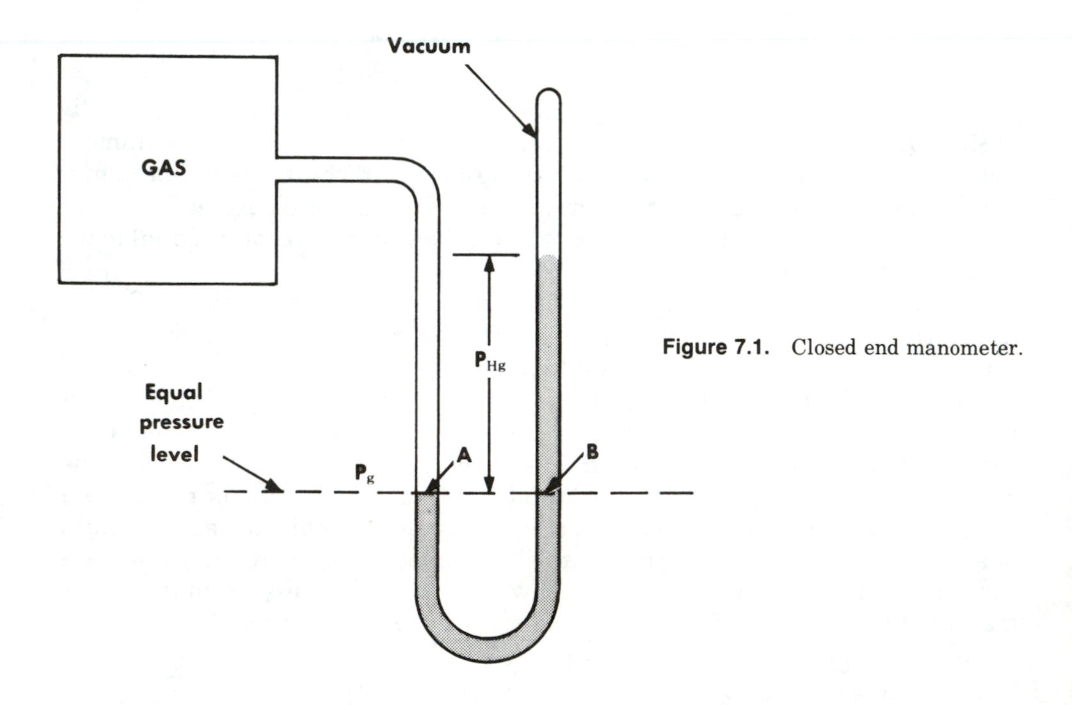

Figure 7.1. Closed end manometer.

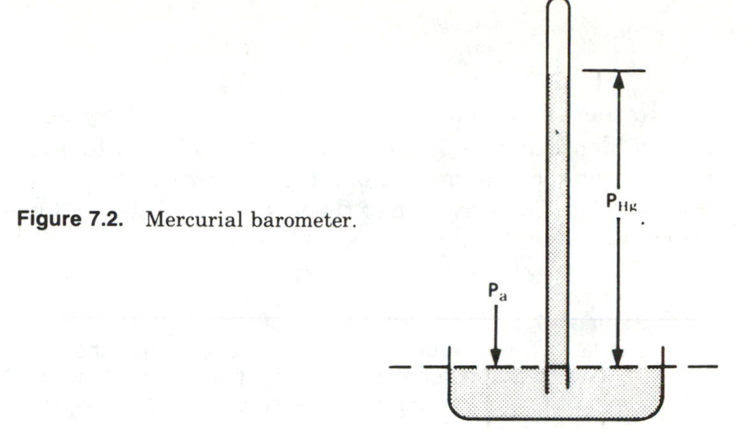

Figure 7.2. Mercurial barometer.

A conventional mercurial barometer is a closed end manometer (Fig. 7.2). At the dotted line drawn through the *lower* liquid surface, pressure inside the tube and outside are equal. Inside the tube, the pressure is P_{Hg}; outside, $P = P_a$, atmospheric pressure. Therefore $P_a = P_{Hg}$, so atmospheric pressure is read directly in inches, mm, or cm of mercury.

In an open end manometer, one leg is exposed to the air. Either of two conditions may exist, as shown in Figure 7.3. In both cases the dotted line drawn through the lower liquid surface establishes the levels of equal pressure in the two legs of the manometer. In Figure 7.3A the pressure at that level in the left leg is the pressure of the gas, P_g. In the right leg at the same level the total pressure is the sum of the pressure of the atmosphere, P_a, and the pressure caused by the mercury column, P_{Hg}. Hence

$$P_g = P_a + P_{Hg}. \tag{7.1}$$

In Figure 7.3B the total pressure in the left leg at the level of the dotted line is the sum of the pressure caused by the mercury column, P_{Hg}, and the pressure of the gas, P_g. In the right leg the total pressure is simply atmospheric pressure, P_a. Hence

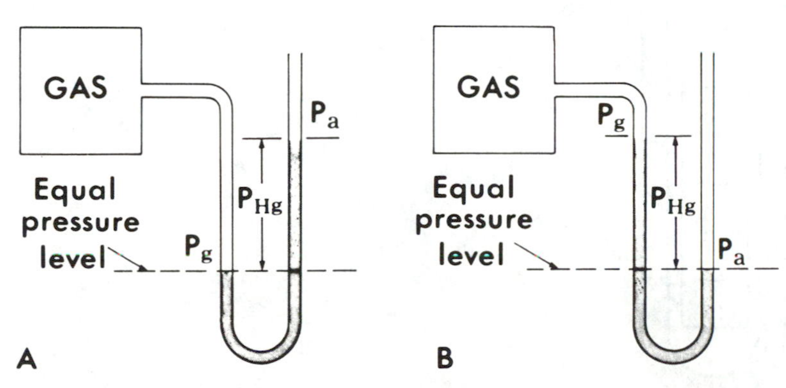

Figure 7.3. Open end manometer.

$$P_g + P_{Hg} = P_a \tag{7.2}$$

$$P_g = P_a - P_{Hg} \tag{7.3}$$

Rather than memorizing two formulas and trying to associate them with nearly identical sketches, the student is advised to analyze each situation by recognizing the equivalence of total pressure at the *lower* liquid surface level in the two legs. The pressure of the gas is found simply by equating pressures at that level.

Example 7.1. An open end mercurial manometer (Fig. 7.4) is connected to a gas tank. The mercury meniscus in the left leg of the manometer is opposite 33.8 cm on the meter stick; in the right leg it is opposite 16.2 cm. Atmospheric pressure is 747 torr. Find the pressure of the gas.

First, by inspection of Figure 7.4, at what level on the meter stick are the pressures in the two legs of the manometer equal, 33.8 or 16.2?

7.1a. 33.8. Always equate pressures at the *lower* liquid surface.
Is gas pressure greater or less than atmospheric pressure?

7.1b. The gas pressure is greater than atmospheric pressure. The diagram shows the gas supporting both the mercury and the atmosphere.

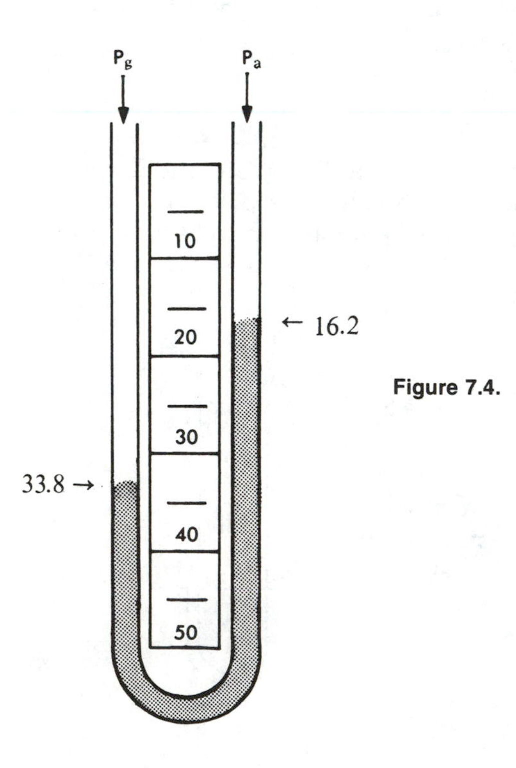

Figure 7.4.

By what amount is P_g greater than P_a, in cm Hg?

7.1c. 17.6 cm Hg. The difference between atmospheric and gas pressures is always equal to the difference in levels of the mercury in the two legs of the manometer. This difference is 33.8−16.2, or 17.6 cm Hg.

From the following, select the number that represents the pressure exerted by the gas in suitable units: 571; 729.4; 764.6; 923.

7.1d. 923 mm Hg, or 923 torr.

If you selected either of the first two numbers, you forgot that gas pressure is greater than atmospheric pressure. If you chose the second or third number, you were trapped by the failure to include units in listing the possible answers above. 764.6 comes from adding 747 *mm* to 17.6 *cm*—and obviously you cannot add mm to cm. Be careful with this; it is a common error. Recognizing that 17.6 cm = 176 mm, the correct calculation is 747 mm + 176 mm = 923 mm. Another correct answer you are likely to have calculated is 92.3 cm Hg—but that number was not among the choices available to you.

Example 7.2. If P_a = 755 mm Hg, find the gas pressure in the demonstration setup in Figure 7.5.

Solve completely.

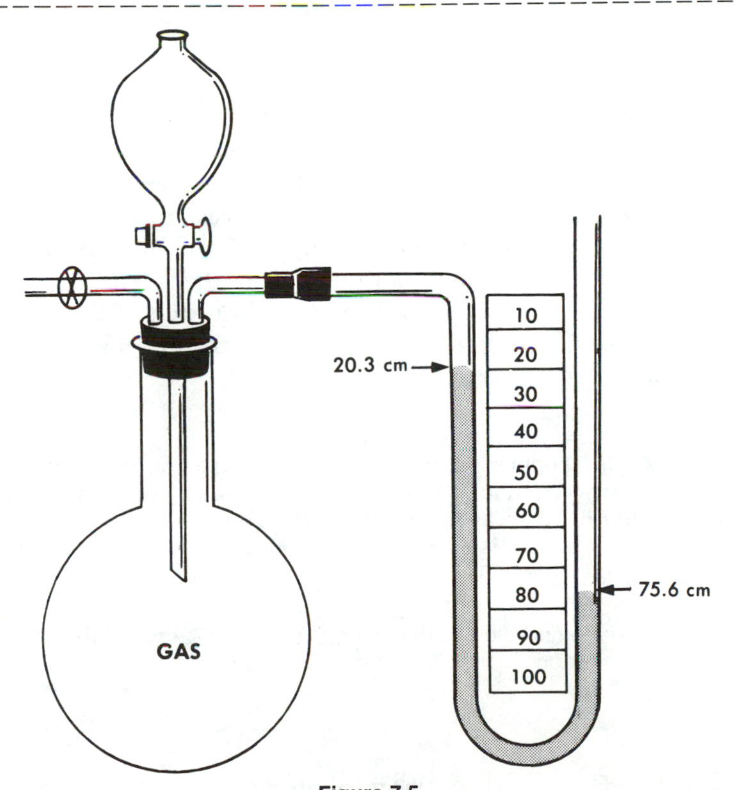

Figure 7.5.
(From Teacher's Guide to Chem Study, 155, Freeman Company).

7.2a. 202 torr.

Solution method: By inspection, the gas pressure is less than atmospheric pressure. P_{Hg} is 75.6 − 20.3 = 55.3 cm = 553 mm. P_g = 755 − 553 = 202 mm Hg = 202 torr.

If the pressure to be measured by an open end manometer is very close to atmospheric pressure, the difference in mercury levels may be too small to be read accurately. Under these circumstances water or some other low density liquid may be used. It is then necessary to convert the observed height of a liquid column to an equivalent height of mercury. We know from the physics involved that liquid pressure at any point is equal to the height of the liquid above that point times the density of the liquid. Thus

$$P = h_l \times D_l.$$

If a mercury column is to produce an equivalent pressure,

$$P = h_{Hg} \times D_{Hg}.$$

The two pressures being equal, the right sides of the two equations may be equated and the resulting equation solved for the height of the equivalent mercury column:

$$h_{Hg} \times D_{Hg} = h_l \times D_l \quad (7.4)$$

$$h_{Hg} = h_l \times \frac{D_l}{D_{Hg}} \quad (7.5)$$

Again, rather than memorizing a formula, the student is advised to realize that the height of a mercury column that exerts a pressure equal to the pressure of an observed column of another liquid may be found by multiplying the observed height by a ratio of liquid densities. In all practical situations the height of the equivalent mercury column will be *less* than that of the column of the other liquid because of the high density (13.6 g/cm^3) of mercury. The ratio, therefore, will always be less than 1.

Example 7.3. In a popular laboratory experiment in general chemistry, a gas is collected in an inverted tube. A typical situation at the end of the experiment is shown in Figure 7.6. If the difference in levels of the liquid surfaces is 172 mm, the specific gravity of the liquid is 1.23, and atmospheric pressure is 763 torr, find the pressure of the gas in torr.

It is first necessary to find the P_{Hg} that is equivalent to 172 mm of the liquid of specific gravity 1.23. Will P_{Hg} be more than 172 or less than 172?

7.3a. Less than 172. The height of the mercury column will always be less than that of a column of lower density liquid.

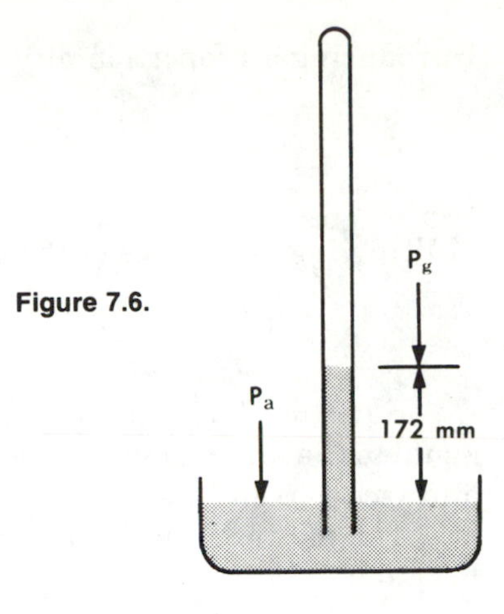

Figure 7.6.

The density of mercury is 13.6 g/cm³. Recall that specific gravity is numerically equal to density in the metric system. Therefore a specific gravity of 1.23 is equivalent to a density of 1.23 g/cm³. By which density ratio, 1.23/13.6 or 13.6/1.23, must 172 be multiplied to yield a product less than 172?

7.3b. 1.23/13.6—a ratio less than one.
Compute the mm Hg that are equivalent to 172 mm of the liquid.

7.3c. 172 mm liquid × 1.23/13.6 = 15.5 mm Hg.
Is the pressure of the gas greater or less than atmospheric pressure?

7.3d. The gas pressure is less than atmospheric pressure. The atmosphere is supporting the gas and the liquid column.
Compute P_g.

7.3e. 748 mm H_g = 748 torr.

$$763 - 15.5 = 748 \text{ mm Hg}$$

7.3 BOYLE'S LAW

It has been determined experimentally that the pressure exerted by a fixed quantity of a confined gas at constant temperature is inversely proportional to its volume. Expressed mathematically,

$$P \propto \frac{1}{V} . \tag{7.6}$$

Introducing a proportionality constant, k,

$$P = k\frac{1}{V}. \tag{7.7}$$

Multiplying both sides of the equation by V,

$$PV = k\ . \tag{7.8}$$

Equation 7.6 is known as Boyle's law. Since the *product* of P and V is always equal to the same number, when either factor *increases* the other must *decrease*, and vice versa. This is the nature of an inverse proportionality.

The mathematical relationship may be extended to show that $P_iV_i = k = P_fV_f$, or

$$P_iV_i = P_fV_f, \tag{7.9}$$

where subscripts *i* and *f* refer respectively to initial and final measurements of pressure and volume. Solving for V_f,

$$V_f = V_i \times \frac{P_i}{P_f}. \tag{7.10}$$

In other words, if the pressure of a confined gas is changed, the final volume may be found by multiplying the initial volume by a *ratio of pressures*—a pressure correction. If final pressure is greater than initial pressure, then, by the inverse proportionality of pressure and volume, final volume must be less than initial volume. Therefore, the pressure correction must be a ratio *less than 1*. Conversely, if final pressure is less than initial pressure, volume must increase and the pressure correction must be more than 1.

This is the key to the "reasoning" method of solving gas law problems. By knowing the inverse character of the pressure-volume relationship you may *reason* whether volume increases or decreases, and select your pressure correction accordingly.

Solve the following example by the reasoning method.

Example 7.4. A certain gas occupies 4.80 liters at 720 torr pressure. Find its volume if the pressure is changed to 750 torr. Temperature remains constant.

From the statement of the problem and from the initial conditions, does pressure increase or decrease?

7.4a. It increases—from 720 torr to 750 torr.
Will this produce an increase or a decrease in volume?

7.4b. A decrease. Pressure and volume are inversely related.

The new volume will be found by applying a pressure correction to the initial volume—by multiplying V_i by a ratio of initial and final pressures. This ratio may be either 720 torr/750 torr or 750 torr/720 torr. Bearing in mind that V_f must be less than V_i, which ratio is correct?

7.4c. 720 torr/750 torr.

Multiplying V_i by a ratio less than 1 makes the product less than V_i. Conversely, a ratio greater than 1 yields a product greater than V_i.

Complete the calculation, $4.80\ l \times 720$ torr/750 torr = . . .

7.4d. 4.61 liters.

This problem, and all others involving the pressure and volume of a confined gas at constant temperature, may be solved by substitution into Equation 7.9 or 7.10. The "formula" method is equally valid and, in its more complete statement, distinctly superior to the reasoning method for some problems. These alternatives will be mentioned throughout the gas law chapters, with comments to guide the student as to the selection of the better method for specific problems. Which method is "better" is, of course, a matter of opinion. As with the matter of indicating states in chemical equations, the student is advised to use the method presented in his chemistry class. In this text the reasoning method will be used except where there is a distinct advantage to using the formula approach. The formula method will be mentioned with each example, however, for those who may prefer it.

Example 7.5. 3000 liters of oxygen, produced at 0.800 atmosphere, is to be compressed into a cylinder of volume 15.0 liters at the same temperature. What pressure, in atmospheres, will the gas exert?

This time it is a new pressure that is sought as volume is changed. Does volume increase or decrease according to the statement of the problem?

7.5a. It decreases—from 3000 liters to 15 liters.

If volume decreases, will pressure increase or decrease?

7.5b. It must increase because pressure and volume are inversely related.

By what volume ratio must P_i be multiplied, $3000\ l/15\ l$ or $15\ l/3000\ l$, to get the new pressure? Set up and solve.

7.5c. 160 atmospheres.

$$0.800 \text{ atm} \times \frac{3000\ \cancel{l}}{15\ \cancel{l}} = 160 \text{ atm}$$

Example 7.5 may also be solved by substitution into Equation 7.10.

7.4 ABSOLUTE TEMPERATURE

The findings from experiments conducted by varying volume and temperature of any confined gas at constant pressure yield a graph such as Figure 7.7, with volume plotted vertically. Experiments relating pressure and temperature at constant volume yield a similar graph, except that pressure is plotted vertically. Both lines, when extrapolated, cross the temperature axis at −273°C. This temperature, at which the volume and pressure of a gas extrapolate to zero, is referred to as *absolute zero*, and it is the zero of an absolute temperature scale. Using degrees the same size as Celsius degrees, the absolute scale is called the Kelvin scale; temperatures on that scale are designated °K. Thus the freezing point of water, 0°C, is 273° above absolute zero, or 273°K. The boiling point of water, 100°C, is 373°K. In general, Kelvin temperature is always 273° greater than the corresponding Celsius temperature. Expressed as an equation,

$$°K = °C + 273 . \tag{7.11}$$

7.5 CHARLES' LAW

If the vertical axis of the graph of volume versus temperature (Fig. 7.7) were drawn at 0°K rather than 0°C, the graph would be a straight line passing through the origin. This is the graph of a direct proportionality. The complete statement of this relationship is known as Charles' Law: *The volume of a fixed quantity of confined gas at constant pressure is directly proportional to the absolute temperature*. Expressed mathematically,

$$V \propto T . \tag{7.12}$$

Introducing a proportionality constant,

$$V = kT . \tag{7.13}$$

By dividing both sides by T,

$$\frac{V}{T} = k . \tag{7.14}$$

If the quotient V/T is to remain constant, an increase in V must be accompanied by an increase in T and vice versa. This is inherent in the nature of a direct proportional relationship.

Following the same approach that was used with Boyle's Law,

$$\frac{V_i}{T_i} = \frac{V_f}{T_f}, \tag{7.15}$$

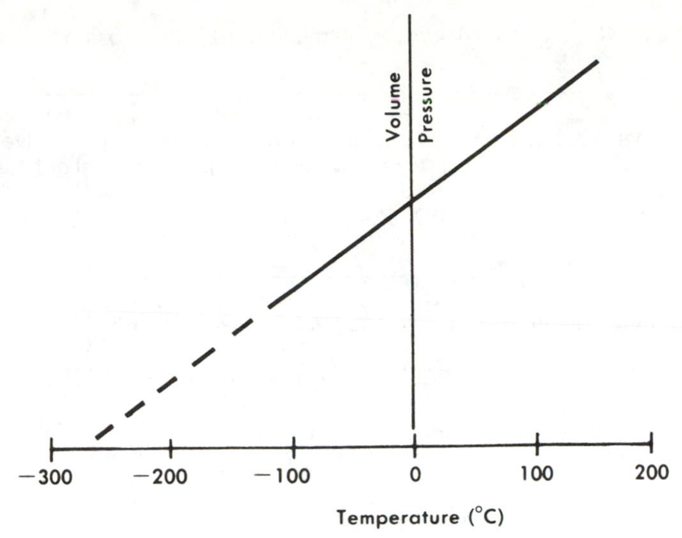

Figure 7.7.

where the subscripts again refer to initial and final measurements of volume and temperature. Solving for V_f,

$$V_f = V_i \times \frac{T_f}{T_i} \; . \qquad (7.16)$$

Thus, if the temperature of a confined gas is changed at constant pressure, the final volume may be found by multiplying the initial volume by a *ratio of temperatures*—a temperature correction. If final temperature is greater than initial temperature, the direct proportionality between volume and temperature predicts that final volume will be greater than initial volume; the temperature correction ratio must be greater than 1. Conversely, a drop in temperature means V_f is less than V_i, and the temperature correction ratio is less than 1.

One extremely important fact must be remembered in working gas law problems involving temperature: *The proportional relationships apply to absolute temperature* rather than the Celsius temperatures, which are frequently given in the statement of the problem. Before solving gas law problems, Celsius temperatures must be converted to °K.

Example 7.6. A 1.20-liter sample of a gas is heated from 15°C to 40°C at constant pressure. What will be the volume at the higher temperature?

First, from the conditions of the problem, tell whether the answer will be more or less than 1.20.

7.6a. More. Gas volume varies directly with temperature. If temperature increases, volume must also increase.

By what ratio, 40/15 or 15/40, would you multiply 1.20 liters to find the new volume?

7.6b. Neither. Volume is proportional to °K, not °C. You must first convert to absolute temperature. (Just think what sort of °C ratio would result if either initial or final temperature were 0° and you failed to convert to °K!)

Find °K values for 15°C and 40°C.

7.6c. Initial temperature = 288°K, and final temperature = 313°K.

$$T_i = 15 + 273 = 288°K. \qquad T_f = 40 + 273 = 313°K.$$

Now select the proper temperature ratio, 288°K/313°K or 313°K/288°K, by which the initial volume, 1.20 liters, must be multiplied to give V_f greater than V_i. Set up and solve.

7.6d. 1.30 liters.

$$1.20\ l \times \frac{313°\text{K}}{288°\text{K}} = 1.30\ l$$

This example may also be solved by substitution into Equation 7.15 or 7.16. The resulting setup and arithmetic are identical.

7.6 GAY-LUSSAC'S LAW

The similarity between volume-temperature and pressure-temperature graphs, shown by their common vertical axis in Figure 7.7, indicates that the pressure-temperature interdependence is the same as the volume-temperature relationship: *The pressure of a fixed quantity of confined gas at constant volume is directly proportional to the absolute temperature:*

$$P \propto T. \tag{7.17}$$

Equation 7.16 corresponds to the equation:

$$P_f = P_i \times \frac{T_f}{T_i}. \tag{7.18}$$

Example 7.7. Air in a steel cylinder is heated from 19°C to 42°C. If the initial pressure was 4.26 atmospheres, what is the final pressure?

First, will the new pressure be greater or less than 4.26 atmospheres?

7.7a. Greater. In this case of direct proportionality, an increase in temperature produces an increase in pressure.

By what ratio must the initial pressure be multiplied to find the new pressure?

7.7b. 315°K/292°K. Certainly you didn't have 41/19! Complete the solution.

7.7c. 4.60 atmospheres.

$$4.26 \text{ atm} \times \frac{315°K}{292°K} = 4.60 \text{ atm}$$

Alternately, Equation 7.18 could have been used on this example. That is,

$$P_f = P_i \times \frac{T_f}{T_i} = 4.26 \text{ atm} \times \frac{315°K}{292°K} = 4.60 \text{ atm.}$$

7.7 COMBINED GAS LAWS

To this point the relationships between any two properties of pressure, volume, or temperature of a fixed quantity of a confined gas have been considered, assuming that the third variable has remained constant. More often than not, however, all three vary. The individual gas laws may be combined to yield

$$\frac{PV}{T} = k \tag{7.19}$$

or

$$\frac{P_iV_i}{T_i} = \frac{P_fV_f}{T_f}. \tag{7.20}$$

Solving for P_f, V_f, and T_f, respectively,

$$P_f = P_i \times \frac{V_i}{V_f} \times \frac{T_f}{T_i} \tag{7.21}$$

$$V_f = V_i \times \frac{P_i}{P_f} \times \frac{T_f}{T_i} \tag{7.22}$$

$$T_f = T_i \times \frac{P_f}{P_i} \times \frac{V_f}{V_i}. \tag{7.23}$$

Examination of these equations shows that problems involving all three variables require the application of two correction factors to a single variable. This can best be illustrated through an example:

Example 7.8. Find the volume of a gas at 800 torr and 40°C if its volume at 720 torr and 15°C is 6.84 liters.

The approach to this problem is to consider first the volume change that is caused by the change in pressure, holding temperature constant momentarily. Then find how the volume changes further because of a change in temperature, holding pressure constant.

Proceed by steps: What will happen to the volume if pressure is changed from 720 torr to 800 torr and temperature remains at 15°C? Will volume increase or decrease?

7.8a. It will decrease. Volume is inversely proportional to pressure.

Fill in the fraction that constitutes the correction factor, 800 torr/720 torr or 720 torr/800 torr.

$$V_f = 6.84\ l \times \_\_\_\_\_\_ \times \_\_\_\_\_\_$$

7.8b. $V_f = 6.84\ l \times \frac{720\ \text{torr}}{800\ \text{torr}} \times \_\_\_\_\_\_$

— l at 800 torr and 15°C

— l at 720 torr and 15°C

The numerical result to this point, if it were to be calculated, would give the volume at P_f and T_i—a volume corrected for pressure but not for temperature. Now predict what further change will be produced by the temperature correction for a change from 15°C to 40°C. Will volume increase or decrease?

7.8c. It will increase. Volume is directly proportional to temperature.

Write the fraction that constitutes the temperature correction factor.

7.8d. 313°K/288°K.

$$\frac{40 + 273}{15 + 273} = \frac{313°K}{288°K}$$

Extend the setup to include this correction factor and complete the problem.

7.8e. 6.69 liters at 800 torr and 40°C.

$$V_f = 6.84\ l \times \frac{720\ \text{torr}}{800\ \text{torr}} \times \frac{313°K}{288°K} = 6.69\ l$$

— l at 800 torr and 15°C

— l at 720 torr and 15°C

Alternately the problem may be solved by substitution into an equation. Rather than try to keep Equations 7.21, 7.22 and 7.23 straight in one's thinking, it is easier to work always with Equation 7.20.

Substituting into Equation 17.20,

$$\frac{(720\ \text{torr})(6.84\ l)}{288°K} = \frac{(800\ \text{torr}) \times V_f}{313°K}.$$

Now solving for V_f,

$$V_f = 6.84\,l \times \frac{720\ \cancel{\text{torr}}}{800\ \cancel{\text{torr}}} \times \frac{313\cancel{°\text{K}}}{288\cancel{°\text{K}}} = 6.69\ l.$$

Example 7.9. 12.8 liters of a certain gas are prepared at 750 torr and −108°C. The gas is then forced into an 855 cm^3 cylinder in which it warms to room temperature, 22°C. Find the pressure of this gas in atmospheres.

This time we are to calculate a final pressure based on changes in volume and temperature. The reasoning process is similar. The volume change is from 12.8 liters to 855 cubic centimeters. Note the difference in units; obviously they must be reconciled. With this in mind, will pressure increase or decrease as a result of the change in volume?

7.9a. It will increase. Volume decreases from 12.8 liters to 0.855 liter (or from 12,800 cm^3 to 855 cm^3), and pressure changes in the opposite direction.

Begin the setup, showing volume correction in liters.

7.9b. $750\ \text{torr} \times \frac{12.8\ \cancel{l}}{0.855\ \cancel{l}} \times$ ———

torr at 0.855 l and −108°C

torr at 12.8 l and −108°C

Will pressure increase or decrease as a result of the temperature change? Extend the setup to include a temperature correction from −108°C to 22°C.

7.9c. $750\ \text{torr} \times \frac{12.8\ \cancel{l}}{0.855\ \cancel{l}} \times \frac{295\cancel{°\text{K}}}{165\cancel{°\text{K}}} \times$ ———

torr at 0.855 l and 22°C.

torr at 0.855 l and −108°C

torr at 12.8 l and −108°C

If you think the problem is completed, you'd better read it again. It specifies an answer in atmospheres. Complete the setup and solve.

7.9d. 26.4 atmospheres at 0.855 liter and 22°C.

$$750\ \cancel{\text{torr}} \times \frac{12.8\ \cancel{l}}{0.855\ \cancel{l}} \times \frac{295\cancel{°\text{K}}}{165\cancel{°\text{K}}} \times \frac{1\ \text{atm}}{760\ \cancel{\text{torr}}} = 26.4\ \text{atm}.$$

Example 7.9 may also be solved by substitution into Equation 7.20.

Problems involving the determination of a final temperature are less frequently encountered. They may be solved by the reasoning approach, but the thought process is more obscure. As a rule external heating or cooling is required to make the gas adjust to a new set of pressure and volume conditions, and the nature of the volume and pressure correction factors is not entirely obvious. These problems are perhaps more readily solved by the formula approach. The following example illustrates the method.

Example 7.10. A gas occupies 4.80 liters at 6.24 atmospheres and 25°C. What will the temperature be in degrees Celsius if it is expanded to 25.0 liters at a pressure of 0.840 atmosphere?

The first three values indicate the initial conditions of volume, pressure, and temperature; the last two values are the final volume and pressure. Equation 7.20 is repeated here:

$$\frac{P_iV_i}{T_i} = \frac{P_fV_f}{T_f}. \qquad (7.20)$$

But the numerical values are not quite ready for direct substitution.
What must first be changed? Make that change. . . .

7.10a. T_i must be changed from °C to °K. 25°C = 25 + 273 = 298°K. Now substitute into Equation 7.19 and solve for T_f, the only unknown.

7.10b. The answer to this point is 209°K, but the problem is not yet finished. . . .

$$\frac{6.24\text{ atm} \times 4.80\,l}{298°\text{K.}} = \frac{0.840\text{ atm} \times 25.0\,l}{T_f}$$

$$T_f = 298°\text{K} \times \frac{25.0\,\not{l}}{4.80\,\not{l}} \times \frac{0.840\,\cancel{\text{atm}}}{6.24\,\cancel{\text{atm}}} = 209°\text{K}$$

Now the final step in the problem. You are asked for the temperature in degrees Celsius. Finish it up. . . .

7.10c. −64°C.

$$°\text{K} = °\text{C} + 273$$

$$°\text{C} = °\text{K} - 273 = 209 - 273 = -64°\text{C}$$

7.8 STANDARD TEMPERATURE AND PRESSURE

Because of the interdependence of temperature, pressure, and volume of a confined gas, it is not possible to specify the *quantity* of gas in volume units without also specifying the temperature and pressure. Volume may be used as a quantity unit, however, if it is referred to arbitrarily established standards of temperature and pressure. The commonly accepted "standard conditions," or "standard temperature and pressure, STP," as they are frequently called and abbreviated, are 0°C (273°K) and 1.00 atmosphere (760 torr). Many gas law problems require the student to convert volume to or from STP. These problems are solved in precisely the same manner as those previously discussed. One example should suffice:

Example 7.11. In a laboratory experiment, 85.3 ml of a gas are collected at 24°C and 733 mm pressure. Find the volume at STP. (Interpreted, what would the volume be at 0°C and 760 mm?)

Set up the problem in its entirety and solve.

7.11a. 75.6 ml at STP.

$$85.3 \text{ ml} \times \frac{733 \text{ torr}}{760 \text{ torr}} \times \frac{273°\text{K}}{297°\text{K}} = 75.6 \text{ ml}$$

7.9 THE EQUATION OF STATE

It has been shown that the pressure exerted by a gas is proportional to the absolute temperature of the gas and inversely proportional to its volume. Pressure also depends upon one other variable: the quantity of gas present. If n is the number of moles of gas molecules,

$$P \propto n \,. \qquad (7.24)$$

Combining this proportionality with $P \propto T$ (Equation 7.17) and $P \propto \frac{1}{V}$ (Equation 7.6),

$$P \propto \frac{nT}{V} \,. \qquad (7.25)$$

Introducing a proportionality constant, R,

$$P = \frac{nRT}{V} \,, \qquad (7.26)$$

which is usually written in the form

$$PV = nRT \,. \qquad (7.27)$$

Equation 7.27 is called an *equation of state*; more specifically, it is the **ideal gas equation.**

The proportionality constant, R, is referred to as the **universal gas constant.** Its value may be found by employing the experimental fact that 1 mole of gas (n = 1) at STP occupies 22.4 liters. Solving Equation 7.27 for R and substituting known values,

$$R = \frac{PV}{nT};$$

$$R = \frac{1 \text{ atm}}{1 \text{ mole}} \times \frac{22.4\ l}{273°K} = 0.0821 \frac{\text{liter-atmospheres}}{\text{mole}-°K}; \tag{7.28}$$

and

$$R = \frac{760 \text{ torr}}{1 \text{ mole}} \times \frac{22.4\ l}{273°K} = 62.4 \frac{\text{liter-torr}}{\text{mole}-°K}. \tag{7.29}$$

The value of R to be used in a specific problem depends upon the units of measurement. With both expressions above, volume must be in liters, temperature in °K, and quantity in moles. If pressure is in atmospheres, the 0.0821 liter-atm/mole−°K value for R is indicated, whereas pressure in torr would require 62.4 liter-torr/mole−°K.

A useful variation of the ideal gas law can be derived as follows: If the weight of any chemical species, g, is divided by the molar weight, MW, the quotient is the number of moles:

$$\frac{g}{MW} = \frac{\text{grams}}{\text{grams/mole}} = \text{moles} = n^*.$$

g/MW may, therefore, be substituted for n in the ideal gas equation:

$$PV = nRT = \frac{g}{MW}RT. \tag{7.30}$$

In solving gas law problems, the recommended procedure is to rearrange the equation algebraically so the desired quantity is on the left and all known variables are on the right, substitute known values, and solve. To illustrate:

Example 7.12. The contents of a 1.85 liter gas cylinder exert a pressure of 3.68 atmospheres at 24°C. How many moles of gas are in the cylinder?

The unknown is moles, or n in the ideal gas equation. Solve Equation 7.27 for n, substitute the given values, and calculate the answer.

---

*In simplifying complex fractions, the denominator of the denominator fraction is moved to the numerator of the main fraction.:

$$\frac{1}{1/a} = \frac{1(a)}{(1/a)(a)} = \frac{a}{1}$$

7.12a. 0.279 mole.

$$n = \frac{PV}{RT} = \frac{3.68\ \text{atm} \times 1.85\ l}{(273+24)\,°K} \times \frac{\text{mole}-°K}{0.0821\ l-\text{atm}} = 0.279\ \text{mole.}$$

0.0821 liter-atmospheres/mole−°K is used for R because pressure is given in atmospheres.

Example 7.13. What is the weight of 18.9 liters of NH_3 at 31°C and 735 torr?

In this problem the mass, g, must be found. Solve Equation 7.30 for g, substitute known values, and compute the answer.

7.13a. 12.5 grams of NH_3.

$$g = \frac{(MW)PV}{RT} = \frac{17.0\ g\ NH_3}{1\ \text{mole}\ NH_3} \times \frac{735\ \text{torr}}{(273+31)\,°K} \times \frac{\text{mole}-°K}{62.4\ l-\text{torr}} \times 18.9\ l$$
$$= 12.5\ g\ NH_3$$

Any of the ideal gas equation variables may be found in the manner illustrated in Examples 7.12 and 7.13. A particularly useful application is the determination of molar weight of an unknown gas from the gas density.

Example 7.14. An unidentified gas has a density of 1.15 grams/liter at 20°C and 748 torr. Find the molar weight of the gas.

The density of a gas is essentially a statement of two variables in Equation 7.30; it gives you the grams and the volume, the latter being 1 liter. This leaves molar weight as the only unknown.

7.14a. 28.1 grams/mole.

$$MW = \frac{gRT}{PV} = \frac{1.15\ g \times (273+20)\,°K}{1\ l \times 748\ \text{torr}} \times \frac{62.4\ l-\text{torr}}{\text{mole}-°K} = 28.1\ g/\text{mole}$$

In a reverse sequence, gas densities may be determined for any gas at any temperature and pressure.

Example 7.15. What is the density of hydrogen at 29°C and 720 torr?

Density is mass per unit volume, i.e., g/V from the ideal gas equation. Solve the equation algebraically for g/V, make the substitutions and calculate the answer.

7.15a. 0.0768 gram/liter.

$$\frac{g}{V} = \frac{P(MW)}{RT} = \frac{720\ \text{torr}}{(273+29)\,°K} \times \frac{2.01\ g\ H_2}{1\ \text{mole}\ H_2} \times \frac{\text{mole}-°K}{62.4\ l-\text{torr}} = 0.0768\ g/l$$

PROBLEMS

Section 7.2

7.1. A mercurial manometer is connected to a gas storage tank. The mercury level difference is 28.4 cm, with the open leg level higher than the level in the closed leg. If atmospheric pressure is 752 torr, what is the pressure of the gas in the tank?

7.2. The water level in the entrance to an underwater sealab is 35.0 feet below the surface of the ocean, as shown in the sketch. The density of sea water is 1.02 g/ml. Find the pressure in the laboratory if the atmospheric pressure is 764 torr.

7.17. A jet eductor is a piece of equipment used to draw gases from other equipment items. When operating on a condenser in a steam plant, the closed leg of a mercury manometer attached to the condenser is 692 mm above the open leg. What pressure is being maintained by the jet eductor if atmospheric pressure is 743 mm Hg?

7.18. A manometer containing a fluid of specific gravity 1.43 is connected to two tanks. The liquid level in the leg connected to tank A is 4.63 inches lower than the level in the leg connected to tank B. If the pressure in tank A is 448.2 torr, what is the pressure in tank B? Answer in torr.

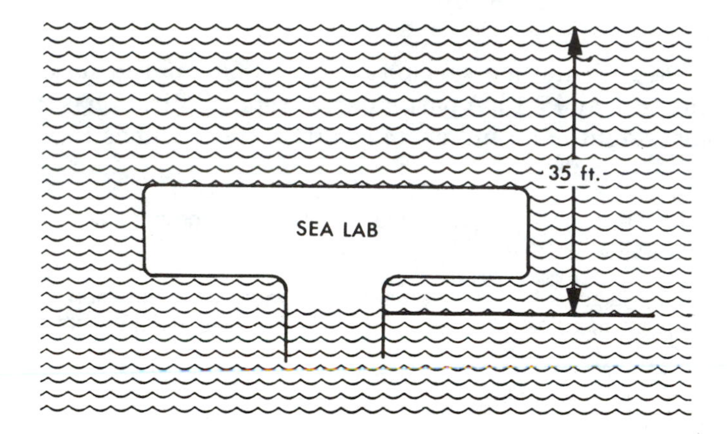

Section 7.3

7.3. A gas is confined in a cylinder with a moveable piston at one end. When the volume of the cylinder is 682 ml the pressure of the gas is 1.33 atmospheres. What will the pressure be if the cylinder volume is reduced to 419 ml?

7.19. A gas in a variable volume tank is exerting a pressure of 88.4 cm of mercury when the volume is 39.2 liters. To what must the volume be changed if the same gas is to exert a pressure of 64.0 cm Hg?

Section 7.5

7.4. A large industrial gas storage tank is designed to deliver fuel to the burners of a boiler at constant pressure, the pressure being maintained by a piston that rises or falls to adjust tank volume. When the plant is shut down at the beginning of a weekend the tank volume is 14.2 cubic meters and the gas temperature is 62°C. What will the volume be for the same quantity of gas on Monday morning when the temperature has dropped to 24°C?

7.20. A spring-loaded closure maintains a constant pressure on a gas system containing a fixed quantity of gas, but allows the volume of the system to adjust to changes in temperature. From a starting point of 12.6 liters at 19°C, to what volume will the system change if the temperature rises to 28°C?

Section 7.6

7.5. A gas in a steel cylinder shows a gauge pressure of 355 pounds per square inch while sitting on a loading dock during a cold winter day when the temperature is −18°C. What pressure will the gauge show when the tank is brought inside and its contents warm up to 23°C?

7.21. A gas storage tank is designed to hold a fixed volume and quantity of gas at 1.74 atmospheres and 35°C. To prevent excessive pressure build-up by overheating, the tank is fitted with a relief valve that opens at 2.00 atmospheres. To what temperature must the gas rise in order to open the valve?

Section 7.7, 7.8

7.6. An industrial process yields 8.42 liters of NO at 35°C and 725 torr. What would be its volume if adjusted to STP?

7.22. A quantity of potassium chlorate is selected to yield, on complete thermal decomposition, 75.0 ml of oxygen, measured at STP. If the actual temperature is 28°C, and the actual pressure is 0.894 atmosphere, what volume of oxygen will result?

7.7. A gasoline station has a compressed air tank having a volume of 0.140 cubic meter. On a certain day the temperature of the compressed air is 33°C, and the pressure gauge indicates 125 pounds per square inch (14.7 psi = 1 atmosphere). To what volume would the contents of the tank expand if opened to an atmospheric pressure of 751 torr and a temperature of 13°C? Answer in cubic meters.

7.23. If 1.62 cubic meters of air at 18°C and 738 torr are compressed into the 0.140 cubic meter tank described in problem 7.7, and the temperature increased to 28°C, what pressure, measured in pounds per square inch, will be seen on the gauge?

7.8. The stack gases of a certain industrial process are discharged to the atmosphere at 135°C and 844 torr. To what volume will 1.00 liter of stack gases change as they adjust to atmospheric conditions of 14°C and 748 torr?

7.24. A collapsible balloon for carrying meteorological testing apparatus aloft is partly filled with 282 liters of helium, measured at 25°C and 756 torr. Assuming the volume of the balloon is free to expand or contract according to changes in pressure and temperature, what volume will it have on reaching an altitude at which the temperature is −58°C and pressure is 0.641 atmosphere?

7.9. Superheated steam at 371°C and 51 atmospheres is delivered from the boilers of an ocean liner to its turbines, where it expands and propels the ship. At one point, one liter of starting steam will have expanded to 153 liters, and its pressure will have been reduced to 131 torr. Assuming compliance with the ideal gas laws, calculate the Celsius temperature at that point in the system.

7.25. In chemical processing plants hot exhaust gases are commonly sent through heat exchangers where they preheat an incoming reactant. In one such system 5.0 liters of exhaust gas at 274°C and 2.60 atmospheres expands to 74.9 liters at 1.12 atmospheres in passing through the heat exchanger. At what temperature is the gas finally exhausted to the atmosphere?

Section 7.9

7.10. A 9.81 liter cylinder contains 23.5 moles of nitrogen at 23°C. What pressure is exerted by the gas? Answer in atmospheres.

7.26. Find the pressure in torr produced by 2.35 grams of carbon dioxide in a 5.00 liter vessel at 18°C.

7.11. A pressure of 850 torr is exerted by 28.6 grams of sulfur dioxide at a temperature of 40°C. Calculate the volume of the vessel holding the gas.

7.12. How many moles of NO_2 are in a 5.24 liter cylinder if the pressure is 1.62 atmospheres at 17°C?

7.13. At what Celsius temperature will argon have a density of 10.3 grams/liter and a pressure of 6.43 atmospheres?

7.14. Calculate the mass of ammonia in a 6.64 liter cylinder if the pressure is 4.76 atmospheres at a temperature of 25°C.

7.15. The density of an unknown gas at 20°C and 749 torr is 1.31 grams/liter. Estimate the molar weight of the gas.

7.16. Air, which is about 21% oxygen (MW = 32) and 79% nitrogen (MW = 28) by volume, behaves as if its "molar weight" is 29. Calculate the density of air at 25°C and 1 atmosphere to two significant figures.

7.27. 20.3 atmospheres is the pressure caused by 6.04 moles of nitric oxide, NO, at a temperature of 18°C. What is the volume of the gas in liters?

7.28. A 721 milliliter hydrogen lecture bottle is left with the valve slightly open. Assuming no air has mixed with the hydrogen, how many moles of hydrogen are left in the bottle after the pressure has become equal to an atmospheric pressure of 752 torr and a temperature of 22°C?

7.29. At what temperature will 0.750 mole of chlorine in a 15.7 liter vessel exert a pressure of 756 torr?

7.30. How many grams of carbon monoxide must be placed into a 40.0 liter tank to develop a pressure of 965 torr at 23°C?

7.31. 0.710 gram of an unidentified gas is placed in a 390 ml cylinder. At 52°C the gas yields a pressure of 520 torr. Find the molar weight of the gas.

7.32. What is the density of neon at 40°C and 1.23 atmospheres?

CHAPTER 8

APPLICATIONS OF THE GAS LAWS

8.1 MOLAR VOLUME

Experimentally it has been found that, in reactions between gases at the same temperature and pressure, the volumes of the different reacting gases are always in the ratio of small whole numbers. This generalization is known as Gay-Lussac's Law of Combining Volumes. From it stems Avogadro's Hypothesis: Equal volumes of different gases at the same temperature and pressure contain the same number of molecules. It follows that, provided the pressure and temperature of a gas are fixed, the volume of the gas is a suitable measure of its quantity—its mass, or the number of gas molecules.

The most frequently adopted "fixed" temperature and pressure used is STP, 0°C and 1.00 atmosphere. It has been found experimentally that at STP one mole of any gas will occupy approximately 22.4 liters. We used this relationship, one mole of any gas equals 22.4 liters at STP, when solving stoichiometry problems involving gases in Section 4.6., page 51.

This value, 22.4 liters per mole, is the molar volume of a *gas at STP*. The student is advised to note carefully the limitations placed on that sentence. 22.4 liters per mole is the molar volume of a *gas*—never a solid or a liquid—if and only if that volume is measured *at STP*—not some other combination of temperature and pressure. Perhaps this point can be emphasized by finding molar volume at another pressure–temperature combination.

Example 8.1. Calculate the molar volume—the volume occupied by 1 mole—of a gas at 25°C and 1 atmosphere.

In essence this is a Charles' Law problem (see page 88). It says 22.4 liters of a fixed quantity of a gas—one mole—are heated from 0°C to 25°C at constant pressure—one atmosphere. Find the new volume.

8.1a. 24.5 liters/mole.

$$\frac{22.4 \text{ liters}}{\text{mole}} \times \frac{(273 + 25)\cancel{^\circ\text{K}}}{273\cancel{^\circ\text{K}}} = 24.5 \text{ liters/mole}$$

The combination 25°C and 1 atmosphere is sometimes referred to as "standard state." You will find in Chapter 12 and elsewhere in this book that much data in chemical handbooks is referenced to these conditions. They are a bit more realistic than STP; not many chemical laboratories have their air conditioner-heater thermostats set at the freezing point of water!

8.2 MOLAR VOLUME PROBLEMS

Density, by definition, is mass per unit volume. In reference to gases, density is usually expressed as grams per liter. Density, molar volume, and molar weight are frequently involved in solving problems with variable quantities of gases. The units of the three quantities are significant:

Density: $\frac{\text{grams}}{\text{liter}}$

Molar weight: $\frac{\text{grams}}{\text{mole}}$

Molar volume: $\frac{\text{liters}}{\text{mole}} = 22.4\ l/\text{mole at STP}$

$= 24.5\ l/\text{mole at 25°C and 1 atmosphere.}$

The use of these concepts can best be illustrated by some examples:

Example 8.2. Find the volume of 0.163 mole of CO_2 at STP.

What is the first entry in the setup of this problem?

8.2a. 0.163 mole.

The initial entry is nearly always the given quantity.

From the discussion in this section, what relationship is known between volume at STP and moles that makes possible the conversion from moles to liters? Complete the setup and solve the problem.

8.2b. 3.65 liters.

$$0.163\ \cancel{\text{moles}} \times \frac{22.4\ l}{1\ \cancel{\text{mole}}} = 3.65\ l$$

If the quantity of gas is given in grams, another step is required:

Example 8.3. What volume would be occupied by 16.0 grams of SO_2 at STP?

In this example, the *identity* of the gas is important. As in other areas of chemistry, it is number of molecules, or moles, that is significant, rather than mass alone. Molar weight—which is unique for each gas—permits conversion from grams to moles. Molar volume makes possible a second conversion from moles to liters. Number of moles is the connecting link between mass and volume.

Beginning with the given quantity, set up the problem to the extent of converting the 16.0 grams of SO_2 to moles.

8.3a. $16.0\ \cancel{g\ SO_2} \times \frac{1\text{ mole } SO_2}{64\ \cancel{g\ SO_2}}$

Now, as in Example 8.1, convert moles of SO_2 to liters. Show the entire setup and solve.

8.3b. 5.60 liters.

$$16.0\ \cancel{g\ SO_2} \times \frac{1\ \cancel{\text{mole } SO_2}}{64\ \cancel{g\ SO_2}} \times \frac{22.4\ l\ SO_2}{1\ \cancel{\text{mole } SO_2}} = 5.60\ l\ SO_2$$

Example 8.4. What is the weight of 13.8 liters of NH_3 at 25°C and 1 atmosphere?

The two-step conversion, a reversal of Example 8.3, should be evident. Start with the given quantity, set up and solve—but watch your molar volume figure.

8.4a. 9.58 grams of NH_3.

$$13.8\ \cancel{l\ NH_3} \times \frac{1\ \cancel{\text{mole } NH_3}}{24.5\ \cancel{l\ NH_3}} \times \frac{17.0\text{ g } NH_3}{1\ \cancel{\text{mole } NH_3}} = 9.58\text{ g } NH_3$$

At 25°C and 1 atmosphere you must use 24.5 liters/mole for molar volume.

Density varies the handling of units somewhat.

Example 8.5. Determine the density of CH_4 at STP.

This time there is no "given quantity." The answer sought is a relationship between grams and volume. The mole is still the connecting link between them. How many grams are equivalent to one mole of CH_4?

8.5a. 16 grams CH_4 = 1 mole CH_4—by calculation of molar weight.
Now, how many liters of CH_4 at STP are equivalent to 1 mole of CH_4?

8.5b. 22.4 liters CH_4 = 1 mole CH_4 —molar volume at STP.
Using the principle that "things equal to the same thing are equal to each other," 16 grams of CH_4 may be equated to 22.4 liters of CH_4 at STP. If 16 grams of CH_4 = 22.4 liters of CH_4, what will be the weight of one liter of CH_4 at STP? Set up and solve.

8.5c. 0.714 gram per liter.

$$\frac{16\text{ grams}}{22.4\ l} = 0.714\text{ g}/l$$

A dimensional analysis approach yields the same result. Dividing molar weight by molar volume,

$$\frac{16 \text{ grams}}{\cancel{\text{mole}}} \times \frac{\cancel{\text{mole}}}{22.4 \, l} = 0.714 \text{ g}/l$$

Example 8.6. An unknown gas is found to have a density of 1.22 grams per liter at STP. Predict its molar weight.

What units are required in the answer to this problem?

8.6a. Grams per mole, the units of molar weight.

The "given quantity" this time is not a pure quantity, but density, a quantity with derived units. This does not alter the procedure, however. Starting with density and the other information at your disposal, set up the problem and solve.

8.6b. 27.3 grams per mole.

$$\frac{1.22 \text{ g}}{1 \, \cancel{l}} \times \frac{22.4 \, \cancel{l}}{1 \text{ mole}} = 27.3 \text{ g/mole}$$

Example 8.7. 4.48 liters of an unknown gas at 25°C and 1 atmosphere weighs 2.93 grams. Estimate its molar weight.

You should be able to work your way through this without leading questions. If you have doubts, don't look ahead to the answer, but rather back to the reasoning in Examples 8.5 and 8.6. These should guide you to the correct setup.

8.7a. Molar weight = 16.0 grams per mole.

$$\frac{2.93 \text{ g}}{4.48 \, \cancel{l}} \times \frac{24.5 \, \cancel{l}}{1 \text{ mole}} = 16.0 \text{ g/mole}$$

The question naturally arises at this point as to how to handle problems at temperatures and pressures other than those for which you know the molar volume. You already know the answer: use the $PV = nRT = \frac{g}{MW} RT$ relationships (Equations 7.27 and 7.28) of the last chapter. Molar volume is simply an easier way to the desired end at those temperatures and pressures for which you happen to know the liters occupied by one mole.

8.3 MOLECULAR FORMULAS

The determination of molar weights by the methods described in Sections 7.9 (page 97) and 8.2 enables us to find molecular formulas from empirical formulas. For example, suppose it has been determined that the empirical formula for a substance is CH. Suppose further that measurements of mass, temperature, pressure, and volume of the substance in the gaseous state show that its molar weight is 78 grams per mole. What is the molecular formula of the substance?

The empirical formula, CH, gives only the ratio of atoms of different elements in the compound. The molecular formula could be CH, C_2H_2, C_3H_3, . . . or C_nH_n, or $(CH)_n$ where n is the number of empirical CH units in the individual molecule. If the molar weight is divided by the molar weight of the empirical formula unit, the result is the number of empirical units per molecule. In this example the molar weight of the empirical formula unit, CH, is 13. Dividing 78 by 13 gives 6 units per molecule. Hence the molecular formula is C_6H_6.

Example 8.8. A pure, unknown liquid is found in the laboratory to contain 14.3% hydrogen and 85.7% carbon by weight. If 29.4 grams of this liquid are vaporized in a 3.60-liter cylinder at 260°C, the pressure is found to be 2.84 atmospheres. Determine the molecular formula of the compound.

The solution to this problem involves three steps: determination of the empirical formula; determination of the molar weight; and determination of the molecular formula.

The empirical formula is found by methods discussed in Chapter 3, page 37. Complete this step.

8.8a. The empirical formula is CH_2.

Assuming a 100 gram sample, the tabulation is as follows:

Element	Grams Element	Moles Element	Mole Ratio
Carbon	85.7	7.14	1
Hydrogen	14.3	14.3	2

Using the ideal gas equation, compute the molar weight of the compound. Solve the equation for MW, substitute and calculate the value.

8.8b. 126 grams/mole.

$$MW = \frac{gRT}{PV} = \frac{29.4\ g}{2.84\ \cancel{atm}} \times \frac{0.0821\ \cancel{l}\text{-}\cancel{atm}}{mole\text{-}\cancel{°K}} \times \frac{(273+260)\cancel{°K}}{3.60\ \cancel{l}} = 126\ grams/mole$$

What is the molar weight of the empirical formula unit, CH_2?

8.8c. 14 grams per mole of empirical formula units.

The empirical formula is CH_2; its molar weight is 14 grams per mole. The molecular formula may be written C_nH_{2n} or $(CH_2)_n$; its molar weight is 126 grams per mole. What value of n will produce a molecular formula having a molar weight of 126? Or, put another way, how many moles of CH_2 empirical formula units at 14 grams per mole, are required to produce one mole of molecular formula units having a molar weight of 126? What it boils down to is how many fourteens in 126?

8.8d. 9. $\frac{126}{14} = 9.$

Write the molecular formula of the compound.

8.8e. C_9H_{18}.

8.4 GAS STOICHIOMETRY

In Chapter 4, page 51, we considered stoichiometry problems in which one of the species was a gas whose volume was measured at STP. At 22.4 liters per mole we could convert from quantity in liters to moles, or moles to quantity in liters, respectively the first and third steps of the stoichiometric pattern shown on page 47. Our present knowledge of the gas laws enables us to go beyond the STP relationship; we are now able to convert between moles and volume measured at any temperature and pressure at which the ideal gas laws are valid.

There are two approaches to converting between moles and volumes at other than STP. One method used the relationship PV = nRT in the form of $V = n \times \frac{RT}{P}$ in going from moles (n) to volume (V), or $n = V \times \frac{P}{RT}$ if liters are to be converted to moles. The second method makes the conversion with 22.4 liters per mole at STP, having adjusted or then adjusting the volume by temperature and pressure corrections. Clearly seen, the second method is probably easier; but without a clear picture of whether the pressure and temperature corrections are *to* or *from* STP the factors frequently appear inverted.

We suggest you choose between these two approaches in this way: If your instructor prefers one, uses only one in the classroom, or directs you to use one rather than the other, that *is* your choice. Disregard the other method altogether. If the choice is left to you, examine both, decide which you prefer, use it exclusively and discard the alternate. Unless you already have considerable maturity in chemical calculations we recommend against learning both methods simultaneously.

Let's begin with method one . . . .

Example 8.9. How many liters of hydrogen at 23°C and 733 torr are released by the reaction between 1.98 grams of sodium and water by the equation $2\ Na + 2\ H_2O \rightarrow H_2 + 2\ NaOH$?

The first two steps of the stoichiometric pattern, converting grams of sodium to moles, and that to moles of hydrogen, are as in other problems. Set up the solution to that point.

8.9a. $1.98\ \cancel{\text{g Na}} \times \frac{1\ \cancel{\text{mole Na}}}{23.0\ \cancel{\text{g Na}}} \times \frac{1\ \text{mole}\ H_2}{2\ \cancel{\text{moles Na}}}$.

Method 1: The ideal gas equation solved for volume, $V = n \times \frac{RT}{P}$, may now be used to complete the problem. Volume is found by multiplying n, the number of moles of H_2 as expressed above, by R and T and dividing by P. Complete the setup and calculate the answer.

8.9b. 1.08 liters H_2.

$$V = 1.98\ \text{g Na} \times \frac{1\ \text{mole Na}}{23.0\ \text{g Na}} \times \frac{1\ \text{mole}\ H_2}{2\ \text{moles Na}} \times \frac{62.4\ l\ \text{torr}}{°\text{K-mole}} \times \frac{296°\text{K}}{733\ \text{torr}} = 1.08\ l\ H_2$$

Method 2: From the setup of moles of H_2 (8.9a) we can convert to liters of H_2 at STP, using 22.4 liters = 1 mole. Extend the problem that far.

8.9c. $1.98\ \text{g Na} \times \frac{1\ \text{mole Na}}{23.0\ \text{g Na}} \times \frac{1\ \text{mole}\ H_2}{2\ \text{moles Na}} \times \frac{22.4\ l\ H_2}{1\ \text{mole}\ H_2}$

We now have the volume of the wanted gaseous species *at STP*. To correct to 733 torr, we must multiply by a pressure correction ratio. It is either $\frac{733\ \text{torr}}{760\ \text{torr}}$ or $\frac{760\ \text{torr}}{733\ \text{torr}}$. Think it through, and extend the setup.

8.9d. $1.98\ \text{g Na} \times \frac{1\ \text{mole Na}}{23.0\ \text{g Na}} \times \frac{1\ \text{mole}\ H_2}{2\ \text{moles Na}} \times \frac{22.4\ l\ H_2}{1\ \text{mole}\ H_2} \times \frac{760\ \text{torr}}{733\ \text{torr}}$

Pressure is *decreasing* from 760 torr to 733 torr. As pressure *decreases*, volume *increases*, so the pressure correction factor must be greater than 1.

Now the temperature correction. Complete the problem.

8.9e. 1.08 liters H_2.

$$1.98\ \text{g Na} \times \frac{1\ \text{mole Na}}{23.0\ \text{g Na}} \times \frac{1\ \text{mole}\ H_2}{2\ \text{moles Na}} \times \frac{22.4\ l\ H_2}{1\ \text{mole}\ H_2} \times \frac{760\ \text{torr}}{733\ \text{torr}} \times \frac{(273+23)°\text{K}}{273°\text{K}} = 1.08\ l\ H_2.$$

Example 8.10. How many grams of ammonia can be formed by the reaction of 7.82 liters of nitrogen at 12.5 atmospheres and 25°C? The equation is $N_2 + 3\ H_2 \rightarrow 2\ NH_3$.

The first step in the stoichiometric pattern is to convert the given quantity, a volume of nitrogen at non-STP conditions in this problem, to moles. Using method 1, the ideal gas equation solved for n yields $n = V \times \frac{P}{RT}$. The data of the problem may be substituted directly into this equation. Complete the setup this far, but do not solve.

8.10a. $n = 7.82\ l\ N_2 \times \frac{12.5\ \text{atm}}{(273+25)°\text{K}} \times \frac{\text{mole-}°\text{K}}{0.0821\ l\text{-atm}}$

Conversion of moles of nitrogen to moles of ammonia, and moles of ammonia to grams of ammonia, is like other stoichiometry problems. Complete the setup and solve.

8.10b. 136 grams NH_3.

$$7.82\ l\ N_2 \times \frac{12.5\ \text{atm}}{(273+25)°\text{K}} \times \frac{\text{mole-}°\text{K}}{0.0821\ l\text{-atm}} \times \frac{2\ \text{moles}\ NH_3}{1\ \text{mole}\ N_2} \times \frac{17.0\ \text{g}\ NH_3}{1\ \text{mole}\ NH_3} = 136\ \text{g}\ NH_3.$$

By method 2, the starting 7.82 liters of nitrogen at 12.5 atmospheres and 25°C is converted by correction factors to its equivalent volume at STP. Set up, but do not solve.

8.10c. $7.82\ l\ N_2 \times \frac{12.5\ \cancel{atm}}{1\ \cancel{atm}} \times \frac{273\ \cancel{°K}}{(273+25)\ \cancel{°K}}$

You now have liters of N_2 at STP. Converting liters to moles, and then to moles and grams of ammonia, is routine. Complete the problem.

8.10d. 136 grams NH_3.

$$7.82\ \cancel{l\ N_2} \times \frac{12.5\ \cancel{atm}}{1\ \cancel{atm}} \times \frac{273\ \cancel{°K}}{298\ \cancel{°K}} \times \frac{1\ \cancel{mole\ N_2}}{22.4\ \cancel{l\ N_2}} \times \frac{2\ \cancel{moles\ NH_3}}{1\ \cancel{mole\ N_2}} \times \frac{17.0\ g\ NH_3}{1\ \cancel{mole\ NH_3}} = 136\ g\ NH_3$$

It was shown in Chapter 4 (page 52) that the mole relationships from the coefficients of a chemical equation are identical to the volume relationships of gaseous species, *provided the gas volumes are measured at the same temperature and pressure.* Example 4.12 considered the burning of 1.39 liters of ethane, measured at STP, by the equation $2\ C_2H_6 + 7\ O_2 \rightarrow 4\ CO_2 + 6\ H_2O$, asking for the volume of oxygen required if it also is measured at STP. The solution of the problem was simply

$$1.39\ \cancel{liters\ C_2H_6} \times \frac{7\ liters\ O_2}{2\ \cancel{liters\ C_2H_6}} = 4.87\ liters\ O_2.$$

Had *both* the ethane and oxygen volumes been measured at 25°C and 750 torr—or any other combination of temperature and pressure—the solution would have been identical.

What happens, though, when both given and wanted species are expressed in volumes, but at *different* temperature-pressure combinations? The most direct approach is to *convert the volume of the given quantity to the temperature and pressure conditions of the wanted species; then solve on the basis of both gas volumes measured at the same temperature and pressure.* The method is illustrated in the example below:

Example 8.11. How many liters of hydrogen at 855 torr and 10°C are required to react with chlorine to produce 3.97 liters of HCl at 740 torr and 29°C? The equation is $H_2(g) + Cl_2(g) \rightarrow 2\ HCl(g)$.

First, find the volume the given species, HCl, would occupy at the conditions at which the wanted species are measured, 855 torr and 10°C. Apply pressure and temperature corrections to the initial 3.97 liters of HCl. Set up, but do not solve.

8.11a.

$$3.97\ l\ HCl \times \frac{740\ \cancel{torr}}{855\ \cancel{torr}} \times \frac{283\ \cancel{°K}}{302\ \cancel{°K}}$$

Now find the volume of hydrogen required to produce the number of liters of HCl represented by the above setup, both gases measured at the same temperature and pressure. The setup requires one additional step—the volume conversion. Solve to an answer.

8.11b. 1.61 liters H_2.

$$3.97\ \cancel{l\ HCl} \times \frac{740\ \cancel{torr}}{855\ \cancel{torr}} \times \frac{283\ \cancel{°K}}{302\ \cancel{°K}} \times \frac{1\ l\ H_2}{2\ \cancel{l\ HCl}} = 1.61\ l\ H_2.$$

8.5 DALTON'S LAW OF PARTIAL PRESSURES

The **partial pressure** of each component in a mixture of gases is the pressure it would exert if it alone occupied the total volume at the same temperature. The partial pressure of each component is independent of the partial pressures of the other components. It follows that the total pressure exerted by the entire mixture is the sum of all the partial pressures. This statement is known as Dalton's Law of Partial Pressures. Expressed mathematically,

$$P = p_1 + p_2 + p_3 + \ldots, \tag{8.1}$$

where P is the total pressure exerted by the mixture and p is the partial pressure of component 1, 2, 3, and so forth.

The direct application of Dalton's Law is illustrated by Example 8.12.

Example 8.12. In a container holding a mixture of nitrogen and oxygen, the partial pressure of the nitrogen is 420 torr, and the partial pressure of the oxygen is 548 torr. What is the total pressure of the mixture?

No comment should be required. Use Equation 8.1.

8.12a. P = 968 torr.

$$P = p_{N_2} + p_{O_2} = 420\ \text{torr} + 548\ \text{torr} = 968\ \text{torr}$$

Example 8.13. The total pressure of hydrogen, helium, and argon is 745 torr in a mixture of these three gases. If p_{He} is 320 torr and p_{Ar} is 405 torr, find p_{H_2}.

8.13a. p_{H_2} = 20 torr.

$$p_{H_2} = P - p_{He} - p_{Ar} = (745 - 320 - 405)\ \text{torr} = 20\ \text{torr}$$

The pressure exerted by a gas is proportional to the number of molecules of gas present, and independent of their identity. If half of the molecules in a mixture of two gases are gas *A*, and the other half are gas *B*, then half of the pressure must be caused by gas *A* and the other half by gas *B*. The partial pressure of each component is half the total pressure. If one-third of the total number of molecules are gas A, then the partial pressure of gas A will be one-third of the total pressure. In general, the partial pressure of any component of a mixture is found by multiplying the total pressure by the fraction of the total number of moles represented by that component.

The chemist expresses this relationship in the following equation:

$$p_A = X_A P. \qquad (8.2)$$

P and p have the same meaning as in Equation 8.1, and the subscript A refers to component A. X is a number fraction, specifically called the **mole fraction. Mole fraction is defined as the number of moles of a component divided by the total number of moles of all components.**

$$X_A = \frac{\text{moles of component } A}{\text{moles } A + \text{moles } B + \text{moles } C + \ldots}$$

$$X_A = \frac{\text{moles } A}{\text{total moles}} \qquad (8.3)$$

Example 8.14. A gaseous mixture contains 8.0 moles of hydrogen, 3.0 moles of helium, and 1.0 mole of methane. Find the partial pressure of each component if the total pressure is 2.4 atmospheres.

Let's do this for the hydrogen first. Using Equation 8.3, find the mole fraction of hydrogen.

8.14a. $X_{H_2} = 0.67$.

$$X_{H_2} = \frac{\text{moles hydrogen}}{\text{total moles}} = \frac{8.0}{8.0 + 3.0 + 1.0} = \frac{8.0}{12.0} = 0.67$$

Mole fractions are generally expressed as decimal fractions, but in use the rational fraction is sometimes more convenient. Using the conventional form of a fraction and Equation 8.2, find the partial pressure of hydrogen.

8.14b. $p_{H_2} = 1.6$ atmospheres.

$$p_{H_2} = \frac{8.0}{12.0} \times 2.4 \text{ atm} = 1.6 \text{ atm}$$

In a similar manner, calculate the partial pressures of helium and methane.

8.14c. $p_{He} = 0.60$ atm; $p_{CH_4} = 0.20$ atm.

$$p_{He} = \frac{3.0}{12.0} \times 2.4 \text{ atm} = 0.60 \text{ atm}$$

$$p_{CH_4} = \frac{1.0}{12.0} \times 2.4 \text{ atm} = 0.20 \text{ atm}$$

The statement of Dalton's Law suggests a way to check your answers. See if the sum of the partial pressures is equal to the total pressure.

8.14d. $p_{H_2} + p_{He} + p_{CH_4} = 1.6 \text{ atm} + 0.60 \text{ atm} + 0.20 \text{ atm} = 2.4 \text{ atm}$

Example 8.15. 30.0 grams of CO_2, 42.0 grams of N_2, and 48.0 grams of SO_2 are mixed in a container in which they exert a total pressure of 960 torr. Find the partial pressure of each gas.

This time the gas quantities are given in grams. They must first be converted to moles. The problem then becomes like Example 8.14. You should be able to take this one all the way.

8.15a. $p_{CO_2} = 223$ torr; $p_{N_2} = 491$ torr; $p_{SO_2} = 246$ torr.

$$30.0 \cancel{\text{g } CO_2} \times \frac{1 \text{ mole } CO_2}{44.0 \cancel{\text{g } CO_2}} = 0.68 \text{ mole } CO_2$$

$$42.0 \cancel{\text{g } N_2} \times \frac{1 \text{ mole } N_2}{28.0 \cancel{\text{g } N_2}} = 1.50 \text{ moles } N_2$$

$$48.0 \cancel{\text{g } SO_2} \times \frac{1 \text{ mole } SO_2}{64.1 \cancel{\text{g } SO_2}} = 0.75 \text{ mole } SO_2$$

$$\text{Total} = 2.93 \text{ moles mixture}$$

$$p_{CO_2} = \frac{0.68}{2.93} \times 960 \text{ torr} = 223 \text{ torr}$$

$$p_{N_2} = \frac{1.50}{2.93} \times 960 \text{ torr} = 491 \text{ torr}$$

$$p_{SO_2} = \frac{0.75}{2.93} \times 960 \text{ torr} = 246 \text{ torr}$$

$$\text{Total} = 960 \text{ torr}$$

Example 8.16. A mixture of hydrocarbons contains three moles of methane, CH_4, four moles of ethane, C_2H_6, and five moles of propane, C_3H_8. The container has a volume of 124 liters and the temperature is 22°C. Find the partial pressure of methane in atmospheres.

The ideal gas equation must be used to solve this problem. There is a long way and a short way. Which will you take? Solve it first, and then look to see.

8.16a. 0.586 atmosphere.

$PV = nRT$. Also, $p_{CH_4}V = n_{CH_4}RT$. In a mixture of gases, each gas is independent of the others and acts as if it alone occupied the space. (Reread the first sentence in this section.)

The gas laws apply to any component of a mixture or to the entire mixture. Therefore the problem should be solved for methane alone, not the entire mixture.

$$p_{CH_4} = \frac{n_{CH_4}RT}{V} = 3\ \text{moles} \times \frac{0.0821\ \ell\text{-atm}}{°\text{K-mole}} \times \frac{295°\text{K}}{124\ \ell} = 0.586\ \text{atm}$$

Alternatively, we could have calculated the total pressure and multiplied by the mole fraction of methane. But that is the long way!

Gases in contact with any liquid will contain some of that liquid in the gaseous or vapor state. The pressure exerted by that vapor is known as its **vapor pressure,** and it is a part of the total pressure exerted by the gas. If the gas is confined above and in contact with the liquid, the liquid and its vapor will reach an equilibrium in which the gas is said to be saturated with the vapor. The saturated vapor pressure of a particular liquid is determined only by the temperature; vapor-pressure-temperature data for various liquids are tabulated in handbooks.

As might be anticipated, the most important liquid that contributes its vapor pressure to total pressures of gases is water. Any time water in contact with a confined gas achieves liquid-vapor equilibrium, the water vapor pressure will be determined solely by the temperature of the system. Table V in the Appendix, page 290, lists these values.

Example 8.17. A student conducts a laboratory experiment in which he collects hydrogen gas by bubbling it through water. If the total pressure of the hydrogen and water vapor with which it is saturated is 752 torr, and the temperature of the system is 26°C, find the partial pressure of the dry hydrogen.

First consult Table V for the water vapor pressure at 26°C.

8.17a. At 26°C the water vapor pressure is 25.2 torr.
If the total pressure is attributable to the water vapor pressure and the partial pressure of hydrogen, find p_{H_2}.

8.17b. p_{H_2} = 727 torr.

$$(752 - 25)\ \text{torr} = 727\ \text{torr}$$

One of the most important types of problems in this area is the determination of the STP volume of a dry gas that has been collected over water.

Example 8.18. 78.6 ml nitrogen are collected over water at 22°C and a total pressure of 728 torr. What volume would the dry nitrogen occupy at STP?

Initially the nitrogen has a volume of 78.6 ml. So does the water vapor: both gases occupy the entire volume. The initial pressure of the nitrogen, however, is not 728 torr. This is the total of the partial pressure of nitrogen and the vapor pressure of water. To find p_{N_2}, the water vapor pressure must be subtracted. The balance of the problem is then solved as any other gas law problem.

What is the partial pressure of the nitrogen?

8.18a. 708 torr.
Water vapor pressure at 22°C is 19.8 torr.

$$(728 - 20)\ \text{torr} = 708\ \text{torr}$$

Now make the necessary corrections to find volume at STP.

8.18b. 67.8 ml dry nitrogen at STP.

$$78.6\ \text{ml}\ N_2 \times \frac{708\ \cancel{\text{torr}}}{760\ \cancel{\text{torr}}} \times \frac{273\cancel{°\text{K}}}{295\cancel{°\text{K}}} = 67.8\ \text{ml}\ N_2$$

In all problems such as Example 8.18, vapor pressure must be subtracted from total pressure in establishing the pressure correction ratio.

The following example, which is frequently encountered in the general chemistry laboratory, incorporates nearly all of the principles that have been discussed in both chapters on gases.

Example 8.19. 84.3 ml of hydrogen are collected in a gas collecting tube as shown in Figure 8.1. Barometric pressure is 756 torr. The system reaches an equilibrium temperature of 24°C. The height of the liquid column in the tube is 214 mm, and the specific gravity of the liquid is 1.13. Find the number of grams of hydrogen collected.

The objective in this problem is to determine the mass of hydrogen. This suggests an ideal gas law approach. Volume, temperature, molar weight, and R are all known. Pressure is not known, but data from which it may be found are given. Pressure of the dry hydrogen

Figure 8.1.

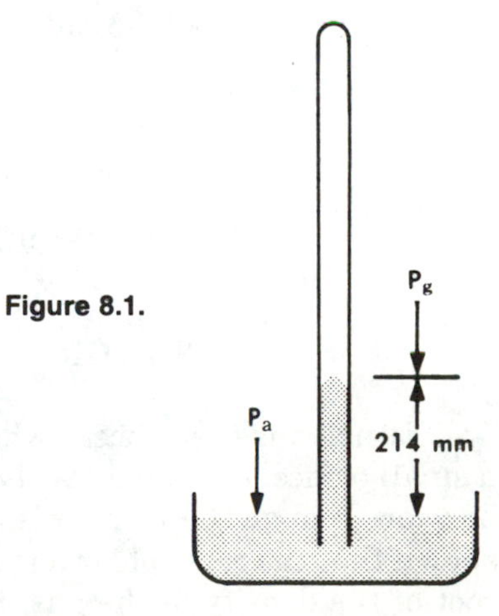

may be found only after determining the total gas pressure. Recall something about levels of equal pressure in liquids and liquid density ratios.

What is the mercury equivalent of a 214 mm column of liquid of specific gravity 1.13?

8.19a. 17.8 mm Hg = 17.8 torr.

$$P_{Hg} = h_l \times \text{density ratio}$$

$$= 214 \text{ mm} \times \frac{1.13}{13.6} = 17.8 \text{ mm Hg}$$

Is the total pressure of the gas more than or less than atmospheric pressure?

8.19b. Less. P_a supports both the liquid column and the gas.
Compute total gas pressure.

8.19c. 738 torr.

$$P_g = P_a - P_{Hg} = (756 - 18) \text{ torr} = 738 \text{ torr}$$

But this pressure includes water vapor. What is the partial pressure of dry hydrogen?

8.19d. 716 torr.

$$P_{H_2} = P_g - p_{H_2O} = (738 - 22) \text{ torr} = 716 \text{ torr}$$

The saturated vapor pressure of water at 24°C is 22.4 torr.

Now solve the ideal gas equation for the weight of hydrogen, g_{H2}, substitute all known values, and calculate the grams of hydrogen produced.

8.19e. 6.58×10^{-3} g H_2.

$$g_{H_2} = \frac{(MW)(p_{H_2})V}{RT}$$

$$= \frac{2.02 \text{ g } H_2}{\cancel{\text{mole}}} \times \frac{716 \cancel{\text{torr}}}{(273+24)\cancel{°\text{K}}} \times \frac{\cancel{\text{mole}}-\cancel{°\text{K}}}{62.4 \cancel{l}-\cancel{\text{torr}}} \times 0.0843 \cancel{l}$$

$$= 6.58 \times 10^{-3} \text{g } H_2$$

8.6 GRAHAM'S LAW OF EFFUSION

Effusion is the process whereby a gas escapes from its container through a small orifice. Diffusion involves the mass transport of a gas through a porous barrier. Though the mechanisms of the two processes differ, experiments have shown that the rates of both processes are inversely proportional to the square root of the density of the gas. For two gases, *A* and *B*:

$$\frac{r_B}{r_A} = \sqrt{\frac{d_A}{d_B}} . \tag{8.4}$$

The symbol r represents effusion or diffusion rate in molecules or moles per unit of time; d represents density. This is a common form of Graham's Law. Inasmuch as the molar weights of gases are proportional to their densities at any given temperature and pressure,* the expression may be written

$$\frac{r_B}{r_A} = \sqrt{\frac{(MW)_A}{(MW)_B}} , \tag{8.5}$$

where MW represents molar weight.

In working diffusion and effusion problems either logic or memorization of a formula and substitution into it may be used. Logic will be used here. Remember that the greater the density or molar weight, the lower will be the rate of the diffusion or effusion.

Example 8.20. Find the ratio of the effusion rate of ammonia to that of carbon dioxide.

To begin, you must know the molar weights of the two gases. Compute them.

8.20a. Ammonia, NH_3, 17.0 grams per mole; carbon dioxide, CO_2, 44 grams per mole.

The expression for effusion rates involves the square root of the ratio of molar weights. Establish that ratio so it has a value greater than one and then take the square root of it. (Making the ratio greater than one minimizes decimal errors in taking square root.)

8.20b. 1.61.

$$\sqrt{\frac{44.0}{17.0}} = \sqrt{2.59} = 1.61$$

Which gas has the higher effusion rate?

8.20c. Ammonia. Gases of lower density or molar weight have higher effusion rates. The relationship is inverse.

Express your result in the form of a ratio of effusion rates of the two gases.

8.20d. $\frac{r_{NH_3}}{r_{CO_2}} = 1.61.$

---

*Solving $PV = \frac{g}{MW} RT$ for density, $\frac{g}{V} = \frac{P}{RT}(MW)$. At a given pressure and temperature, $\frac{P}{RT}$ is constant. Therefore, $\frac{g}{V} \propto MW$.

If we can use molar weights to predict relative effusion rates, we should be able to use measured effusion rates to predict unknown molar weights.

Example 8.21. A certain number of moles of oxygen escaped from an effusion apparatus in 70.0 seconds. The oxygen was then flushed from the apparatus and replaced with an unknown gas at the same initial temperature and pressure. It took 55.5 seconds for the same number of moles of unknown gas to effuse. Calculate the molar weight of the unknown gas.

Let's predict at the outset which gas has the *higher* molar weight. This will be the gas with the *lower* effusion rate. Consider the problem carefully. Which will it be, oxygen or the unknown gas?

8.21a. Oxygen has the *higher* molar weight. It takes *longer* for a given number of moles of oxygen to escape, so its effusion rate is slower.

Now what is the ratio of effusion rates? In this problem we have times, not rates. Without pursuing the algebra involved, for a given quantity the time ratio is inversely proportional to the rate ratio. (Think about it: If you go from here to there in *half* the *time* it takes your brother, how does your *speed* compare to his?) So a time ratio is an *inverse* rate ratio. Set up this ratio and equate it to the square root of the molar weight ratio, substituting the molar weight of oxygen in its proper place. If you make the rate ratio greater than one, and consider your decision as to which gas has the larger molar weight, there should be no question about whether the oxygen goes into the numerator or denominator.

8.21b.

$$\frac{\text{rate unknown}}{\text{rate } O_2} = \frac{\text{time } O_2}{\text{time unknown}} = \frac{70.0 \text{ sec}}{55.5 \text{ sec}} = \sqrt{\frac{\text{MW oxygen}}{\text{MW unknown}}} = \sqrt{\frac{32.0}{\text{MW unknown}}}$$

With the rate ratio greater than one, the molar weight ratio must also be greater than one. Inasmuch as oxygen has the larger molar weight, it must be in the numerator.

Now solve the equation for the unknown molar weight. Suggestion: Compute the decimal value of the ratio, square both sides of the equation, and then solve for the unknown.

8.21c. Molar weight of unknown = 20.1 grams per mole.

$$\frac{70.0}{55.5} = 1.26; \qquad 1.26^2 = \frac{32.0}{\text{MW of unknown}}$$

$$\text{Molar weight} = 20.1$$

One more example—set up and solve.

Example 8.22. 35 seconds are required for the effusion of a given number of moles of ethane, C_2H_6, and 41 seconds for the same quantity of an unknown gas. Both gases are initially at the same conditions. Estimate the molar weight of the unknown.

8.22a. 41.1 grams per mole.

Rate ratio = 41/35. 41/35 squared equals 1.37. The unknown takes longer to effuse and therefore has a higher molar weight than ethane.

$$1.37 = \frac{MW_u}{MW_{ethane}} = \frac{MW_u}{30}$$

$$MW_u = 41.1 \text{ g/mole}$$

PROBLEMS

Section 8.2

8.1. What volume will be occupied by 28.4 grams of propane, C_3H_8, at STP?

8.2. How many grams of nitrogen will be in a 1.50 liter flask at STP?

8.3. Find the STP density of neon.

8.4. If the density of an unknown gas at STP is 1.63 grams per liter, what is the molar weight of the gas?

8.5. When filled with an unknown gas at STP, a 2.10 liter vessel weighs 2.63 grams more than it does when evacuated. Find the molar weight of the gas.

8.24. Calculate the STP volume of 3.07 grams of argon.

8.25. Calculate the mass of 162 liters of chlorine, measured at STP.

8.26. Calculate the density of nitric oxide, NO, at STP.

8.27. The STP density of an unidentified gas is 2.14 grams per liter. Calculate the molar weight of the gas.

8.28. A student evacuates a gas-weighing bottle and finds its mass to be 135.821 grams. She then fills the bottle with an unknown gas, adjusts the temperature to 0°C and the pressure to 1.00 atmosphere and weighs it again at 136.201 grams. She then fills the bottle with water and finds its mass to be 385.42 grams. Find the molar weight of the gas.

Section 8.3

8.6. Just above its boiling point at 445°C, sulfur appears to be a mixture of polyatomic molecules. Above 1000°C, however, there is but one structure. Determine the formula of molecular sulfur if its vapor density is 0.625 gram per liter at 1.10 atmospheres and 1100°C.

8.7. An organic compound has the following percentage composition: 55.8% carbon, 7.0% hydrogen and 37.2% oxygen. 3.26 grams of the compound occupies 1.47 liters at 160°C and 0.914 atmosphere. Find the molecular formula of the compound.

8.8. Two compounds both contain 7.69% hydrogen and 92.31% carbon by weight. 4.35 grams of one, a gas at room conditions, occu-

8.29. Phosphorus vapor apparently consists of polyatomic molecules, the number of atoms in the molecule depending on the temperature. Measured as 790 torr, the vapor density is 2.74 grams per liter at 300°C and 0.617 gram per liter at 1000°C. Determine the molecular formulas at the two temperatures.

8.30. If 0.271 gram of the liquid rocket propellant hydrazine is vaporized at 748 torr and 180°C it occupies 320 milliliters. Analysis of the sample shows that it contains 0.0334 gram of hydrogen, and the balance is nitrogen. Calculate the molecular formula of hydrazine.

8.31. Two organic liquids are analyzed, and both are found to be 85.7% carbon and 14.3% hydrogen by weight. At 750 torr and 150°C,

pies 4.16 liters at 22°C and 738 torr. The other, a liquid at room conditions, has a vapor density of 1.88 grams per liter at 195°C and 702 torr. Determine the molecular formula of each compound.

both are vapors. At these conditions an 800 ml gas-weighing bottle holds 1.60 grams of compound A, whereas the corresponding mass of compound B is 2.22 grams. Find the molecular formulas of the two compounds.

Section 8.4. The reactions described in most of the problems in this section are based on industrial processes. Temperatures and pressures are not necessarily as they are in real systems, primarily because of the complexities introduced by the gaseous mixtures used in actual practice. We shall use these examples to illustrate the principles of gas stoichiometry, leaving to the chemical engineer the task of modifying them to where they are practical and profitable.

8.9. Considering natural gas in a laboratory burner to be pure methane, CH_4, calculate the number of grams of carbon dioxide that would result from the complete burning of 35.0 liters of methane, measured at 749 torr and 22°C.

8.10. Dolomite, used in the manufacture of refractory brick for lining very high temperature furnaces, is processed through a rotary kiln in which carbon dioxide is driven off: $CaCO_3 \cdot MgCO_3 \rightarrow CaO \cdot MgO + 2\ CO_2$. For each kilogram of dolomite processed, how many liters of carbon dioxide escape to the atmosphere at 225°C and 825 torr?

8.11. Nitrogen dioxide is used in the chamber process for manufacturing sulfuric acid. It is made by direct reaction of oxygen with nitric oxide: $2\ NO(g) + O_2(g) \rightarrow 2\ NO_2(g)$. How many liters of nitrogen dioxide will be produced by the reaction of 155 liters of oxygen, both gases being measured at atmospheric conditions, 24°C and 752 torr?

8.12. Carbon monoxide is the gaseous reactant in a blast furnace that reduces iron ore to iron. It is produced by the reaction of coke with oxygen from preheated air. How many liters of atmospheric oxygen at an effective pressure of 160 torr and 23°C are required to produce 500 liters of carbon monoxide at 440 torr and 1700°C? The equation is $C + \frac{1}{2} O_2 \rightarrow CO$.

8.13. A commercial fuel known as producer gas comes from coke by way of the following reaction: $2\ C(s) + O_2(g) + 4\ N_2(g) \rightarrow 2\ CO(g) + 4\ N_2(g)$. The oxygen-nitrogen combination among the reactants represents the two major constituents of air in their appropriate volume

8.32. How many liters of hydrogen, measured at 0.940 atmosphere and 32°C, will result from the electrolytic decomposition of 10.0 milliliters of water?

8.33. The reaction chamber in a modified Haber process for making ammonia by direct combination of its elements is operated at 550°C and 250 atmospheres. How many grams of ammonia will be produced by the reaction of 75.0 liters of nitrogen if introduced at the temperature and pressure of the chamber?

8.34. One of the methods for making sodium sulfate, used largely in the production of kraft paper for grocery bags, involves passing air and sulfur dioxide from a furnace over lumps of salt. The equation is $4\ NaCl + 2\ SO_2 + 2\ H_2O + O_2 \rightarrow 2\ Na_2SO_4 + 4\ HCl$. (Note the hydrochloric acid, an important by-product of the process.) If 2000 cubic feet of oxygen at 400°C and 835 torr react in a given period of time, how many cubic feet of sulfur dioxide react if measured at the same conditions?

8.35. In the natural oxidation of hydrogen sulfide released by decaying organic matter, the following reaction occurs: $2\ H_2S + 3\ O_2 \rightarrow 2\ SO_2 + 2\ H_2O$. How many milliliters of oxygen at 4.52 atmospheres and 18°C are required to react with 2.09 liters of hydrogen sulfide measured at 31°C and 0.923 atmosphere in a laboratory reproduction of the reaction?

8.36. A major cause of smoke pollution is the incomplete combustion of coal and some of the complex ingredients in coal tar. The burning of naphthalene vapor (moth balls), $C_{10}H_8$, is an example. Shown with inert nitrogen to get a more accurate picture of air volume proportion, whereas the carbon monoxide-nitrogen represent the major constituents of producer gas. In terms of gaseous "molecules," the equation could be written 2 C(s) + 5 Air(g) $\rightarrow$ 6 PG(g). For the reaction consuming 1000 liters of air at 747 torr and 19°C, how many liters of producer gas, measured at 804 torr and 225°C, will be formed?

8.14. How much energy will be released by the complete combustion of 7.23 liters of methane, measured at 1.35 atmospheres and 7°C? The thermochemical equation is $CH_4(g) + 2\,O_2(g) \rightarrow CO_2(g) + 2\,H_2O(l)$; $\Delta H = -210.8$ kcal.

Section 8.5

8.15. 24.0 grams of NH_3, 21.0 grams of CO_2 and 34.0 grams of N_2 exert a total pressure of 642 torr. Find the partial pressure of each component.

8.16. A 4.60 liter cylinder contains 1.6 moles of helium, 0.34 mole of neon, and 1.2 moles of argon at 27°C. Find the partial pressure of neon in atmospheres.

8.17. A chimney gas has the following composition by volume: 9.5% CO_2; 0.2% CO; 9.6% O_2; 80.7% N_2. If the gas escapes to the atmosphere at 120°C and 791 torr, calculate the grams of poisonous carbon monoxide released per 1000 liters of gas up the chimney.

8.18. In a laboratory experiment a student collects oxygen by water displacement, i.e., by bubbling the gas into an overturned bottle filled with water. If the total pressure of the gas is 786 torr and the temperature of the system is 21°C, determine the partial pressure of the oxygen.

8.19. In a water displacement collection of hydrogen at 32°C, a student collects 80.4 ml of gas at a total pressure of 755 torr. Calculate the volume that the hydrogen would occupy when dry at STP.

8.20. A student collects 2.63 liters of oxygen over water at a temperature of 29°C and a total pressure of 769 torr. If the oxygen came from the complete thermal decomposition of potassium chlorate, $KClO_3$, how many grams of starting material were used? The equation is $2\,KClO_3 \rightarrow 2\,KCl + 3\,O_2$.

required, the equation is $C_{10}H_8(g) + 12\,O_2(g) + 48\,N_2(g) \rightarrow 10\,CO_2(g) + 4\,H_2O(g) + 48\,N_2(g)$. How many liters of air at 764 torr and 16°C are required to burn one liter of naphthalene vapor, measured at 815 torr and 243°C? "Air" volume includes both oxygen and nitrogen.

8.37. $\Delta H = -1376$ kcal for the burning of butane (bottled gas) by the reaction $2\,C_4H_{10}(g) + 13\,O_2(g) \rightarrow 8\,CO_2(g) + 10\,H_2O(l)$. How many liters of butane, measured at 3.86 atmospheres and 3°C, must be burned to release 15,000 kcal?

8.38. Equal weights of methane, CH_4, ethane, C_2H_6, and propane, C_3H_8 produce a total pressure of 800 torr. What is the partial pressure of each gas?

8.39. A mixture of 37.0 grams of oxygen and 63.0 grams of nitrogen—37.0% oxygen by weight—is forced into a volume of 8.34 liters at 24°C. Calculate the partial pressure of oxygen in atmospheres.

8.40. The gaseous mixture from a sulfur burner is 0.8% SO_3, 7.8% SO_2, 12.2% O_2 and 79.2% N_2 by volume. At a total pressure of 1.12 atmospheres and 86°C, find the grams of SO_2 released in 25.0 liters of mixed gas.

8.41. Air, which we shall assume is 21.0% oxygen and 79.0% nitrogen by volume, is bubbled through water and collected by water displacement at 28°C. If the total pressure of the mixture is 739 torr, calculate the partial pressure of the oxygen.

8.42. How many grams of nitrogen are collected by water displacement if the volume is 133 milliliters at 22°C and a total pressure of 776 torr?

8.43. An experiment is being designed for use in a general chemistry laboratory. Students are to collect a maximum of 95.0 milliliters of hydrogen by water displacement at 26°C and a total pressure of 1 atmosphere. What is the maximum mass of magnesium that may be used in the reaction if the equation is $Mg + 2\,HCl \rightarrow H_2 + MgCl_2$?

Section 8.6

8.21. Find the ratio of the effusion rate of CH_4 to that of C_2H_6.

8.22. 62 seconds are required for the escape of a specified number of moles of nitrogen from an effusion apparatus. If the escape of the same number of moles of an unknown gas requires 51 seconds, find the molar weight of the unknown.

8.23. 45 seconds are required for a given number of moles of hydrogen to effuse through a small opening. How long will it take for the same number of moles of oxygen to escape if the starting conditions are the same?

8.44. Under comparable starting conditions, which gas, CO or CO_2, will effuse more rapidly? By what ratio?

8.45. Helium is used as a standard in a molecular weight determination by effusion. 23 seconds are required for a given number of moles of helium to escape. An unknown gas from identical starting conditions requires 63 seconds. Find the molecular weight of the unknown gas.

8.46. A vial of H_2S was broken at one end of a large college laboratory during a weekend when the ventilation system was not in operation. It took a student working at the other end of the laboratory 77 seconds to notice the foul odor. How long would the odor detection have taken if the gas had been ammonia instead of hydrogen sulfide, assuming comparable starting partial pressures?

CHAPTER 9

SOLUTION CONCENTRATIONS

A solution is a homogeneous mixture of two or more substances. The dissolved substance, called the solute, is ordinarily present in small quantity compared to the medium in which it is dissolved, which is named the solvent. This is generally true of aqueous solutions, where the solvent is water; but it is not necessarily true for all solutions involving gases, liquids, and solids which, theoretically at least, may function in either capacity. This chapter will be confined to solutions of a solid, liquid, or gas in a large quantity of liquid solvent.

Concentration of a solution expresses the quantity of solute in a given quantity of solvent or total solution.

9.1 PERCENTAGE CONCENTRATION

If a solution concentration is given in per cent, assume it to be weight per cent unless specifically stated otherwise. Thus percentage concentration may be defined as follows:

$$\% \text{ concentration} = \frac{\text{grams solute}}{\text{grams solution}} \times 100 \qquad (9.1)$$

$$= \frac{\text{grams solute}}{\text{grams solute} + \text{grams solvent}} \times 100. \qquad (9.2)$$

Example 9.1. 80 grams of a solution when evaporated to dryness yield 24 grams of salt. What was the percentage of salt in the solution?

Use the definition, set up, and solve.

9.1a. 30%.

$$\frac{\text{g solute}}{\text{g solution}} \times 100 = \frac{24}{80} \times 100 = 30\%$$

Example 9.2. 25 grams of sugar are dissolved in 100 grams of water. What is the per cent of sugar in the solution?

9.2a. 20%.

$$\frac{\text{g solute}}{\text{g solute} + \text{g solvent}} \times 100 = \frac{25}{25 + 100} \times 100 = \frac{25}{125} \times 100 = 20\%$$

Example 9.3. How many grams of solute are in 420 grams of 15% solution?

You may approach this as a typical percentage problem, or from a dimensional analysis viewpoint, where percentage is grams of solute per 100 grams of solution. On this basis, 15 grams of solute = 100 grams of solution.

9.3a. 63 grams of solute.

$$420 \cancel{\text{ g solution}} \times \frac{15 \text{ g solute}}{100 \cancel{\text{ g solution}}} = 63 \text{ g solute;}$$

or,

$$0.15 \times 420 = 63 \text{ g solute .}$$

Sometimes it is necessary to prepare a solution of a given percentage concentration which contains a certain quantity of solute. The algebra is a bit more complex, but the problem involves only direct substitution into Equation 9.2.

Example 9.4. How many grams of water would you use to dissolve 150 grams of potassium hydroxide to produce a 25% solution?

9.4a. 450 grams of water.

$$\frac{\text{g solute}}{\text{g solute} + \text{g solvent}} \times 100 = \%$$

Let y = g solvent = g H_2O:

$$\frac{150 \text{ g}}{150 \text{ g} + y} \times 100 = 25$$

$$y = 450 \text{ g } H_2O$$

This problem can be reasoned out in another way too: If 150 grams represent 25 per cent of the solution, the weight of the entire solution may be determined as follows: Let the weight of the solution be X:

$$25\% \text{ of } X = 150 \text{ g}$$

$$0.25X = 150 \text{ g}$$

$$X = \frac{150}{0.25} = 600 \text{ g}$$

$$600 \text{ g solution} - 150 \text{ g solute} = 450 \text{ g solvent}$$

Example 9.5. How many grams of NaCl would you dissolve in 60 grams of water to produce a 20% solution?

There are several approaches to this problem too. The plug-in is one; reasoning is another. Take your choice.

9.5a. 15 grams of NaCl.
By plug-in: Let z = grams of NaCl

$$\frac{z}{z+60} \times 100 = 20$$

$$z = 15 \text{ g NaCl}$$

By reasoning: In a 20% solution 100 − 20 = 80 is the percentage *solvent*.

$$0.80 \times (\text{g solution}) = 60 \text{ g}$$

$$\text{g solution} = \frac{60 \text{ g}}{0.80} = 75 \text{ g}$$

$$75 \text{ g solution} - 60 \text{ g solvent} = 15 \text{ g solute}$$

9.2 MOLE FRACTION

The concept of mole fraction was introduced in the partial pressure section of Chapter 8. By definition,

$$X = \frac{\text{moles solute}}{\text{moles solute} + \text{moles solvent}}, \qquad (9.3)$$

where X is the mole fraction of solute.

Example 9.6. 23 grams of methanol, CH_3OH, are dissolved in 45 grams of water. Find the mole fraction of both methanol and water in the solution.

The definition shows clearly that the moles of both solute and solvent must be computed. Find these first.

9.6a. 0.72 mole methanol; 2.50 moles water.

$$23 \cancel{\text{g } CH_3OH} \times \frac{1 \text{ mole } CH_3OH}{32 \cancel{\text{g } CH_3OH}} = 0.72 \text{ mole } CH_3OH$$

$$45 \cancel{\text{g } H_2O} \times \frac{1 \text{ mole } H_2O}{18 \cancel{\text{g } H_2O}} = 2.50 \text{ moles } H_2O$$

Now insert the molar quantities into Equation 9.3 and solve for both mole fractions.

9.6b. $X_{methanol} = 0.224$; $X_{water} = 0.776$.

Total moles = 2.50 + 0.72 = 3.22.

$$X_{methanol} = \frac{0.72}{3.22} = 0.224 \qquad X_{water} = \frac{2.50}{3.22} = 0.776$$

What do you notice about the sum of the mole fractions?

By the nature of the definition, the sum of the mole fractions of any mixture must be 1. The following algebra shows why:

$$\frac{a}{a+b} + \frac{b}{a+b} = \frac{a+b}{a+b} = 1.$$

In Example 9.6,

$$\frac{0.72}{3.22} + \frac{2.50}{3.22} = \frac{0.72 + 2.50}{3.22} = \frac{3.22}{3.22} = 1.$$

9.3 MOLALITY

Molality is a concentration unit employed to express the relationship between concentration and many physical properties of solutions. **Molality is defined as the number of moles of solute per kilogram of solvent.** Mathematically this may be expressed as

$$m = \frac{\text{moles of solute}}{\text{kilograms of solvent}}, \qquad (9.4)$$

where m is the symbol for molality. In essence, the molality of a solution is the *number of moles of solute that are equivalent to one kilogram of solvent.* In a 1.5 molal solution, for example, 1.5 moles of solute = 1 kilogram of solvent.

The quantity of solvent in molality problems is usually given in grams, or in volume units that may be converted to grams by multiplying by density. If the solvent is water, with its density of 1.00 gram/milliliter, the volume in milliliters is the same as the mass in grams. Molality, however, is based on *kilograms* of solvent. Converting grams to kilograms is done so often, therefore, that it is worth developing the ability to do it quickly and mentally, without making the change a part of the problem setup. For example, suppose a solution is prepared in 75 grams of solvent. How many kilograms is this? The setup would be

$$75 \text{ \sout{grams}} \times \frac{1 \text{ kilogram}}{1000 \text{ \sout{grams}}} = 0.075 \text{ kilograms.}$$

The conversion is simply division of the number of grams by 1000: **to convert grams to kilograms, move the decimal three places to the left.** This will be done without further comment in the examples that follow.

Example 9.7. What is the molality of a solution made by dissolving 24.0 grams of KNO_3 in 56.0 ml of water?

First, how many kilograms of solvent are there?

9.7a. 0.0560 kg water.

$$56.0 \text{ ml} = 56.0 \text{ g} = 0.0560 \text{ kg}$$

In this problem we seek a concentration in moles of solute per kilogram of solvent. From the data of the problem the concentration can be expressed in terms of grams of solute per kilogram of solvent. Set up, but do not solve, this concentration expression: grams solute/kilogram solvent.

9.7b. $\dfrac{24.0 \text{ g } KNO_3}{0.0560 \text{ kg } H_2O}$

Completion of the problem now requires the familiar gram to mole conversion, yielding moles of solute per kilogram of solvent. Finish the setup and solve.

9.7c. 4.24 m.

$$\frac{24.0 \cancel{\text{ g } KNO_3}}{0.0560 \text{ kg } H_2O} \times \frac{1 \text{ mole } KNO_3}{101 \cancel{\text{ g } KNO_3}} = 4.24 \text{ m}$$

Molality problems may require computation of the quantity of either solvent or solute. The next two examples illustrate this.

Example 9.8. How many grams of water must be used in dissolving 150 grams of KOH to produce a 5.95 molal solution?

There is a given quantity in this problem. What is it?

9.8a. 150 grams of KOH. 5.95 m is a concentration for any quantity of the solution; it is not a given quantity on which other quantities will be based.

Molality is concerned with moles of solute, rather than grams. Set up, but do not solve, the conversion of 150 grams of KOH to moles.

9.8b. $150 \cancel{\text{ g KOH}} \times \dfrac{1 \text{ mole KOH}}{56.1 \cancel{\text{ g KOH}}}$

The molal concentration furnishes an equivalence that is useful in this problem. Write that equivalence.

9.8c. 5.95 moles solute = 1 kg solvent.

From the equivalence given, a typical dimensional analysis conversion factor emerges by which the setup for 9.8b may be extended, yielding kilograms of solvent. Remembering that the original problem asks for *grams* of water, complete both the setup and the solution.

9.8d. 449 grams of water.

$$150\ \cancel{\text{g KOH}} \times \frac{1\ \cancel{\text{mole KOH}}}{56.1\ \cancel{\text{g KOH}}} \times \frac{1\ \cancel{\text{kg}}\ H_2O}{5.95\ \cancel{\text{moles KOH}}} \times \frac{1000\ \text{g}}{1\ \cancel{\text{kg}}} = 449\ \text{g}\ H_2O$$

Example 9.9. How many grams of NaCl must be added to 60.0 grams of water to produce a solution that is 4.27 m?

This problem is in essence like Example 9.8 and is solved with the same sort of reasoning. Starting with the given quantity, using the equivalence character of molality, and remembering that the solvent quantity must be expressed in kilograms, set up and solve the entire problem.

9.9a. 15.0 grams NaCl.

$$0.0600\ \cancel{\text{kg water}} \times \frac{4.27\ \cancel{\text{moles NaCl}}}{1\ \cancel{\text{kg water}}} \times \frac{58.5\ \text{g NaCl}}{1\ \cancel{\text{mole NaCl}}} = 15.0\ \text{g NaCl}$$

9.4 INTERCONVERSION OF CONCENTRATION UNITS

If Examples 9.7, 9.8 and 9.9 appeared familiar, there is a reason. They involve the same quantities of solute and solvent as Examples 9.1, 9.4, and 9.5. This shows that if the identities and quantities of both solvent and solute are known, any concentration value thus far considered may be calculated from the basic definitions. The following example illustrates the procedures involved:

Example 9.10. Determine the mole fraction of solute and the molality of a 30.0% sugar solution. The molar weight of sugar is 342 grams per mole.

First, thinking of percentage as parts per hundred, how many grams of sugar and how many grams of water are in 100 grams of solution?

9.10a. 30.0 grams of sugar; and 100.0 − 30.0 = 70.0 grams of water.

The process of converting this information into molality is now the same as in Example 9.7. Set up and solve.

9.10b. 1.25 m.

$$\frac{30\ \cancel{\text{g sugar}}}{0.0700\ \text{kg water}} \times \frac{1\ \text{mole sugar}}{342\ \cancel{\text{g sugar}}} = 1.25\ \text{m}$$

Converting 30.0 grams of sugar and 70.0 grams of water to mole fraction is identical to Example 9.6. Set up and solve.

9.10c. $X_{sugar} = 0.0221$.

$$\text{moles sugar} = 30.0\,\cancel{\text{g sugar}} \times \frac{1 \text{ mole sugar}}{342\,\cancel{\text{g sugar}}} = 0.0877$$

$$\text{moles water} = 70.0\,\cancel{\text{g water}} \times \frac{1 \text{ mole water}}{18.0\,\cancel{\text{g water}}} = 3.88$$

$$\text{Total moles} = 3.97$$

$$X_{sugar} = \frac{\text{moles sugar}}{\text{total moles}} = \frac{0.0877}{3.97} = 0.0221$$

9.5 MOLARITY

All concentration units considered to this point have involved mass or moles of solvent and solute, both of which are determined by weighing. But, with liquid solutions, quantities are more readily measured by volume. It is logical, then, that solution concentrations also be expressed in terms of volume of solution. The most widely used concentration unit of this nature is **molarity.** Abbreviated M, **molarity is defined as moles of solute per liter of solution.** The units appear in the mathematical expression of molarity:

$$M = \frac{\text{moles solute}}{\text{liters solution}}. \tag{9.5}$$

In essence, molarity is the *number of moles of solute that are equivalent to one liter of solution.* In a 1.5 molar solution, for example, 1.5 moles of solute = 1 liter of solution.

Example 9.11. 46.2 grams of sodium hydroxide are dissolved in water to give a final volume of 350 ml. What is the molarity of the solution?

Note the wording of the problem. It did not say the solute was dissolved in 350 ml of water, but in a final 350 ml of solution. The amount of water present is *not* specified. Actually it must be somewhat less than 350 ml, but there is no way of knowing from the information given exactly how great it is.

Observe that in molarity problems volumes are often given in milliliters, whereas the definition involves liters. This is similar to the gram-kilogram relationship in molality. It is advisable to convert volume units to liters, which requires that the decimal point be moved three places to the left if the given volume is in milliliters.

In this example you have two given quantities, 46.2 g NaOH and a final solution volume of 350 ml. Put them together in such a way as to give you a concentration in grams NaOH per liter solution.

9.11a. $$\frac{46.2 \text{ grams NaOH}}{0.350 \text{ liter solution}}.$$

You should require no further guidance on the conversion of grams to moles. Complete the setup and solve.

9.11b. 3.30 M NaOH.

$$\frac{46.2\ \text{g NaOH}}{0.350\,l\ \text{solution}} \times \frac{1\ \text{mole NaOH}}{40.0\ \text{g NaOH}} = 3.30\ \text{moles NaOH}/l\ \text{solution}$$

Example 9.12. How would you prepare 1500 ml of 0.10 M NaCl?

For all practical purposes this question is asking how many grams of NaCl are in 1.50 liters of 0.10 M NaCl. What is the given quantity?

9.12a. 1500 ml or 1.50 *l*.
The molarity is a concentration unit, valid for any quantity.
Proceed with the setup and solve.

9.12b. 8.78 grams of NaCl, dissolved in 1.50 liters of solution.

$$1.50\ l \times \frac{0.10\ \text{mole NaCl}}{1\ l} \times \frac{58.5\ \text{g NaCl}}{1\ \text{mole NaCl}} = 8.78\ \text{g NaCl}$$

Note again the wording: You would not use 1.50 liters of water, but just enough water to make up 1.50 liters of solution.

It is sometimes necessary to convert a known number of moles of a solute to volume of solution. The following is typical:

Example 9.13. How many milliliters of 0.77 M H_2SO_4 would have to be poured out to obtain 0.50 mole?

From the given quantity, the problem is solved by a one-or two-step conversion. Set up and solve.

9.13a. 650 ml.

$$0.50\ \text{mole}\ H_2SO_4 \times \frac{1000\ \text{ml}}{0.77\ \text{mole}\ H_2SO_4} = 650\ \text{ml}$$

The solution shown above is the one-step setup. Notice that it expresses molarity in terms of moles/1000 ml, which is obviously equivalent to moles/liter. This eliminates the conversion from liters to milliliters for the final answer.

Example 9.14. How many milliliters of 0.54 M $AgNO_3$ would you require to obtain 0.34 gram of solute?

This problem is like Example 9.13, except that the preliminary conversion of grams of solute to moles is required. Set up and solve.

9.14a. 3.70 ml.

$$0.34\ \cancel{\text{g AgNO}_3} \times \frac{1\ \cancel{\text{mole AgNO}_3}}{170\ \cancel{\text{g AgNO}_3}} \times \frac{1000\text{ ml solution}}{0.54\ \cancel{\text{mole AgNO}_3}} = 3.70\text{ ml solution}$$

Perhaps the easiest and one of the most important problems involving molarity is the conversion of volume of solution into moles of solute.

Example 9.15. How many moles of solute are there in 125 ml of a 0.864 M solution?

Set up and solve.

9.15a. 0.108 mole.

$$0.125\ \cancel{\text{liter}} \times \frac{0.864\text{ mole}}{1\ \cancel{\text{liter}}} = 0.108\text{ mole;}$$

or,

$$125\ \cancel{\text{ml}} \times \frac{0.864\text{ mole}}{1000\ \cancel{\text{ml}}} = 0.108\text{ mole.}$$

Do not allow the brevity of this example to tempt you to think it is unimportant. *It is VERY important.*

9.6 THE CHEMICAL EQUIVALENT

The concept of the **chemical equivalent** is found at different levels of study; several appear in the beginning chemistry course. One of these, based on oxidation-reduction reactions, will be deferred until that general subject is studied. At this point consideration will be restricted to conventional acid-base chemistry and chemical equivalents associated therewith. In this limited context **one equivalent of any substance is that quantity that is chemically equal to one mole of replaceable hydrogen ions in an acid-base reaction.** Notice the reliance of the equivalent on a specific chemical reaction. Only when the reaction is clearly identified is all ambiguity removed from the concept of the equivalent.

It is convenient to relate the equivalent to a more familiar quantity, the mole. This is most easily done with common acids. One mole of hydrogen chloride, for example, ionizes in water to yield one mole of replaceable hydrogen:

$$HCl(g) \rightarrow H^+(aq) + Cl^-(aq).$$

As a consequence, one mole of HCl contains one equivalent of HCl. Sulfuric acid, on the other hand, contains two replaceable hydrogens per mole and, therefore, two equivalents per mole:

$$H_2SO_4(l) \rightarrow 2H^+(aq) + SO_4^{2-}(aq).$$

The importance of the word "replaceable" becomes apparent in the case of acetic acid, the formula of which is commonly written $HC_2H_3O_2$. There is but one mole of replaceable hydrogen in a mole of acetic acid, despite the presence of four hydrogen atoms in the molecule. The three hydrogen atoms bonded to one of the carbon atoms do not participate in a normal acid-base reaction; only one hydrogen ionizes:

$$HC_2H_3O_2 \rightarrow H^+ + C_2H_3O_2^-.$$

Consequently there is one equivalent in one mole of acetic acid.

The definition of an equivalent refers to a mole of replaceable hydrogen ions *in an acid-base reaction*. The same acid in different reactions may yield a different number of moles of hydrogen ion per mole of acid. Phosphoric acid is an example. In some reactions the molecule yields a single hydrogen ion, leaving the dihydrogen phosphate ion:

$$H_3PO_4(l) \rightarrow H^+(aq) + H_2PO_4^-(aq).$$

There is one equivalent per mole of acid in this reaction. In another reaction two moles of hydrogen ion separate from one mole of acid molecules, leaving the monohydrogen phosphate ion:

$$H_3PO_4(l) \rightarrow 2\ H^+(aq) + HPO_4^{2-}(aq).$$

There are two equivalents per mole of acid in this reaction. The third hydrogen is also ionizable in the presence of a very strong base:

$$H_3PO_4(l) \rightarrow 3\ H^+(aq) + PO_4^{3-}(aq).$$

In this reaction the phosphate ion is left, and there are three equivalents per mole of acid.

In all acids considered so far, each mole of replaced hydrogen is able to react with one mole of sodium hydroxide. If the mole of hydrogen represents one equivalent, then the amount of NaOH chemically equal to it must also be one equivalent. This, in fact, leads to a secondary definition of the equivalent as *that quantity of a base that yields one mole of hydroxide ions in a chemical reaction.* Thus, NaOH and KOH have one equivalent per mole; $Mg(OH)_2$, $Ca(OH)_2$, and $Ba(OH)_2$ have two equivalents per mole; and $Al(OH)_3$ has three equivalents per mole. The number of equivalents per mole is equal to the number of moles of hydroxide ions in one mole of the base.

In summary, the number of equivalents per mole of an acid is equal to the number of moles of replaceable hydrogens per mole; and the number of equivalents per mole of a base is equal to the number of moles hydroxide ions per mole. Using these criteria, complete the following exercise:

Example 9.16. For each compound listed, write the number of equivalents per mole: HNO_3; $Sr(OH)_2$; $HC_7H_5O_2$.

9.16a.

HNO_3: One equivalent/mole, based on one mole H^+ per mole of acid.
$Sr(OH)_2$: Two equivalents/mole, based on two moles OH^- per mole of base.
$HC_7H_5O_2$: One equivalent/mole, based on one mole H^+ per mole of acid. This is comparable to the acetic acid example above.

9.7 EQUIVALENT WEIGHT

The term **equivalent weight** or "gram-equivalent weight," as it is called in some texts, is analogous to molar weight and is associated with molar weight through the number of equivalents per mole of a compound. **Equivalent weight may be defined as the number of grams of a substance in one equivalent.** As usual, the units are suggested by the definition, grams/equivalent. Finding equivalent weight is straightforward if the concept of equivalents per mole is understood. The following examples are drawn from compounds in Example 9.16:

Example 9.17. What is the equivalent weight of $Sr(OH)_2$?

As a first step, determine the molar weight of strontium hydroxide.

9.17a. 122 g/mole.
The molar weight may now be converted to grams per equivalent (g/eq), knowing the number of equivalents per mole, found in Example 9.16. Set up and solve.

9.17b. 61 grams per equivalent.

$$\frac{122 \text{ g } Sr(OH)_2}{1 \cancel{\text{ mole } Sr(OH)_2}} \times \frac{1 \cancel{\text{ mole } Sr(OH)_2}}{2 \text{ eq } Sr(OH)_2} = 61 \text{ g/eq}$$

Interconversion between grams and equivalents is straightforward:

Example 9.18. How many equivalents are there in 126 grams of HCl?

This type of problem can be done in two steps by either of two processes. For the first, convert grams to moles, and then moles to equivalents. For the second, determine the equivalent weight of the solute, and then use it to convert grams to equivalents. The first approach is more consistent with the techniques developed in this book, but the choice is strictly optional. Choose your method, set up, and solve completely.

9.18a. 3.45 equivalents of HCl.
First method:

$$126 \cancel{\text{ g HCl}} \times \frac{1 \cancel{\text{ mole HCl}}}{36.5 \cancel{\text{ g HCl}}} \times \frac{1 \text{ eq HCl}}{1 \cancel{\text{ mole HCl}}} = 3.45 \text{ eq HCl}$$

Second method:

$$\frac{36.5 \text{ g HCl}}{1 \cancel{\text{mole HCl}}} \times \frac{1 \cancel{\text{mole HCl}}}{1 \text{ eq HCl}} = 36.5 \text{ g HCl/eq}$$

$$126 \cancel{\text{g HCl}} \times \frac{1 \text{ eq HCl}}{36.5 \cancel{\text{g HCl}}} = 3.45 \text{ eq HCl}$$

The second method, involving the calculation of an intermediate answer, is the more common thought process. The single setup advantage of the first method is apparent. Furthermore, the two steps in the first method can be combined into one. The purpose is to convert grams into equivalents. Observe the three-part equality that is present in the two steps of the problem: 36.5 g HCl = 1 mole HCl = 1 eq HCl. Why not eliminate the mole as a connecting link and convert directly from grams to equivalents? What is involved is equating the molar weight with the number of equivalents per mole. This approach leads directly to the second of the two steps immediately above.

$$126 \cancel{\text{g HCl}} \times \frac{1 \text{ eq HCl}}{36.5 \cancel{\text{g HCl}}} = 3.45 \text{ eq HCl}$$

Example 9.19. How many grams of H_2SO_4 are in 1.37 equivalents?

This is the reverse of the previous problem. Set up and solve.

9.19a. 67.2 g H_2SO_4.

$$1.37 \cancel{\text{eq } H_2SO_4} \times \frac{98.1 \text{ g } H_2SO_4}{2 \cancel{\text{eq } H_2SO_4}} = 67.2 \text{ g } H_2SO_4$$

A final and important generalization about chemical equivalents remains to be made. It is best put forth in the form of a chemical equation:

$$3\ H_2SO_4 + 2\ Al(OH)_3 \rightarrow Al_2(SO_4)_3 + 6\ H_2O.$$

Question: How many *moles* of H_2SO_4 and $Al(OH)_3$ are represented in this equation?

Answer: Three and two, respectively. The coefficients in the equation indicate the numbers of moles involved in the reaction.

Question: How many *equivalents* of H_2SO_4 and $Al(OH)_3$ are represented in this equation?

Answer: Six of each.

$$3 \cancel{\text{moles } H_2SO_4} \times \frac{2 \text{ eq } H_2SO_4}{1 \cancel{\text{mole } H_2SO_4}} = 6 \text{ eq } H_2SO_4$$

$$2 \cancel{\text{moles } Al(OH)_3} \times \frac{3 \text{ eq } Al(OH)_3}{1 \cancel{\text{mole } Al(OH)_3}} = 6 \text{ eq } Al(OH)_3$$

In certain areas of quantitative analysis the concept of equivalents and equivalent weights is extended to salts, such as aluminum sulfate. Had it been done here, it would have been found that the equation involves six equivalents of aluminum sulfate too.

This illustrates the essence of chemical equivalence. *Regardless of the different numbers of grams or moles of the various species in a chemical equation, the number of equivalents of each species is the same. If you find the number of equivalents of one species in the reaction, you have found the number of equivalents of every species.*

9.8 NORMALITY

Another way of expressing solution concentration is in terms of **normality, designated N and defined as the number of equivalents of solute per liter of solution.** Again the units are implied in the definition and appear in the mathematical statement of that definition:

$$N = \frac{\text{equivalents}}{\text{liter}} = \text{eq}/l. \qquad (9.6)$$

Any time normality, N, appears in a problem, units of equivalents/liter should be applied. In essence, normality is the number of equivalents of solute that is "equivalent" to one liter of solution. In a 1.5 normal solution, for example, 1.5 equivalents of solute = 1 liter of solution.

Normality problems are similar to molarity problems, except that they contain an additional step relating moles to equivalents.

Example 9.20. 28.9 grams of H_2SO_4 are dissolved in 450 ml solution. What is the normality of the solution?

This problem starts out like Example 9.11. Begin by expressing the concentration in the units of grams per liter. Set up but do not solve.

9.20a. $\frac{28.9 \text{ g } H_2SO_4}{0.450\ l \text{ solution}}$

At this point you have a concentration in grams/liter. The answer you require is in equivalents/liter. The necessary conversion is from grams to equivalents. Set up completely and solve.

9.20b. 1.31 N H_2SO_4.

$$\frac{28.9 \cancel{\text{g } H_2SO_4}}{0.450\ l \text{ solution}} \times \frac{2 \text{ eq } H_2SO_4}{98.1 \cancel{\text{g } H_2SO_4}} = 1.31 \text{ eq}/l = 1.31 \text{ N}$$

Example 9.21. How would you prepare 750 milliliters of 0.20 N NaOH?

First, write the equivalence represented by 0.20 N NaOH.

9.21a. 0.20 equivalent of NaOH = 1 liter of solution.
The question in this example actually asks how many grams of NaOH are in 750 ml of a 0.20 N solution of NaOH. The given quantity, volume, may be converted to equivalents through normality as indicated above, and the equivalents converted to grams. Set up and solve.

9.21b. 6.0 grams of NaOH.

$$0.750\ \not{l} \times \frac{0.20\ \text{eq NaOH}}{1\ \not{l}} \times \frac{40\ \text{g NaOH}}{1\ \text{eq NaOH}} = 6.0\ \text{g NaOH}$$

9.9 INTERCONVERSION BETWEEN MOLARITY AND NORMALITY

If the solution concentration is given in either molarity or normality it can be converted to the other by the moles-equivalents relationship.

Example 9.22. What is the normality of 0.655 M H_2SO_4?

Express the concentration in terms of moles per liter, set up, and solve.

9.22a. 1.31 N.

$$\frac{0.655\ \text{mole}\ H_2SO_4}{1\ l} \times \frac{2\ \text{eq}\ H_2SO_4}{1\ \text{mole}\ H_2SO_4} = 1.31\ \text{eq}/l = 1.31\ \text{N}$$

If the problem looks familiar, it should. It is a repetition of Example 9.20 in slightly different terms. But you have not yet converted from normality to molarity. Example 9.23 gives you the opportunity to do so.

Example 9.23. Find the molarity of 1.22 N NaOH.

Whether you "see" the answer or not, it is suggested you set up and solve the problem so that you will be able to do it for more complicated examples.

9.23a. 1.22 M NaOH.

$$\frac{1.22\ \text{eq NaOH}}{1\ l} \times \frac{1\ \text{mole NaOH}}{1\ \text{eq NaOH}} = 1.22\ \text{moles NaOH}/l = 1.22\ \text{M}$$

PROBLEMS

Section 9.1

9.1. 135 grams of solution contain 18.5 grams of dissolved salt. What is the percentage of salt in the solution?

9.2. How many grams of solute are in 65.0 grams of a 13.0% solution?

9.3. How many grams of water would you use to dissolve 20.0 grams of potassium nitrate to produce a 12.0% solution?

9.4. The density of 16.0% ammonium sulfate solution is 1.09 grams per cubic centimeter. How many grams of solute are in 400 milliliters of the solution?

9.5. Calculate the grams of boric acid you would dissolve in 100 milliliters of water to produce a 4.00% solution.

9.36. Calculate the percentage concentration of a solution prepared by dissolving 5.29 grams of silver nitrate in 81.0 grams of water.

9.37. How many grams of ammonium nitrate must be weighed out to make 440 grams of a 58.0% solution?

9.38. You wish to prepare an 8.00% solution containing 4.50 grams of NaOH. How many milliliters of water would you use?

9.39. Calculate the grams of sodium bicarbonate (baking soda) and milliliters of water required to prepare 1.00 liter of 5.50% solution if its specific gravity is 1.04.

9.40. A laboratory technician is to prepare about 250 ml of a 7.20% saline (salt) solution. He measures out 250 ml of water. How many grams of salt should he use?

Section 9.2

9.6. Calculate the mole fraction of both components in a solution prepared by dissolving 12.0 grams of acetic acid, $HC_2H_3O_2$, in 150 ml of water.

9.41. A solution is prepared by combining 14.5 grams of ethylene glycol (anti-freeze), $C_2H_6O_2$, in 65.0 ml of water. Find the mole fraction of both solute and solvent.

Section 9.3

9.7. If 20.0 grams of sugar, $C_{12}H_{22}O_{11}$, are dissolved in 100 ml of water, what is the molality of the solution?

9.8. If you are to prepare a 4.00 m solution of urea in water, how many grams of urea, $CO(NH_2)_2$, would you dissolve in 80.0 ml of water?

9.9. Into what volume of water must 90.9 grams of acetic acid, $HC_2H_3O_2$, be dissolved to produce a 1.40 molal solution?

9.42. Calculate the molal concentration of a solution of 26.9 grams of naphthalene, $C_{10}H_8$, in 125 grams of benzene, C_6H_6.

9.43. Diethylamine, $(CH_3CH_2)_2NH$, is highly soluble in ethanol, C_2H_5OH. Calculate the number of grams of diethylamine that would be dissolved in 400 grams of ethanol to produce a 4.50 m solution.

9.44. How many grams of methyl ethyl ketone, C_4H_8O, a solvent popularly known as MEK used to cement plastics, must be dissolved in 100 ml of benzene, C_6H_6, specific gravity 0.879, to yield a 0.720 molal solution?

Section 9.4

9.10. A solution contains equal weights of two alcohols, CH_3OH and C_2H_5OH. Calculate (a) the percentage concentration of each component; (b) the mole fraction of each component; and (c) the molality, considering C_2H_5OH as the solvent.

9.11. What are the mole fraction and percentage concentration of a 1.80 molal solution of ammonium chloride, NH_4Cl?

9.12. Find the mole fraction of each component and molality of a 25.0% solution of formic acid, HCOOH, in water.

9.45. 45.0 grams of levulose, $C_6H_{12}O_6$, are dissolved in 125 grams of water. Find (a) the percentage concentration of the levulose, (b) the mole fraction of solute and solvent, and (c) the molality of the solution.

9.46. The mole fraction of a solution of urea, $CO(NH_2)_2$, in water is 0.120. Calculate the molality and percentage concentration.

9.47. Calculate the percentage concentration and mole fraction of each component in 4.12 m $CH_2OHCHOHCH_2OH$ (glycerol) in water.

Section 9.5

9.13. 600 milliliters of a solution contains 23.5 grams of sodium sulfate, Na_2SO_4. What is the molarity of the solution?

9.14. What is the molarity of a solution prepared by dissolving 120 grams of $Na_2S_2O_3 \cdot 5\ H_2O$ in water and diluting to 1250 milliliters?

9.15. How many grams of sodium carbonate, Na_2CO_3, are required to prepare 400 ml 0.800 molar solution?

9.16. What number of grams of acetic acid must be dissolved in water and diluted to 750 ml to yield 0.600 M $HC_2H_3O_2$?

9.17. A reaction requires 0.0150 moles of hydrochloric acid. How many milliliters of 0.850 M HCl would you use?

9.18. Calculate the milliliters of concentrated ammonia solution, which is 15 molar, that contain 75.0 grams of NH_3.

9.19. How many moles of solute are in 65.0 ml 2.20 M NaOH?

9.20. 29.3 ml 0.482 M H_2SO_4 are used to titrate a base of unknown concentration. How many moles of sulfuric acid react?

9.21. 100 ml concentrated HCl, which is 12 molar, are diluted to 2000 ml. Calculate the molarity of the diluted solution.

9.48. Calculate the molarity of the solution prepared by dissolving 3.59 grams of KI in water and diluting to 50.0 ml.

9.49. 15.0 grams of anhydrous nickel chloride, $NiCl_2$, are dissolved in water and diluted to 75.0 ml. 25.0 grams of nickel chloride hexahydrate, $NiCl_2 \cdot 6H_2O$, are also dissolved in water and diluted to 75.0 ml. Identify the solution with the higher molar concentration and calculate its molarity.

9.50. How many grams of silver nitrate must be used in the preparation of 250 ml of 0.125 M $AgNO_3$?

9.51. You are to prepare 2.50 liters of 2.25 M KOH. Calculate the mass of solute required.

9.52. What volume of concentrated sulfuric acid, which is 18 molar in concentration, is required to obtain 2.8 moles of the acid?

9.53. 0.150 M NaCl is to be the source of 8.33 grams of dissolved solute. What volume of solution is needed?

9.54. Calculate the moles of silver nitrate in 18.6 ml of 0.204 molar solution.

9.55. Find the number of moles of potassium permanganate in 25.0 ml 0.0904 molar solution.

9.56. What molarity results if 45.0 ml 17 M $HC_2H_3O_2$ are diluted to 1.2 liters?

9.22. How many milliliters of concentrated ammonia, 15 M NH_3, would you dilute to 500 ml to produce 6 M NH_3?

9.57. You are to prepare 750 ml 0.50 M HNO_3 from concentrated nitric acid, which has a molarity of 16. How many milliliters of the concentrated acid are required?

Section 9.6

9.23. How many equivalents are in one mole of each of the following species: HF; HSO_3^-; $H_2C_2O_4$ in $H_2C_2O_4 \rightarrow 2\ H^+ + C_2O_4^{2-}$?

9.24. Determine the number of equivalents in one mole of $Ni(OH)_2$; LiOH; $Zn(OH)_2$.

9.58. State the number of equivalents in one mole of HNO_2; HCO_3^-; H_3PO_4 in $H_3PO_4 \rightarrow 2\ H^+ + HPO_4^{2-}$.

9.59. How many equivalents are in 1 mole of each of the following: $Fe(OH)_3$; RbOH; $Cu(OH)_2$?

Section 9.7

9.25. Calculate the equivalent weight of each species in Problem 9.23.

9.26. Find the equivalent weight of each base in Problem 9.24.

9.27. How many grams of acetic acid, $HC_2H_3O_2$ are in 0.196 equivalents; how many grams of $Ca(OH)_2$ in 0.045 equivalent?

9.28. How many equivalents are in (a) 150 grams of NaOH; (b) 42.0 grams of formic acid in $HCOOH \rightarrow H^+ + HCOO^-$?

9.60. Calculate the equivalent weight of each species in Problem 9.58.

9.61. State the equivalent weight of the three bases in Problem 9.59.

9.62. Find the weight of 0.515 equivalent of CsOH; calculate the grams in 1.20 equivalents of H_3PO_4 in $H_3PO_4 \rightarrow 3\ H^+ + PO_4^{3-}$.

9.63. Determine the number of equivalents in (a) 0.165 gram of $Ba(OH)_2$; (b) 13.4 grams of NaH_2PO_4 in $H_2PO_4^- \rightarrow H^+ + HPO_4^{2-}$.

Section 9.8

9.29. Calculate the normality of a solution prepared by dissolving 17.2 grams of $HC_2H_3O_2$ in 300 ml. solution.

9.30. 9.79 grams of $NaHCO_3$ are dissolved in 500 milliliters of solution. What is the normality in the reaction

$$HCO_3^- + H^+ \rightarrow H_2O + CO_2?$$

9.31. How many grams of KOH must be used to prepare 600 milliliters of 2.00 N KOH?

9.32. Calculate the grams of oxalic acid dihydrate, $H_2C_2O_4 \cdot 2H_2O$, required to prepare 250 ml 0.500 N $H_2C_2O_4$ for the reaction

$$H_2C_2O_4 + 2\ OH^- \rightarrow C_2O_4^{2-} + 2\ HOH.$$

9.64. What normality results from dissolving 8.09 grams of KOH in water and diluting to 250 milliliters?

9.65. 3.48 grams of $H_2C_2O_4$ are dissolved in water, diluted to 100 ml, and used in a reaction in which it ionizes as follows:

$$H_2C_2O_4 \rightarrow H^+ + HC_2O_4^-.$$

What is the normality of the solution?

9.66. Sodium hydrogen sulfate is used as an acid in the reaction $HSO_4^- \rightarrow H^+ + SO_4^{2-}$. How many grams of sodium hydrogen sulfate must be dissolved in 500 ml of solution to produce 0.200 N $NaHSO_4$?

9.67. Sodium carbonate decahydrate, $Na_2CO_3 \cdot 10\ H_2O$, is used as a base in the reaction $CO_3^{2-} + H^+ \rightarrow HCO_3^-$. Calculate the grams of the hydrate needed to prepare 100 ml 0.400 N Na_2CO_3.

9.33. How many milliliters of 12 M HCl would you use to prepare one liter of 0.50 N HCl?

9.68. Calculate the volume of 18 M H_2SO_4 required to prepare 3.0 liters 3.0 N H_2SO_4 for reactions in which the sulfuric acid is completely ionized.

Section 9.9

9.34. What is the normality of 0.620 M $HC_2H_3O_2$? of 0.051 M $Ba(OH)_2$?

9.69. What is the normality of 5.95 M NaOH? What is the normality of 0.464 M H_3PO_4 in the reaction

$$H_3PO_4 + 2\ NaOH \rightarrow 2\ HOH + Na_2HPO_4?$$

9.35. What is the molar concentration of 2.15 N HNO_3 in an acid-base reaction? What is the molarity of 0.025 N $Sr(OH)_2$?

9.70. Calculate the molarity of 0.622 N LiOH. Find the molarity of 0.806 N NaH_2PO_4 in the reaction $H_2PO_4^- + 2\ OH^- \rightarrow 2\ HOH + PO_4^{3-}$.

CHAPTER 10

SOLUTION STOICHIOMETRY

10.1 EQUATIONS FOR SOLUTION STOICHIOMETRY

Some lead(II) nitrate is dissolved in water. In a separate beaker some calcium chloride is dissolved in water. The solutions are combined. A precipitate of lead(II) chloride forms. What chemical equation describes the reaction that has taken place?

In answering this question it is necessary to think of what happened in each step of the process. As lead(II) nitrate dissolves it releases the lead(II) and nitrate ions that form the solid:

$$Pb(NO_3)_2(s) \rightarrow Pb^{2+}(aq) + 2\ NO_3^-(aq). \qquad (10.1)$$

The same thing occurs with calcium chloride:

$$CaCl_2(s) \rightarrow Ca^{2+}(aq) + 2\ Cl^-(aq). \qquad (10.2)$$

When the two solutions are combined it is not, therefore, a matter of combining lead(II) nitrate with calcium chloride. It is, rather, that a pair of ions in one beaker is being combined with a pair of ions in another beaker. And out of the mixture of four ions, the lead(II) from one beaker precipitates with the chloride of the other as solid lead chloride:

$$Pb^{2+}(aq) + 2\ Cl^-\ (aq) \rightarrow PbCl_2(s). \qquad (10.3)$$

Equation 10.3 is called a **net ionic equation.** It describes *precisely* what takes place in a reaction *between two ionic solutions*—and no more. It does not consider the source of the ions. A more complete picture may be obtained by adding Equations 10.1, 10.2, and 10.3:

$$Pb(NO_3)_2(s) \rightarrow \cancel{Pb^{2+}(aq)} + 2\ NO_3^-(aq) \qquad (10.1)$$

$$CaCl_2(s) \rightarrow Ca^{2+}(aq) + 2\ \cancel{Cl^-(aq)} \qquad (10.2)$$

$$\cancel{Pb^{2+}(aq)} + 2\ \cancel{Cl^-(aq)} \rightarrow PbCl_2(s) \qquad (10.3)$$

$$Pb(NO_3)_2(s) + CaCl_2(s) \rightarrow PbCl_2(s) + Ca^{++}(aq) + 2\ NO_3^-(aq) \qquad (10.4)$$

If the initial lead(II) nitrate and calcium chloride are in stoichiometrically equal quantities (neither in excess), and the solution resulting from the combination of the two solutions is evaporated to dryness, $Ca(NO_3)_2$ will crystallize. The final overall reaction equation then becomes

$$Pb(NO_3)_2(s) + CaCl_2(s) \rightarrow PbCl_2(s) + Ca(NO_3)_2(s). \quad (10.5)$$

Each of the above equations is an accurate equation for what it purports to describe. Either the net ionic equation (10.3) or the overall equation (10.5) is usually the basis for solution stoichiometry problems. The net ionic equation has the advantage that it accurately describes the chemical change taking place in the reaction (usually identified in the problem), whereas the overall equation does not. On the other hand, users of net ionic equations must take into account that ionic molarities are not always the same as the molarities of the salt solutions from which the ions come, a fact which adds one or two steps to each problem. (The $CaCl_2$ solution described is an example. The Cl^- molarity is twice the molarity of the $CaCl_2$.) The overall equation does not have this drawback and may therefore be easier to use.

In this text, solution stoichiometry problems will be based on overall equations. The student should be aware of the dual approach in this area, however. Once again, he is advised to make his method conform to the method presented in his chemistry class.

10.2 MOLES AND MOLARITY

On page 47 of this book you were introduced to a three-step pattern for solving stoichiometry problems. The first of these was a conversion of a given quantity to moles of reaction species; the third was the conversion of moles of a reaction species to the quantity units required. You have applied these steps to quantities expressed in grams and, in the case of gases, volumes at known temperatures and pressures. In Examples 9.13 and 9.15, pages 128 and 129, you converted from moles of a solute to volume of a solution of known molarity, and from volume of such a solution to moles of solute. In other words, volume of a solution becomes a third quantity unit which may be related to moles in a stoichiometric context. In this section we shall expand our stoichiometric skills to include solutions of known molarity.

Example 10.1. How many grams of barium chromate can be precipitated by adding excess barium chloride solution to 50.0 ml of 0.469 M K_2CrO_4? The overall equation is $K_2CrO_4(aq) + BaCl_2(aq) \rightarrow BaCrO_4(s) + 2\ KCl(aq)$.

With barium chloride in excess, the calculation must be based on the number of moles of potassium chromate. Given the volume in milliliters (or liters, if you wish), and the molarity in moles per liter, you should be able to set up the first step of the stoichiometry pattern, conversion of the given quantity to moles. Refer to Example 9.15 if necessary.

10.1a.

$$0.0500\ \cancel{l} \times \frac{0.469 \text{ mole } K_2CrO_4}{\cancel{l}}$$

This step illustrates a useful generalization: the volume times the molarity equals the number of moles: V × M = moles.

The remaining steps of the stoichiometric pattern should be straightforward. Complete the problem.

10.1b. 5.93 g $BaCrO_4$.

$$0.0500\ \cancel{l} \times \frac{0.469\ \cancel{\text{mole } K_2CrO_4}}{1\ \cancel{l}} \times \frac{1\ \cancel{\text{mole } BaCrO_4}}{1\ \cancel{\text{mole } K_2CrO_4}} \times \frac{253 \text{ g } BaCrO_4}{1\ \cancel{\text{mole } BaCrO_4}} = 5.93 \text{ g } BaCrO_4$$

Example 10.2. How many grams of silver chromate, Ag_2CrO_4, will precipitate when 150 ml of 0.500 M $AgNO_3$ are added to 100 ml of 0.400 M K_2CrO_4?

The overall equation, please. . . .

10.2a. $2\ AgNO_3 + K_2CrO_4 \rightarrow Ag_2CrO_4 + 2\ KNO_3$.

This is an excess stoichiometry problem, so you must determine the number of moles of each reactant and then decide which is the limiting species. Do this next.

10.2b.

$$0.150\ \cancel{l} \times \frac{0.500 \text{ mole } AgNO_3}{1\ \cancel{l}} = 0.0750 \text{ mole } AgNO_3$$

$$0.100\ \cancel{l} \times \frac{0.400 \text{ mole } K_2CrO_4}{1\ \cancel{l}} = 0.0400 \text{ mole } K_2CrO_4$$

From the equation, two moles of silver nitrate are required for every mole of potassium chromate. From the quantities present, which reactant is the limiting species?

10.2c. $AgNO_3$. 0.0400 mole of potassium chromate requires 0.0800 mole of silver nitrate, which is more than is present. Therefore, silver nitrate is the limiting species.

The problem may now be completed in the usual fashion, based on moles of silver nitrate.

10.2d. 12.5 g Ag_2CrO_4.

$$0.0750\ \cancel{\text{mole } AgNO_3} \times \frac{1\ \cancel{\text{mole } Ag_2CrO_4}}{2\ \cancel{\text{moles } AgNO_3}} \times \frac{332 \text{ g } Ag_2CrO_4}{1\ \cancel{\text{mole } Ag_2CrO_4}} = 12.5 \text{ g } Ag_2CrO_4$$

Example 10.3. How many milliliters of 0.280 M $Ba(NO_3)_2$ are required to precipitate as $BaSO_4$ all the sulfate ion from 25.0 ml of 0.350 M $Al_2(SO_4)_3$? The overall equation is $Al_2(SO_4)_3 + 3\ Ba(NO_3)_2 \rightarrow 3\ BaSO_4 + 2\ Al(NO_3)_3$.

Using the techniques developed so far, take this setup to *moles* of $Ba(NO_3)_2$, but do not solve.

10.3a. $0.0250\ \cancel{l} \times \frac{0.350\ \cancel{\text{mole } Al_2(SO_4)_3}}{1\ \cancel{l}} \times \frac{3\text{ moles } Ba(NO_3)_2}{1\ \cancel{\text{mole } Al_2(SO_4)_3}}$

From the number of moles of barium nitrate we must find the volume of the solution. Do we have an equivalence between solution volume and moles of barium nitrate? The answer, of course, is yes. What is it?

10.3b. 0.280 mole $Ba(NO_3)_2$ = 1 liter solution = 1000 ml solution.

From this the conversion from moles of barium nitrate to milliliters of solution should be apparent. Check Example 9.13 if necessary.

10.3c. 93.8 ml of barium nitrate solution.

$$0.0250\ \cancel{l} \times \frac{0.350\ \cancel{\text{mole } Al_2(SO_4)_3}}{\cancel{l}} \times \frac{3\ \cancel{\text{moles } Ba(NO_3)_2}}{1\ \cancel{\text{mole } Al_2(SO_4)_3}} \times \frac{1000\text{ ml}}{0.280\ \cancel{\text{mole } Ba(NO_3)_2}} = 93.8\text{ ml}$$

Example 10.4. How many grams of NaOH are required to neutralize 45.0 ml of 0.750 M HCl?

First, the overall equation . . .

10.4a. $NaOH + HCl \rightarrow H_2O + NaCl$.

You should need no help on this one. Set up completely and solve.

10.4b. 1.35 g NaOH.

$$0.0450\ \cancel{l} \times \frac{0.750\ \cancel{\text{mole HCl}}}{1\ \cancel{l}} \times \frac{1\ \cancel{\text{mole NaOH}}}{1\ \cancel{\text{mole HCl}}} \times \frac{40.0\text{ g NaOH}}{1\ \cancel{\text{mole NaOH}}} = 1.35\text{ g NaOH}$$

Example 10.5. How many liters of hydrogen, measured at 718 torr and 28°C, can be liberated from 25.0 ml of 0.250 M HCl, using excess magnesium? The equation is $Mg + 2\ HCl \rightarrow H_2 + MgCl_2$.

Let's take this setup to moles of hydrogen as the first step.

10.5.a $$0.0250\ \cancel{l} \times \frac{0.250\ \cancel{\text{mole HCl}}}{1\ \cancel{l}} \times \frac{1\ \text{mole H}_2}{2\ \cancel{\text{moles HCl}}}$$

A bit of gas law refresher is in order for the next step. A similar problem appears as Example 8.9, page 106. The ideal gas equation, PV = *n* RT, when solved for volume becomes V = *n*RT/P. The above setup gives the number of moles, *n*, in the equation. Multiplication by RT/P should yield the volume of hydrogen. Or you may prefer to use molar volume at STP, followed by pressure and temperature corrections.

10.5b. 0.0817 liter H_2

$$0.0250\ \cancel{l\ \text{solution}} \times \frac{0.250\ \cancel{\text{mole HCl}}}{1\ \cancel{l\ \text{solution}}} \times \frac{1\ \cancel{\text{mole}}\ \text{H}_2}{2\ \cancel{\text{moles HCl}}} \times \frac{62.4\ l\text{-}\cancel{\text{torr}}}{\cancel{{}^\circ\text{K-mole}}} \times \frac{(273+28)\cancel{{}^\circ\text{K}}}{718\ \cancel{\text{torr}}} = 0.0817\ l\ \text{H}_2$$

Alternately,

$$0.0250\ \cancel{l\ \text{solution}} \times \frac{0.250\ \cancel{\text{mole HCl}}}{1\ \cancel{l\ \text{solution}}} \times \frac{1\ \cancel{\text{mole H}_2}}{2\ \cancel{\text{moles HCl}}} \times \frac{22.4\ l\ \text{H}_2}{1\ \cancel{\text{mole H}_2}} \times \frac{(273+28)\cancel{{}^\circ\text{K}}}{273\cancel{{}^\circ\text{K}}} \times \frac{760\ \cancel{\text{torr}}}{718\ \cancel{\text{torr}}} = 0.0817\ l\ \text{H}_2$$

10.3 EQUIVALENTS AND NORMALITY

Problems of this section are based on two principles: the number of equivalents of all species in a chemical reaction is the same, and volume × normality = number of equivalents. From these statements it follows that, in a reaction between two solutions, those volumes that will *just* react with each other can be related by the equation

$$V_1N_1 = V_2N_2. \quad (10.6)$$

The mathematical statement simply declares the equality of the number of equivalents of the two species.

Titration in analytical chemistry involves placing a measured volume of one solution in a flask along with a suitable indicator to signal completion of the reaction. Small increments of a second solution are introduced through a buret. Eventually the number of equivalents of titrant added from the buret will just equal the number of equivalents of reagent in the flask. The next drop of buret solution puts that solution in excess. It promptly reacts with the indicator, changing its color, and thereby signals the equivalence point. From initial and final buret readings, the volume of solution added is readily determined. If the normality of either solution is known, the normality of the other is found by direct substitution into Equation 10.6.

Example 10.6. A 25.0 ml sample of a base solution of unknown concentration is titrated with 0.452 N HCl. What is the normality of the base if 18.4 ml of acid are required for neutralization?

Set up completely and solve.

10.6a. $N_{base} = 0.333$.
Substituting into $V_1N_1 = V_2N_2$, Equation 10.6,

$$25.0 \text{ ml} \times N_{base} = 18.4 \text{ ml} \times 0.452; \qquad N_{base} = 0.333$$

Example 10.7. What volume of 0.755 N H_2SO_4 is required to just neutralize 10.0 ml of 0.493 N NaOH?

Again it is a straight substitution. Set up and solve.

10.7a. $V_{acid} = 6.53$ ml.

$$V_{acid} \times 0.755 = 10.0 \text{ ml} \times 0.493 \qquad V_{acid} = 6.53 \text{ ml}$$

Titration techniques may be used also to determine the equivalent weight of an unknown substance. If a solid acid, for example, is weighed, dissolved, and then titrated with a base of known normality, the volume of the base multiplied by its normality gives the number of equivalents of base added. This must be equal to the number of equivalents of acid present. Dividing the number of grams of acid by the number of equivalents gives the number of grams per equivalent, i.e., the equivalent weight of the acid.

Example 10.8. 0.304 gram of an unknown solid acid, when dissolved, requires 16.2 ml of 0.224 N NaOH for neutralization. Find the equivalent weight of the acid.

First, how many equivalents of base are involved in the reaction?

10.8a. 0.00363 equivalent of base.

$$0.0162 \not{l} \times \frac{0.224 \text{ equivalent}}{\not{l}} = 0.00363 \text{ equivalent}$$

Now how many equivalents of acid are there?

10.8b. 0.00363 equivalent of acid. The number of equivalents of *all* species in a chemical reaction is the same.

Equivalent weight is measured in units of grams per equivalent. You have the number of equivalents in a known weight. Put them together.

10.8c. 83.7 grams per equivalent.

$$\frac{0.304 \text{ g}}{0.00363 \text{ eq}} = 83.7 \text{ g/eq}$$

PROBLEMS

Section 10.2

10.1. How many grams of AgCl can be precipitated from 50.0 milliliters of 0.855 M $AgNO_3$?

10.2. How many grams of barium flouride can be precipitated from 40.0 ml 0.436 M NaF by the addition of excess barium nitrate solution?

10.3. 25.0 ml 0.350 M NaOH are added to 45.0 ml 0.125 M $CuSO_4$. How many grams of copper(II) hydroxide will precipitate?

10.4. What volume of 0.415 M $AgNO_3$ will be required to precipitate as AgBr all the bromide ion in 35.0 ml 0.128 M $CaBr_2$?

10.5. What volume of 0.496 M HCl is required to neutralize 20.0 ml 0.809 M NaOH?

10.6. How many milliliters of 0.715 M HCl are required to neutralize 1.24 grams of sodium carbonate? The reaction proceeds to the formation of carbon dioxide and water.

10.7. What minimum number of grams of oxalic acid, $H_2C_2O_4 \cdot 2H_2O$, would you specify for a student experiment involving a titration of no fewer than 15.0 ml 0.100 M NaOH? Both oxalic acid hydrogens are replaceable in this reaction.

10.8. How many liters of chlorine, measured at 740 torr and 26°C, could be recovered from 50.0 ml 1.20 M HCl by the reaction MnO_2 + 4 HCl → $MnCl_2$ + 2 H_2O + Cl_2, assuming complete conversion of reactants to products?

10.9. An analytical procedure for the determination of the chloride ion concentration in a solution involves the precipitation of silver chloride: $Ag^+ + Cl^- \rightarrow AgCl$. What is the molarity of the chloride ion if 16.8 ml 0.629 M $AgNO_3$ are required to precipitate all of the chloride ion in a 25.0 ml sample of a solution of unknown concentration?

10.17. Calculate the grams of magnesium hydroxide that will precipitate from 25.0 ml 0.611 M $MgCl_2$ by the addition of excess sodium hydroxide solution.

10.18. The iron(III) content of a solution may be determined by precipitating it as $Fe(OH)_3$, and then decomposing the hydroxide to Fe_2O_3 by heat. How many grams of iron(III) oxide may be precipitated from 35.0 ml 0.345 M $Fe(NO_3)_3$?

10.19. 25.0 ml 0.235 M $Mg(NO_3)_2$ are combined with 30.0 ml 0.260 M KOH. How many grams of magnesium hydroxide will precipitate?

10.20. Calculate the volume of 0.0955 M KIO_3 that must be added to 20.0 ml 0.142 M $Cu(NO_3)_2$ to precipitate the Cu^{2+} ion as copper(II) iodate.

10.21. Calculate the milliliters of 0.585 M KOH that are required to neutralize 50.0 ml 0.109 M H_2SO_4.

10.22. What volume of 0.842 M NaOH would react with 2.14 grams of sulfamic acid, NH_2SO_3H, a solid acid with a single replaceable hydrogen?

10.23. A student is to titrate solid maleic acid, $H_2C_4H_2O_4$ (two replaceable hydrogens), with 0.500 M NaOH. What is the maximum number of grams of maleic acid to be used if the titration is not to exceed 25.0 milliliters?

10.24. How many milliliters of 1.50 M NaOH must react with aluminum to yield 2.00 liters of hydrogen, measured at 22°C and 710 torr, by the reaction 2 Al + 6 NaOH → 2 Na_3AlO_3 + 3 H_2, assuming complete conversion of reactants to products?

10.25. Calculate the hydroxide ion concentration in a 20.0 ml sample of an unknown if 14.3 ml 0.248 M H_2SO_4 are required in a neutralization titration.

10.10. 60.0 ml 0.322 M KI are combined with 20.0 ml 0.530 M $Pb(NO_3)_2$. (a) How many grams of PbI_2 will precipitate? (b) What is the final molarity of the K^+ ion? (c) What is the final molarity of the NO_3^- ion? (d) What is the final molarity of the Pb^{2+} or I^- ion, whichever is in excess?

10.26. If 25.0 ml 0.339 M NaOH and 35.0 ml 0.238 M H_2SO_4 are combined, what will be the final concentration of sodium ion, sulfate ion, and either hydrogen or hydroxide ion, whichever is in excess? Also, how many grams of sodium sulfate could be recovered if the resulting solution is evaporated to dryness?

Section 10.3

10.11. What is the normality of an acid if 12.8 ml are required to titrate 15.0 ml 0.882 N NaOH?

10.27. Calculate the normality of a solution of sodium carbonate if a 25.0 ml sample requires 22.8 ml 0.405 N H_2SO_4 in a titration.

10.12. 28.4 ml 0.424 N $AgNO_3$ are required to titrate the chloride ion in a 25.0 ml sample of nickel chloride solution. Find the normality of the nickel chloride. (Compare the solution of this problem with that of Problem 10.9.)

10.28. 28.9 ml of 0.402 N NaOH are required to titrate 50.0 ml of a solution of tartaric acid ($H_2C_4H_4O_6$) of unknown concentration. Find the normality of the acid.

10.13. Using a particular acid-base indicator, 20.0 ml of a phosphoric acid solution requires 32.6 ml 0.208 N NaOH in a titration reaction. Find the normality of the acid.

10.29. Repeating the titration described in Problem 10.13, but with a different indicator, it is found that only 16.3 ml 0.208 N NaOH are required for 20.0 ml of the phosphoric acid solution. Calculate the normality of the acid and account for the difference in the answers in Problem 10.13 and this problem.

10.14. 3.29 grams of oxalic acid dihydrate, $H_2C_2O_4 \cdot 2H_2O$, are dissolved in water and diluted to 500.0 ml. A 25.0 ml sample of this solution is then titrated with sodium hydroxide solution, 30.1 ml being required for neutralization. Calculate the normality of (a) the oxalic acid and (b) the base for the reaction of both replaceable hydrogens in the oxalic acid.

10.30. Potassium hydrogen phthalate, $KHC_8H_4O_4$, is a commonly used primary standard against which to standardize bases of unknown concentration. If 2.14 grams of this solid are dissolved in water and diluted to 100.0 ml, and 10.0 ml of the solution require 12.9 ml sodium hydroxide solution for neutralization, calculate the normality of the base. (There is one replaceable hydrogen in $KHC_8H_4O_4$.)

10.15. 0.305 grams of sulfamic acid, NH_2SO_3H, are dissolved in water and the resulting solution titrated with a sodium hydroxide solution of unknown concentration. If neutralization of the one replaceable hydrogen requires 26.4 ml of the sodium hydroxide solution, calculate its normality.

10.31. 22.4 milliliters of a base are required to titrate a solution containing 440 milligrams of oxalic acid dihydrate, $H_2C_2O_4 \cdot 2H_2O$. Find the normality of the base if the acid is diprotic (two replaceable hydrogens).

10.16. 15.6 ml 0.562 N NaOH are required to titrate a solution prepared by dissolving 0.631 grams of an unknown acid. What is the equivalent weight of the acid?

10.32. 1.21 grams of an organic compound that functions as a base in reaction with sulfuric acid are dissolved in water and titrated with 0.294 N H_2SO_4. What is the equivalent weight of the base if 30.7 ml of acid are required in the titration?

CHAPTER 11

COLLIGATIVE PROPERTIES OF SOLUTIONS

As an "impure" substance, consisting of two or more components, a solution lacks the fixed physical properties of a pure substance. Some physical properties of solutions depend upon the *concentration* of the component particles, and are independent of the identity or character of the particles. Such properties are known as **colligative properties.** Vapor pressure, boiling point, and freezing point are three such properties.

The character of the solute, i.e., whether it is volatile or nonvolatile, and whether it is an electrolyte or nonelectrolyte, affects colligative properties. In Sections 11.1–11.3 we shall consider only nonvolatile nonelectrolytes. The effect of these other variables will be explained in Section 11.4.

11.1 RAOULT'S LAW

It has been shown experimentally that, at a given temperature, the vapor pressure of a solution of a nonvolatile solute is directly proportional to the *mole fraction* of the *solvent* particles. This relationship, an example of Raoult's Law, may be expressed mathematically as follows:

$$P_n = X_1 P_1^\circ, \tag{11.1}$$

where P_n is the vapor pressure of the solution, X_1 the mole fraction of solvent in that solution, and P_1° the vapor pressure of pure solvent at the same temperature. The solution always has a lower vapor pressure than pure solvent, as indicated by Equation 11.1, in which the mole fraction of solvent, X_1, must always be less than 1.

A useful relationship can be derived by subtracting both sides of Equation 11.1 from the identity, $P_1^\circ = P_1^\circ$:

$$P_1^\circ - P_n = P_1^\circ - X_1 P_1^\circ. \tag{11.2}$$

The left side of Equation 11.2 is the difference between the vapor pressure of the pure solvent and the vapor pressure of the solution, or ΔP. Substituting this expression and factoring the right side of the equation gives:

$$\Delta P = P_1^\circ(1 - X_1). \qquad (11.3)$$

The sum of the mole fractions of solvent and solute is 1:

$$X_1 + X_2 = 1 \qquad (11.4)$$

where X_2 is the mole fraction of solute. Rearranging,

$$X_2 = 1 - X_1. \qquad (11.5)$$

Substituting X_2 for its equivalent, $1 - X_1$, in Equation 11.3,

$$\Delta P = X_2 P_1^\circ. \qquad (11.6)$$

Hence, it is clear that the *difference* in vapor pressure between a pure solvent and a solution is proportional to the *mole fraction* of *solute particles*.

Solution vapor pressure problems may be solved by substitution into either Equation 11.1 or 11.6.

Example 11.1. Find the vapor pressure of a solution of 20.0 grams of sucrose, $C_{12}H_{22}O_{11}$, in 40.0 grams of water at 34°C.

Solve the problem first by direct substitution into Equation 11.1. The vapor pressure of water at 34° is 40.00 torr. By the method of the previous chapter, calculate the mole fraction of *water* (solvent) in the solution.

11.1a. 0.974.

$$20.0\ \cancel{g\ C_{12}H_{22}O_{11}} \times \frac{1\ \text{mole}\ C_{12}H_{22}O_{11}}{342\ \cancel{g\ C_{12}H_{22}O_{11}}} = 0.0585\ \text{mole}\ C_{12}H_{22}O_{11}$$

$$40.00\ \cancel{g\ H_2O} \times \frac{1\ \text{mole}\ H_2O}{18.0\ \cancel{g\ H_2O}} = 2.22\ \text{moles}\ H_2O$$

$$0.0584 + 2.22 = 2.28\ \text{Total moles}$$

$$X_{H_2O} = \frac{2.22}{2.28} = 0.974$$

Now substitute into Equation 11.1 and solve for P_n.

11.1b. 39.0 torr.

$$P_n = X_1 P_1^\circ = 0.974 \times 40.00 = 39.0\ \text{torr}$$

Now approach the same problem by finding the vapor pressure difference, ΔP, between the solution and solvent as in Equation 11.6.

11.1c. 1.02 torr.

$$\text{Mole fraction of } C_{12}H_{22}O_{11} = \frac{0.0585}{2.28}$$

$$\Delta P = X_2P_1^\circ = \frac{0.0585}{2.28} \times 40.00 = 1.03 \text{ torr}$$

The vapor pressure of the solution is now found by subtracting the vapor pressure difference from the vapor pressure of the solvent. Complete the problem.

11.1d. 38.98 torr.

$$P_n = P_1^\circ - \Delta P = 40.00 - 1.03 = 38.98 \text{ torr}$$

Note the difference between the answers in 11.1b and 11.1d. In problems where the mole fraction of the solvent is close to unity, it is often possible to obtain an additional significant figure by using Equation 11.6 rather than 11.1.

11.2 BOILING AND FREEZING POINTS

The boiling point of a solution of a nonvolatile solute is higher than the boiling point of the pure solvent. The difference between the boiling points of the solution and the solvent is known as the *boiling point elevation* and is designated ΔT_B. By contrast, the freezing point of a solution is ordinarily lower than the freezing point of the pure solvent. The difference between the temperatures is called the *freezing point depression* and is designated ΔT_F. Boiling point elevation and freezing point depression are colligative properties, and they are proportional to the *molal concentration* (m) of the *solute* particles. Converting these proportionalities into equations by means of proportionality constants gives

$$\Delta T_B = K_B m \qquad (11.7)$$

$$\Delta T_F = K_F m, \qquad (11.8)$$

where K_B and K_F are the molal boiling point constant and the molal freezing point constant, respectively.*

The units of K_B or K_F found by solving either Equation 11.7 or 11.8 are

$$\frac{°C}{\text{moles solute/kg solvent}}, \quad \text{or} \quad \frac{°C\text{–kg solvent}}{\text{moles solute}}.$$

These are, at best, awkward units to handle. Rather than be enslaved by dimensional analysis because it is usually a helpful tool, the student is advised

*Equations 11.7 and 11.8 are valid only for dilute solutions in which interaction between solute and solvent particles is negligible, which we shall assume to be the case for solutes that are nonelectrolytes.

TABLE 11.1 MOLAL FREEZING AND BOILING POINT CONSTANTS

Solvent	Freezing Point (°C)	Molal Freezing Point Constant	Boiling Point (°C)	Molal Boiling Point Constant
Water	0.0	− 1.86	100.0	0.52
Acetic Acid	16.6	− 3.90	118.5	2.93
Benzene	5.50	− 5.10	80.1	2.53
Cyclohexane	6.5	−20.2	81	2.79
Phenol	43	− 3.56	182	7.40
Naphthalene	80.2	− 6.9		

to leave it freely if a substitute-and-solve-algebraically approach is less cumbersome. This will be done here, using units only for expressing results or if they contribute to the understanding of the problem.

Molal boiling and freezing point constants are properties of the solvent: for solutions in a given solvent, ΔT_B and ΔT_F depend only on the molal concentration of solute particles, irrespective of their identity. For water, K_B is 0.52, and K_F is 1.86. Thus an aqueous solution that is one molal in solute particle concentration should freeze at −1.86°C and boil at 100.52°C. Other molal boiling and freezing point constants are given in Table 11.1.

The method for solving boiling point problems is identical to that for freezing point problems. These methods are illustrated in the following examples:

Example 11.2. Determine the boiling and freezing points of a solution of 10.0 grams of urea, $CO(NH_2)_2$, in 200 grams of water.

Both Equations 11.7 and 11.8 require the molality of the solution. Determine this.

11.2a. m = 0.833 mole per kg water.

$$\frac{10.0\ \cancel{g\ CO(NH_2)_2}}{0.200\ kg\ H_2O} \times \frac{1\ mole\ CO(NH_2)_2}{60.0\ \cancel{g\ CO(NH_2)_2}} = 0.833\ mole\ CO(NH_2)_2/kg\ H_2O$$

Equations 11.7 and 11.8 call for multiplication of the molality by the molal boiling and freezing point constants. Look these up in the table and calculate the ΔT values.

11.2b. $\Delta T_B = 0.43°C$; $\Delta T_F = -1.55°C$

$$\Delta T_B = K_B m = 0.52 \times 0.833 = 0.43°C$$

$$\Delta T_F = K_F m = -1.86 \times 0.833 = -1.55°C.$$

In calculating ΔT_F, the negative sign of K_F is frequently not used in a strictly algebraic sense. The magnitude of the freezing point change is determined using the absolute value of K_F. Realizing that the freezing point of a solution is always lower than that of the pure solvent, the solution's freezing point is found by subtracting the change from the freezing point of the pure solvent.

Complete the problem by stating the boiling and freezing temperatures of the solution.

11.2c. $T_B = 100.43°C$; $T_F = -1.55°C$.

The molal boiling and freezing point constants of a solvent may be determined by Equations 11.7 or 11.8, based on the experimental measurement of ΔT with a solution of known molality.

Example 11.3. A certain solvent freezes at 76.3°C. If 11.0 grams of naphthalene, $C_{10}H_8$, are dissolved in 120 grams of the solvent, the freezing temperature is 73.5°C. Calculate K_F for the solvent.

To find K_F from Equation 11.8, ΔT_F and molality must be known. Calculate ΔT_F first.

11.3a. 2.8°C.

$$76.3° - 73.5° = 2.8°$$

Now determine the molality from the data in the problem.

11.3b. 0.716 m.

$$\frac{11.0 \text{ g } C_{10}H_8}{0.120 \text{ kg solvent}} \times \frac{1 \text{ mole } C_{10}H_8}{128 \text{ g } C_{10}H_8} = 0.716 \text{ mole } C_{10}H_8/\text{kg solvent}$$

With both molality and ΔT known, direct substitution into Equation 11.8 yields K_F.

11.3c. 3.9.

$$K_F = \frac{\Delta T_F}{m} = \frac{2.8}{0.716} = 3.9.$$

Probably the most familiar application of a colligative property calculation occurs in connection with the use of antifreeze in automobiles. . . .

Example 11.4. Calculate the number of grams of ethylene glycol, $C_2H_6O_2$, per kilogram of water that will prevent freezing down to −20.0°F (−28.9°C).

This time we have data from which ΔT_F may be calculated, and we seek a concentration. Solve first for molality by rearranging Equation 11.8 and substituting.

11.4a. m = 15.5 moles $C_2H_6O_2$/kg water.

The pure solvent freezes at 0.0°C, and the solution at −28.9°C. Therefore, ΔT is −28.9°C.

Solving Equation 11.8 for molality and substituting,

$$m = \frac{\Delta T_B}{K_B} = \frac{-28.9}{-1.86} = 15.5 \text{ moles } C_2H_6O_2\text{/kilogram water.}$$

The only thing remaining is to convert the *moles* of solute per kilogram to *grams* of solute per kilogram.

11.4b. 961 grams $C_2H_6O_2$/kg water.

$$\frac{15.5 \cancel{\text{ moles } C_2H_6O_2}}{\text{kg water}} \times \frac{62.0 \text{ g } C_2H_6O_2}{\cancel{\text{mole } C_2H_6O_2}} = 961 \text{ g } C_2H_6O_2\text{/kg water}$$

11.3 MOLAR WEIGHT DETERMINATION FROM ΔT_F AND ΔT_B

ΔT_B and ΔT_F may be used for the experimental determination of the approximate molar weights of unknown solutes. A weighed quantity of unknown solute is dissolved in a weighed quantity of a solvent of known freezing point and freezing point constant. The concentration can be expressed in grams solute/kg solvent. The freezing point of the solution is determined experimentally. Using Equation 11.8, the molality of the solution may be determined. This expresses the concentration of the solution in moles of solute per kilogram of solvent. The quantity of solute in a kilogram of solvent is now known in both *grams* and *moles*. Division of one by the other yields molar weight:

$$\frac{\text{g solute}/\cancel{\text{kg solvent}}}{\text{moles solute}/\cancel{\text{kg solvent}}} = \text{g solute/mole solute.}$$

Example 11.5. 25.1 grams of a solute are dissolved in 150 grams of benzene. The freezing point of the solution is $-1.92°$C. Pure benzene freezes at 5.50°C, and its molal freezing point constant is 5.10. Find the approximate molar weight of the solute.

First, from the data of the problem, determine the concentration of the solution in grams of solute per kilogram of solvent.

11.5a. 167 g solute/kg benzene.

$$\frac{25.1 \text{ g solute}}{0.150 \text{ kg benzene}} = 167 \text{ g solute/kg benzene}$$

Now the molality must be found. The first step is to determine the freezing point depression, ΔT_F.

11.5b. 7.42°C.

$$\Delta T_F = 5.50 - (-1.92) = 7.42°C$$

Now, using this and the information in the problem, substitute into Equation 11.8 and find the molality.

11.5c. m = 1.45.

$$m = \frac{\Delta T_F}{K_F} = \frac{7.42}{5.10} = 1.45 \text{ moles solute/kg benzene}$$

Finally, divide one concentration by the other to obtain the molar weight.

11.5d. 115 g/mole.

$$\frac{167 \text{ g solute/}\cancel{\text{kg benzene}}}{1.45 \text{ moles solute/}\cancel{\text{kg benzene}}} = 115 \text{ g solute/mole solute}$$

Another approach to this type of problem may be developed by modifying the defining equation for molality,

$$m = \frac{\text{moles solute}}{\text{kilogram solvent}} . \tag{9.4}$$

As was shown on page 124, the number of moles of a substance is equal to grams/molar weight. Substituting into Equation 9.4 yields

$$m = \frac{\text{grams solute/MW solute}}{\text{kilograms solvent}} . \tag{11.9}$$

The use of Equation 11.9 will be illustrated in the next example:

Example 11.6. 35.0 grams of a nonelectrolyte are dissolved in 100 ml water, producing a solution that boils at 102.16°C. Determine the approximate molar weight of the solute.

First determine the molality by Equation 11.7. You have the temperature from which to calculate ΔT_B, and the boiling point constant is known.

11.6a. m = 4.15 moles solute/kilogram solvent.

$$m = \frac{\Delta T_B}{K_B} = \frac{102.16 - 100.00}{0.52} = 4.15 \text{ moles solute/kg solvent}$$

Now Equation 11.9: you have everything but MW. Substitute and solve.

11.6b. 84 grams/mole.

$$4.15 = \frac{35.0/\text{MW}}{0.100}$$

$$\text{MW} = \frac{35.0}{4.15 \times 0.100} = 84.3 \text{ grams/mole}$$

The answer has been expressed in two significant figures because of the two significant figures in K_B in step 11.6a.

Example 11.6 could be solved equally well, of course, by the method used for Example 11.5. As in other situations where alternative methods have been presented, we recommend first that you adopt the method preferred by your instructor. If the choice is left to you, make the choice, adopt the approach that appeals to you, and discard the other.

11.4 THE EFFECT OF THE NATURE OF THE SOLUTE ON ΔT_F AND ΔT_B

In both discussion and example problems to this point qualifying adjectives have been used to limit the type of solute involved. In considering Raoult's Law, only nonvolatile solutes were mentioned. For so-called "ideal" solutions, if both solute and solvent are volatile, the vapor pressure of a solution is the sum of the partial vapor pressure of the solute plus that of the solvent. That is:

$$P_n = X_1 P_1^\circ + X_2 P_2^\circ$$

where P_2° is the vapor pressure of pure solute.

In describing boiling and freezing point changes, only nonelectrolytes were mentioned as solutes. What happens if the solute is an electrolyte that dissociates into two or more particles in dissolving? Colligative properties concern only the *number* of solute particles, be they ionic or molecular. For example, NaCl dissociates into ions on dissolving: $NaCl \rightarrow Na^+ + Cl^-$. One mole of NaCl yields one mole of Na^+ ions and one mole of Cl^- ions, or two moles of ionic particles. A 0.1 molal solution of NaCl is therefore 0.2 molal in particles, and thus produces approximately twice the freezing point depression or boiling point elevation of a 0.1 molal solution of a nonelectrolyte. Similarly, a 0.1 molal solution of $CaCl_2$ is 0.3 molal in particles; or 0.1 molal Na_3PO_4 is 0.4 molal in particles.

Equations 11.7 and 11.8 may be modified to allow for the dissociation of electrolytes by inserting an "ion multiplier," i, that is equal to the number of moles of ions resulting from the dissociation of one mole of solute.

$$\Delta T_B = iK_B m \qquad (11.10)$$

$$\Delta T_F = iK_F m \qquad (11.11)$$

For NaCl, $i = 2$; for $CaCl_2$, $i = 3$; and for Na_3PO_4, $i = 4$.

Example 11.7. Calculate the freezing point of 0.20 m Na_2SO_4.

First, what is the value of i for a solution of Na_2SO_4?

11.7a. $i = 3$.

One mole of Na_2SO_4 yields three moles of ions on dissociation:

$$Na_2SO_4 \rightarrow 2\ Na^+ + SO_4^{2-}.$$

From here the problem is solved by direct substitution into the appropriate equation. . . .

11.7b. $T_F = -1.1°C$.

$$\Delta T_F = iK_Fm = 3 \times 1.86 \times 0.20 = 1.1°C$$

The solution freezes 1.1°C below the freezing point of water: $T_F = 0.0 - 1.1 = -1.1°C$.

Even if dissociation of the solute is considered, most solutions do not conform precisely with calculated values. This is because the solute and/or solvent particles interact with each other or with themselves, which reduces the "effective" concentration to a value below that calculated from the amounts of solute and solvent present in the solution. These interactions are electrical in character. Solutions that actually possess the physical properties calculated as in this chapter are called ideal solutions. The more dilute the solution, the weaker the interactions between particles and, therefore, the closer the approach to ideal character. For precise work the departure from ideal solutions must be accounted for in colligative property problems. These corrections, however, are beyond the scope of this book.

11.5 OSMOTIC PRESSURE

Figure 11.1 illustrates the phenomenon of osmotic pressure. Two compartments of a container are separated by a membrane through which a solvent will pass. A dilute solution is prepared, using solute that cannot pass through the membrane. If pure solvent is placed in the left compartment and the solution in the right, the rate of movement of solvent particles through the membrane will be greater from the pure solvent to the solution (left to right) than from the solution to the solvent (right to left), as shown in part *a* of the illustration. If pressure is applied to the solution side, as in part *b*, the solvent molecules may be "pushed" back into the solution side at the same rate at which they move from it. The pressure required to balance the rates is called the **osmotic pressure.**

Osmotic pressure is a colligative property which is proportional to the molar concentration of the solution. (Some texts indicate the proportionality is with the molal concentration. The distinction is academic in practical situa-

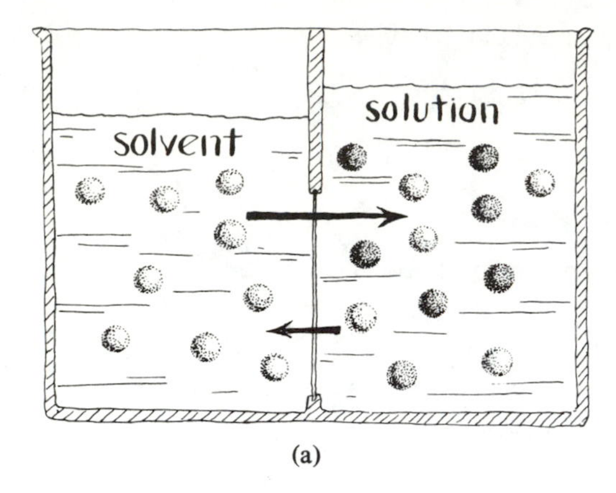

(a)

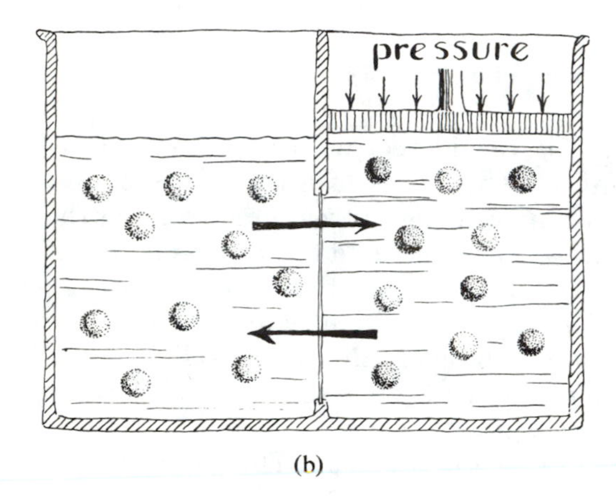

(b)

Figure 11.1. Osmotic pressure.

tions: at the dilute concentrations at which the relationship is valid, molarity and molality are essentially the same.) Thus

$$\pi \propto M, \tag{11.12}$$

where π is the osmotic pressure. The proportionality constant that converts this relationship into an equation is the product of the gas constant and the absolute temperature:

$$\pi = MRT. \tag{11.13}$$

Inasmuch as molarity is the number of moles divided by the volume, n/V, Equation 11.13 is of the same form as the ideal gas equation: $\pi = (n/V)RT$, or $\pi V = nRT$. The units are the same as in gas law problems.

One form of the osmotic pressure relationship that is of particular interest is the variation

$$\pi V = \frac{g}{MW}RT, \tag{11.14}$$

where grams/molar weight is substituted for the number of moles, as in Equation 7.30, page 82. By direct substitution into this equation the osmotic pressure of solutions may be predicted and the molar weights of large, complex molecules estimated.

Example 11.8. Predict the osmotic pressure at 25°C of a solution of 35.0 grams of sugar, $C_{12}H_{22}O_{11}$, in a liter of solution.

To begin, solve Equation 11.14 for π.

11.8a.

$$\pi = \frac{gRT}{V(MW)}.$$

Compute the molar weight of sugar and the temperature in °K. Substitute these and other required values and solve for π in atmospheres. Recall the gas laws in selecting a value for R.

11.8b. 2.50 atmospheres.

$$\frac{35.0\ \text{g sugar}}{1\ \ell\ \text{solution}} \times \frac{0.0821\ \ell\text{-atm}}{°\text{K-mole}} \times 298°\text{K} \times \frac{1\ \text{mole}}{342\ \text{g sugar}} = 2.50\ \text{atm}$$

Example 11.9. The osmotic pressure of an aqueous solution of 3.85 grams of a protein in 250 ml of solution is 10.4 torr at 30°C. Estimate the molar weight of the protein.

As a first step, solve Equation 11.14 for molar weight.

11.9a.

$$MW = \frac{gRT}{\pi V}.$$

Make the proper substitutions and solve.

11.9b. 28,000 g/mole.

$$MW = \frac{3.85\ \text{g}}{0.250\ \ell} \times \frac{62.4\ \ell\text{-torr}}{°\text{K-mole}} \times \frac{303°\text{K}}{10.4\ \text{torr}} = 28{,}000\ \text{g/mole}$$

PROBLEMS

Sections 11.1 and 11.2

11.1. Glucose, $C_6H_{12}O_6$, is a nonvolatile nonelectrolyte. For an aqueous solution of 50.0 grams of glucose dissolved in 100 grams of water, calculate the (a) vapor pressure at 25°C; (b) boiling point; and (c) freezing point.

11.2. Calculate (a) the vapor pressure at 20°C; (b) the boiling point; and (c) the freezing point of a 10.0% aqueous solution of urea, $CO(NH_2)_2$.

11.3. 4.34 grams of paradichlorobenzene, $C_6H_4Cl_2$, are dissolved in 65.0 grams of naphthalene, $C_{10}H_8$. Calculate the freezing point of the solution, using data from Table 11.1.

11.4. Calculate the freezing and boiling points of a solution prepared by dissolving 40.0 grams of pentanedioic (glutaric) acid, $HOOC(CH_2)_3COOH$, in 150 grams of water.

11.5. The boiling point of a solution of 1.40 grams of urea, NH_2CONH_2, in 16.3 grams of a certain solvent is 3.92°C higher than the boiling point of the pure solvent. Calculate the molal boiling point constant for the solvent.

11.6. How many kilograms of glycerol, $C_3H_8O_3$, must be dissolved in 8.50 kg of water to make a solution that freezes at −32°C?

11.7. Calculate the molal concentration of an aqueous solution that boils at 100.84°C.

11.15. 20.0 grams of analine, $C_6H_5NH_2$, are dissolved in 120 grams of water. Assuming the solute to be a nonvolatile nonelectrolyte, compute (a) the vapor pressure at 28°C; (b) the boiling point; and (c) the freezing point of the solution.

11.16. Determine (a) the vapor pressure at 23°C; (b) the boiling point; and (c) the freezing point of 0.65 m $C_3H_8O_3$, an aqueous solution of glycerol.

11.17. Calculate the freezing point of a solution of 1.99 grams of naphthalene, $C_{10}H_8$, in 32.0 grams of benzene, C_6H_6, using data from Table 11.1.

11.18. Calculate the boiling and freezing points of a solution of 75.0 grams of ethylene glycol, $C_2H_6O_2$ (antifreeze), in 200.0 grams of water.

11.19. Calculate the molal freezing point constant of an unknown solvent if its normal freezing point is 28.7°C, and a solution of 11.4 grams of ethanol, C_2H_5OH, in 200 grams of the solvent freezes at 21.5°C.

11.20. How many kilograms of ethylene glycol, $C_2H_6O_2$, must be dissolved in 12 liters of water in an automobile cooling system to prevent freezing at any temperature above −13°F (−25°C)?

11.21. What is the molality of a solution of an unknown solute in acetic acid if it freezes at 13.4°C?

Section 11.3

11.8. If a solution prepared by dissolving 26.0 grams of an unknown solute in 380 grams of water freezes at −1.18°C, what is the approximate molar weight of the solute?

11.9. A solution of 12.0 grams of an unknown nonelectrolyte dissolved in 80.0 grams of naphthalene, $C_{10}H_8$, freezes at 71.3°C. Using the data of Table 11.1, estimate the molar weight of the solute.

11.22. A solution of 15.1 grams of an unknown solute in 600 grams of water boils at 100.28°C. What is the approximate molar weight of the unknown?

11.23. When 11.0 grams of an unknown solute are dissolved in 90.0 grams of phenol, the freezing point depression is 9.6°C. Estimate the molar weight of the solute.

Section 11.4

11.10. Estimate the freezing point of 0.10 m KBr.

11.11. A solution prepared by dissolving 9.41 grams of $NaHSO_3$ in 1.00 kg of water freezes at −0.33°C. From these data show that $NaHSO_3 \rightarrow Na^+ + HSO_3^-$ is a correct expression for the ionization of $NaHSO_3$, whereas $NaHSO_3 \rightarrow Na^+ + H^+ + SO_3^{2-}$ is not.

11.24. Calculate the theoretical boiling point of 0.090 m $(NH_4)_3PO_4$.

11.25. 0.010 m H_3PO_4 has a freezing point depression of 0.030°C. Assuming the only ionization of H_3PO_4 is described by the equation $H_3PO_4 \rightarrow H^+ + H_2PO_4^-$, estimate the percent ionization, i.e., the percentage of the original 0.010 mole of H_3PO_4 in a kilogram of water that ionizes by the equation shown.

Section 11.5

11.12. Find the osmotic pressure in mm Hg (torr) of a solution of 4.00 grams of $C_{12}H_{22}O_{11}$ in 500 ml at 20°C.

11.13. A 3.0% sugar solution has a density of 1.01 g/ml and is 0.089 molar. At 25°C, what height of solution can be supported by osmotic pressure in a device such as the one illustrated in Figure 11.2? Answer in meters.

11.14. Predict the vapor pressure at 25°C, the freezing point, and the boiling point of sea water, which is essentially a 3.5% sodium chloride solution.

11.26. Predict the osmotic pressure at 28°C of a solution prepared by dissolving 3.75 grams of urea, $CO(NH_2)_2$, in water and diluting to 250 ml.

11.27. A solution of 1.82 grams of an organic compound in 500 ml of water exerts an osmotic pressure of 32.4 torr at 18°C. Find the molar weight of the compound.

11.28. Assuming the molarity and molality of sea water are identical (see Problem 11.14), what osmotic pressure might be developed by sea water against fresh water?

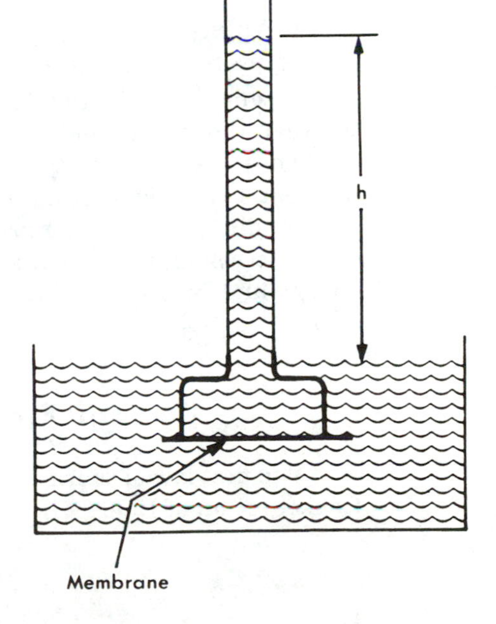

Figure 11.2.

CHAPTER 12

CHEMICAL THERMODYNAMICS

Chemical thermodynamics is the study of energy changes that accompany chemical reactions. This energy may appear in several forms, such as heat, light, or work. These energy relationships have been investigated in the laboratory; the results of experimental observations have been organized into a set of concepts including enthalpy, entropy and free energy; they have been summarized in certain "laws of thermodynamics"; and through the application of these laws one can predict whether or not it is possible for a reaction to occur spontaneously.

12.1 THE SYSTEM AND THE SURROUNDINGS

In the study of energy relationships we confine our investigation to a specific region of the universe we call a **system.** The system is ordinarily "closed" in the sense that, for the duration of the study, there is no passage of matter between the system and the rest of the universe, called the **surroundings.**

The isolation of matter within the system does not extend to energy. Energy may enter the system from the surroundings, or leave the system and go to the surroundings. Within the present chapter we shall limit our consideration of energy passing between the system and its surroundings to two types, heat energy and "pressure-volume" work.

12.2 THE FIRST LAW OF THERMODYNAMICS

The first law of thermodynamics is a mathematical statement of the law of Conservation of Energy in the form of the equation

$$\Delta E = Q - W. \tag{12.1}$$

Here ΔE represents the change in **internal energy** of the system, i.e., the change in kinetic and potential energies of the molecules making up the system; Q is

heat flow, the heat energy transferred between the system and its surroundings; and W is the **work** done by the system on the surroundings.

The internal energy of a system cannot be measured directly, but heat flow and work are measurable, and from them the *change* in internal energy can be calculated. On page 57 we saw that the heat absorbed by a sample of matter that does not change its physical state, i.e., solid, liquid or gas, is found by the equations:

$$Q = m \times c \times \Delta T, \tag{5.1}$$

where m is mass in grams, c is specific heat in calories/gram-°C, and ΔT is the temperature change of the sample; or

$$Q = n \times C \times \Delta T, \tag{5.2}$$

where n is the number of moles and C is the molar heat capacity in calories/mole-°C. If heat flow is accompanied by a change of physical state at constant temperature it is calculated from

$$Q = m \times \text{latent heat}, \tag{5.3}$$

where latent heat is the calories transferred as one gram of substance changes state; or

$$Q = n \times \Delta H_{\text{change of state}} \tag{5.4}$$

where ΔH, with its appropriate subscript, is the molar heat of fusion, vaporization, condensation or solidification, as the case may be, measured in calories/mole.

The term *work* includes all forms of energy except heat and light. In this chapter we shall restrict our consideration to "presure-volume" work that appears when a gas expands against a resisting pressure, as when gases push the piston in the cylinder of an automobile, thereby converting the heat flow accompanying a chemical reaction into mechanical work. If the expansion is against a constant opposing pressure, pressure-volume work is given by the expression

$$W = P\Delta V. \tag{12.2}$$

When pressure is measured in atmospheres and volume in liters, the unit of $P\Delta V$ work is the liter-atmosphere. This unit may be converted into the more familiar calorie or kilocalorie by the relationship

$$1.00 \text{ liter-atmosphere} = 24.2 \text{ calories} = 0.0242 \text{ kilocalories}. \tag{12.3}$$

Returning to the first law relationship, $\Delta E = Q - W$, a system initially has a definite amount of internal energy, E. If heat flows into the system from the surroundings, Q is positive thereby increasing E; if heat leaves the system, Q is

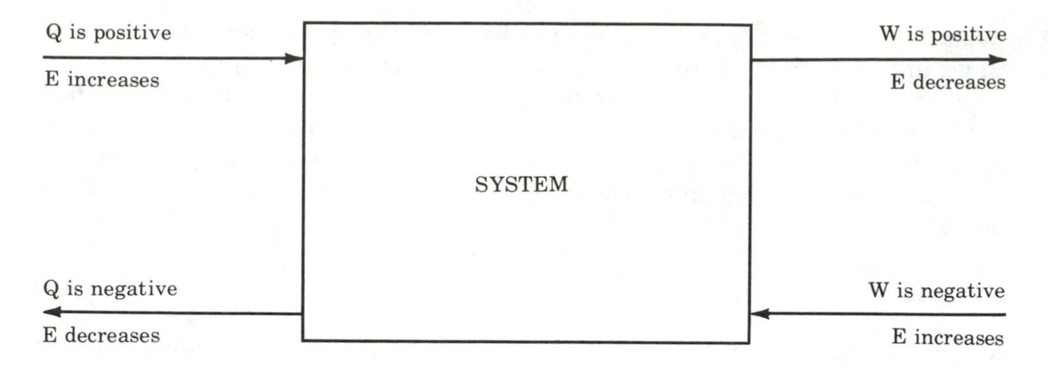

negative thereby decreasing E. Work done *by the system on the surroundings* is considered to be positive, and energy is *lost* to the system. It is subtracted, therefore, in Equation 12.1. Conversely, if the surroundings do work on (add energy to) the system, W is negative. Proper substitution of Q and W, including their algebraic signs, into Equation 12.1 will yield a correct value for ΔE. This is diagramed in Figure 12.1.

We are now ready to apply the first law of Thermodynamics.

Example 12.1. In a certain exothermic reaction (see page 70 if you do not recall the meaning of this term), 5.62 kcal of heat are transferred from the system to the surroundings. The expansion of gases in the reaction raises a piston, doing 0.46 kcal of work on the surroundings. Calculate ΔE.

First, from the statement of the problem, determine if Q is a positive or negative number.

12.1a. Q is a negative number. In an exothermic reaction, in which heat is *released,* or lost by the system to the surroundings, internal energy is reduced. Q is positive only when the system absorbs energy from the surroundings.

Now how about the sign of W?

12.1b. W is positive. By pushing the piston back, the system is doing work on the surroundings, effectively transferring some of its energy to the surroundings. From the standpoint of the *system*, it is doing *work*, and that is positive.

Now substitute the appropriate values into Equation 12.1 and solve.

12.1c. $\Delta E = -6.08$ kcal.

$$\Delta E = Q - W = -5.62 - 0.46 = -6.08 \text{ kcal}$$

Example 12.2. Calculate the change in internal energy for the change in state:

$$H_2O\ (l,\ 100°C,\ 1.00\ \text{atm}) \rightarrow H_2O\ (g,\ 100°C,\ 1.00\ \text{atm})$$

The molar heat of vaporization of water is 9.72 kcal/mole. Assume the gas laws apply to steam at its normal boiling point.

To use Equation 12.1 we must find both Q and W. The first is easy; what is the value of Q?

12.2a. Q = 9.72 kcal.
The molar heat of vaporization is the heat that must be absorbed by one mole of water at 100°C to convert it to steam at the same temperature. Because the water *absorbs* heat (increase in energy), Q is positive.

Work is calculated from Equation 12.2 and is equal to $P\Delta V$. It is the work done by the steam by pushing back against a constant resisting pressure of one atmosphere. One atmosphere is, therefore, the P factor in $P\Delta V$. We must yet determine ΔV, which, recalling that Δ always means final value minus initial values, is:

V (1 mole steam, 100°C, 1.00 atm) − V (1 mole liquid water, 100°C, 1.00 atm)

Using the ideal gas law (see page 95 if necessary), calculate the volume of 1 mole of steam at its normal boiling point.

12.2b. V = 30.6 liters.

$$V = \frac{nRT}{P} = \frac{1.00\ \cancel{\text{mole}}}{1.00\ \cancel{\text{atm}}} \times \frac{0.0821\ \text{liter-}\cancel{\text{atm}}}{\cancel{\text{mole}}\text{-}\cancel{°\text{K}}} \times (273+100)\cancel{°\text{K}} = 30.6\ \text{liters}$$

Now find the initial volume of one mole of liquid water at 100°C. Assume the density of water is 1.0 gram/milliliter at all temperatures in the liquid phase.

12.2c. V = 18 ml, or 0.018 liter.

$$1.00\ \cancel{\text{mole}} \times \frac{18.0\ \cancel{\text{grams}}}{1\ \cancel{\text{mole}}} \times \frac{1\ \cancel{\text{ml}}}{1.0\ \cancel{\text{gram}}} \times \frac{1\ \text{liter}}{1000\ \cancel{\text{ml}}} = 0.018\ \text{liter}$$

You are now ready to find ΔV.

12.2d. ΔV = 30.6 liters.

$$\Delta V = V_{final} - V_{initial} = 30.6 - 0.018 = 30.6\ \text{liters}$$

As this example illustrates, the volume of a liquid is almost always negligible compared to the volume of a vapor in a change of state at ordinary temperatures and pressures. The same is true of the volume of a solid. *Unless stated otherwise, we shall assume that for a process involving gases, volumes of liquids and solids may be neglected in calculating ΔV.*

Now calculate W in liter-atmospheres.

12.2e. W = 30.6 liter-atmospheres.

$$W = P\Delta V = 1.00\ \text{atm} \times 30.6\ \text{liters} = 30.6\ \text{liter-atmospheres.}$$

In order to apply Equation 12.1, Q and W must be expressed in the same units. Since Q is in kcal, let us convert W to kcal.

12.2f. W = 0.741 kcal.

$$W = P\Delta V = 30.6\ \cancel{\text{liter-atm}} \times \frac{0.0242\ \text{kcal}}{1\ \cancel{\text{liter-atm}}} = 0.741\ \text{kcal}$$

Finally you are ready to calculate ΔE.

12.2g. ΔE = 8.98 kcal.

$$\Delta E = Q - W = 9.72 - 0.741 = 8.98\ \text{kcal}$$

Notice that not all of the heat added to water to vaporize it (9.72 kcal) goes to increase internal energy. In this instance 0.741 kcal leave the system in the form of work. Only 8.98 kcal remain as internal energy. We shall later return to this observation and, using it as a base, identify a new thermodynamic function.

Example 12.3. Calculate the change in internal energy when one mole of ice melts at 0°C and a constant pressure of 1 atmosphere. The molar heat of fusion is 1.44 kcal/mole. The density of ice at 0°C is 0.918 g/cm^3; of water at 0°C, 1.00 g/ml.

First, what is Q, the heat flow required to melt one mole of ice at 0°C?

12.3a. Q = 1.44 kcal.
The molar heat of fusion is by definition, the amount of heat that must be absorbed by one mole of a solid to change it to a liquid at its melting point.

To find W you must again calculate $P\Delta V$. This time ΔV is the difference in volume between one mole of water as liquid and as solid. From the density of liquid water, what volume is occupied by one mole? Express your answer in liters.

12.3b. V = 0.0180 liter, as in 12.2c.
Now the volume of 1 mole of ice in liters. . .

12.3c. V = 0.0196 liter.

$$V = 1\ \cancel{\text{mole}}\ \text{ice} \times \frac{18.0\ \cancel{\text{g}}\ \text{ice}}{1\ \cancel{\text{mole}}\ \text{ice}} \times \frac{1\ \cancel{\text{cm}^3}}{0.918\ \cancel{\text{g}}} \times \frac{1\ \text{liter}}{1000\ \cancel{\text{cm}^3}} = 0.0196\ \text{liter}$$

Now complete your calculation of $P\Delta V$. Be sure to express your answer in kcal so it may be subtracted from the value of Q without further unit adjustment.

12.3d. W = -3.87×10^{-5} kcal.

$$W = P\Delta V = 1.00\ \cancel{\text{atm}} \times (0.0180 - 0.0196)\ \cancel{\text{liter}} \times \frac{0.0242\ \text{kcal}}{\cancel{\text{liter-atm}}} = -3.87 \times 10^{-5}\ \text{kcal}$$

Notice that W has a negative sign, the result of a decrease in volume from the initial to the final conditions, making ΔV negative.

Now solve for ΔE.

12.3e. $\Delta E = 1.44$ kcal.

$$\Delta E = Q - W = 1.44 - (-0.0000387) = 1.44 \text{ kcal}$$

In this example we see that ΔV is so small that the work term is negligible. This is typical: in a change in which all reactants and products are solids and/or liquids, $P\Delta V$ is usually negligible compared to Q. Consequently the relationship $\Delta E = Q - W = Q - P\Delta V$ becomes $\Delta E \approx Q$.

Even reactions involving gases as reactants and/or products may be conducted under conditions in which volume remains constant. A reaction that occurs in a rigid container, such as a bomb calorimeter, is an example. Under these circumstances $\Delta V = 0$ and $\Delta E = Q_V$, where the subscript is used to indicate constant volume.

12.3 STATE FUNCTIONS

Internal energy is an example of a **state function.** At any given instant a system has a certain quantity of internal energy. How the system obtained this internal energy, the amount of internal energy it had a few seconds ago or the amount it will have a few microseconds hence, has no bearing on the internal energy *now*. This is the character of a state function: it identifies the instantaneous value of some property of the system without regard for how it reached that value. Pressure, volume and temperature are other familiar state functions that are important to our understanding of thermodynamics.

A change in a state function is independent of the process by which the change occurs, commonly referred to as the path of the change. Herein lies the importance of the concept of state functions. We can calculate quite readily the changes in some state functions over paths that cannot be duplicated physically, whereas determining the same values experimentally is not always feasible. Furthermore, if it is possible to determine the value of a change in a state function by any process, real or theoretical, we may be confident that the same value will hold for any other process.

Heat flow, Q, and work, W, are not state functions, but may have quite different values for a given change in state, depending on the path followed. The only restrictions on Q and W for a given *change* in internal energy are that the *difference* $Q - W$ must be the same for any path, much as $9 - 4$ is the same as $7 - 2$.

12.4 ENTHALPY

In Example 12.2, in which we considered the change of one mole of water to a vapor at its normal boiling point, we found that Q has a value of 9.72 kcal, while ΔE is 8.98 kcal, the difference being converted into $P\Delta V$ work. It would be convenient to have a single thermodynamic function that identifies the heat flow in a process conducted at constant pressure, the condition that usually exists when we carry out a reaction in the laboratory. **Enthalpy** is such a function. Identified by the symbol H, enthalpy is a state function that is defined by the equation

$$H = E + PV. \tag{12.4}$$

Like internal energy, which is not measurable on an absolute scale, enthalpy cannot be measured directly. But the *change* in enthalpy of a system is measurable through calorimetric methods. This change is represented as ΔH. Accordingly,

$$\Delta H = \Delta E + \Delta(PV). \tag{12.5}$$

When pressure is constant, i.e., equal to and balanced by a constant opposing external pressure—and for this introduction to chemical thermodynamics we shall consider only constant pressure changes—Equation 12.5 becomes

$$\Delta H = \Delta E + P\Delta V. \tag{12.6}$$

The first law of thermodynamics states that $\Delta E = Q - W$; at constant pressure this becomes $\Delta E = Q_P - P\Delta V$. Substituting into Equation 12.6:

$$\Delta H = Q_P - P\Delta V + P\Delta V$$

$$\Delta H = Q_P. \tag{12.7}$$

By measuring the heat flow at constant pressure, we therefore measure ΔH of a chemical and/or physical change. We have already used this fact in Examples 12.2 and 12.3, where molar heat of vaporization, ΔH_{vap}, and molar heat of fusion, ΔH_{fus}, were equated to heat flow at atmospheric pressure.

Let us interrupt our development for a moment to make an important point. Equation 12.7 equates ΔH to Q_P. You must remember that this equation is valid only if two conditions are met: (1) the change occurs at constant pressure and (2) the only work is $P\Delta V$ work. These restrictions will be in effect throughout the remainder of this chapter; but you will find in your more advanced consideration of thermodynamics that this simplification will not always be permissible.

The pressure-volume term of Equation 12.5 is important in a constant pressure chemical or physical change only when there is a change in volume. As we have seen, volume changes are significant only when gases are involved. This provides an alternate approach to finding the value of $\Delta(PV)$. The ideal gas law tells us $PV = nRT$. Therefore, $\Delta(PV) = \Delta(nRT)$. We can substitute this into Equation 12.5:

$$\Delta H = \Delta E + \Delta(PV) = \Delta E + \Delta(nRT).$$

In an isothermal (constant temperature) process, n is the only variable in the second term. The equation, therefore, becomes

$$\Delta H = \Delta E + RT\Delta n_g \tag{12.8}$$

where Δn_g is the change in number of moles *of gas* in the reaction, $n_{final} - n_{initial}$.

It is also worth noting that, if $\Delta(PV) = P\Delta V = W$ at constant pressure, and if $\Delta(PV) = \Delta(nRT) = RT\Delta n_g$ at constant temperature, then *at constant temperature* **and** *constant pressure*

$$W_{TP} = RT\Delta n_g. \tag{12.9}$$

Before either Equation 12.8 or 12.9 can be applied to yield work in calories or kilocalories, we must establish a value for R in the proper units. The familiar 0.0821 liter-atmospheres per mole-degree Kelvin may be converted to calories per mole-degree Kelvin by Equation 12.3 (page 161):

$$R = \frac{0.0821 \text{ }\cancel{\text{liter-atm}}}{\text{mole-°K}} \times \frac{24.2 \text{ cal}}{1 \text{ }\cancel{\text{liter-atm}}} = 1.99 \frac{\text{cal}}{\text{mole-°K}}$$

$$R = 1.99 \frac{\text{cal}}{\text{mole-°K}} = 0.00199 \frac{\text{kcal}}{\text{mole-°K}}. \tag{12.10}$$

Before tackling an example problem, it is helpful to gather together all the equations encountered for changes involving only pressure–volume work at constant pressure:

$$\Delta E = Q - W \tag{12.1}$$

$$Q_{\text{change of state}} = n \times \Delta H_{\text{change of state}} \tag{5.2}$$

$$W = P\Delta V \tag{12.2}$$

$$\Delta H = \Delta E + P\Delta V \tag{12.6}$$

$$\Delta H = Q_P \tag{12.7}$$

$$\Delta H = \Delta E + RT\Delta n_{gas} \text{ (constant temperature)} \tag{12.8}$$

$$W = RT\Delta n_{gas} \text{ (constant temperature)} \tag{12.9}$$

Example 12.4. The molar heat of vaporization, ΔH_{vap}, of carbon tetrachloride, CCl_4, is 7.15 kcal/mole at its boiling point of 76.8°C. If 50.0 grams of CCl_4 are vaporized at 1.00 atmosphere and the boiling point, calculate Q, W, ΔE and ΔH.

Heat of vaporization, by definition, is the heat that must be absorbed to vaporize *one mole* of a substance at its boiling point. The problem calls for vaporizing only 50.0 grams. Calculate Q for the vaporization of 50.0 grams of CCl_4.

12.4a. Q = 2.32 kcal.

$$Q = n \times \Delta H_{vap} = 50.0 \cancel{\text{ g } CCl_4} \times \frac{1 \cancel{\text{ mole } CCl_4}}{154 \cancel{\text{ g } CCl_4}} \times \frac{7.15 \text{ kcal}}{1 \cancel{\text{ mole } CCl_4}} = 2.32 \text{ kcal}$$

The given quantity, 50.0 grams of CCl_4, is first changed to moles, and then multiplied by the molar heat of vaporization, according to Equation 5.2.

Next calculate W. There are two correct methods. One involves Equation 12.2 as it was used in Example 12.2. Since this change occurs at both constant pressure and constant temperature we can also apply Equation 12.9, $W = RT\Delta n_{gas}$. Let's use the second method. In doing so, what value will you use for Δn_{gas}? The equation for the change of state may help you to see the answer to this question: $CCl_4(l) \rightarrow CCl_4(g)$.

12.4b. Δn_{gas} = 1 mole gas per mole CCl_4 vaporized.

Δn is the change in the number of moles *of gas* in the reaction. The equation shows one mole of gas on the right side, and none on the left. Therefore, $\Delta n_{gas} = 1 - 0 = 1$.

Δn established, the rest of the calculation for W is straightforward. Be sure you use a proper value for R.

12.4c. W = 226 cal = 0.226 kcal.

$$W = RT\Delta n_{gas}$$

$$= \frac{1.99 \text{ cal}}{\text{mole-°K}} \times (273 + 77)\text{°K} \times 50.0 \text{ g } CCl_4 \times \frac{1 \text{ mole gas}}{154 \text{ g } CCl_4}$$

$$= 226 \text{ cal} = 0.226 \text{ kcal}$$

Alternately, assuming $PV = \frac{gRT}{MW}$ is valid for CCl_4 vapor at its boiling point and one atmosphere pressure, the volume of the vapor is

$$V = \frac{gRT}{(MW)P} = \frac{50.0 \text{ g } CCl_4}{1 \text{ atm}} \times \frac{1 \text{ mole } CCl_4}{154 \text{ g } CCl_4} \times \frac{0.0821 \text{ liter-atm}}{\text{mole-°K}} \times 350\text{°K}$$

$$= 9.33 \text{ liters}$$

Assuming the volume of the liquid is negligible compared to the volume of the vapor, ΔV = 9.33 liters. Thus,

$$W = P\Delta V = 1 \text{ atm} \times 9.33 \text{ liters} \times \frac{24.2 \text{ cal}}{1 \text{ liter-atm}} = 226 \text{ cal} = 0.226 \text{ kcal.}$$

The first method is surely the easier of the two.

With both Q and W known, ΔE follows readily. . . .

12.4d. ΔE = 2.09 kcal.

$$\Delta E = Q - W = 2.32 - 0.226 = 2.09 \text{ kcal}$$

ΔH is even easier. It requires *no* calculation. . . .

12.4e. ΔH = 2.32 kcal.

As noted previously, at constant pressure $\Delta H = Q_p$. Heat flow was calculated in the first step of the problem.

You may recall that enthalpy was introduced in Chapter 6 of this book (page 70), where it was equated to the heat flow for a reaction at constant temperature and pressure. At that time you learned that, by the process of enthalpy summation, you can calculate ΔH for any reaction if you have values for the "standard heat of formation" of every compound in the reaction equation. Standard heat of formation, ΔH_f°, is defined as the ΔH for the reaction in which one mole of a compound is produced from its elements, both products and reactants being in their standard state. All elemental substances in their natural states are assigned a ΔH_f° of zero for purposes of these calculations. Values of ΔH_f° are included in Table III, page 285.

The superscript "o" designates standard state, which is taken to be one atmosphere and the temperature at which the change occurs. Standard state heats of formation are generally tabulated at 25°C. We shall follow the common practice of using super zero to identify both 1 atmosphere and 25°C, *unless temperature is specified otherwise.*

To find ΔH for any reaction, the sum of the heats of formation of the products is subtracted from the sum of the heats of formation of the reactants. Expressed mathematically,

$$\Delta H = \Delta H^\circ = \sum \Delta H_f^\circ \text{ (products)} - \sum \Delta H_f^\circ \text{ (reactants)}, \qquad (6.13)$$

ΔH has been shown equal to ΔH° because it is an experimental fact that ΔH is essentially independent of pressure and its variation with temperature is negligible.

The use of Equation 6.13 in the determination of ΔH is a forecast of things to come in this chapter. A review example is in order. . . .

Example 12.5. Using ΔH_f° values from Table III (page 285), calculate ΔH for the reaction

$$CH_4(g) + 2\ Cl_2(g) \rightarrow CH_2Cl_2(g) + 2\ HCl(g).$$

Refer to page 74 if you need a refresher for handling this problem.

12.5a. $\Delta H = -47$ kcal.

$$\Delta H = \sum \Delta H_f^\circ \text{ (products)} - \sum \Delta H_f^\circ \text{ (reactants)}$$
$$= -21 + 2(-22.06) - (-17.89) = -47 \text{ kcal}$$

12.5 ENTROPY

One of the functions of thermodynamics is to develop a criterion for predicting whether or not a chemical change can occur spontaneously. There are two "driving forces" that tend to promote spontaneous change. The first is a *minimization of energy*, manifested as a negative value for ΔH. In the temperatures and pressures of our normal environment, a large majority of spontaneous changes have negative ΔH values. But the exceptions are too numerous to disregard, and at elevated temperatures the exceptions are even more common. Clearly a negative ΔH contributes to spontaneous change, but it cannot stand alone as a predictor of change.

The second driving force for spontaneous change is a tendency toward the most random arrangement of things in the universe. This may be described as a maximization of **entropy**, a function associated with disorder or randomness. Like internal energy and enthalpy, entropy is a state function. A precise thermodynamic definition of entropy is mathematical in character, involving calculus, and therefore beyond the scope of this book. Accordingly, our consideration of entropy will be rather empirical in nature.

Unlike internal energy and enthalpy, entropy values may be stated on an absolute scale based on the fact that a "perfect crystal" at 0°K has an entropy of zero. Entropy increases as temperature rises, and absolute values may be calculated. A list of absolute entropies at standard state (1 atmosphere) and 25°C, designated $S°$ and recorded in calories/mole–°K, is included in Table III (page 285).

Just as $\Delta H°$ for a reaction may be determined from known $\Delta H_f°$ values, so $\Delta S°$ may be found from known $S°$ values:

$$\Delta S° = \sum S° \text{ (products)} - \sum S° \text{ (reactants)}. \tag{12.11}$$

Except at low temperatures, entropy *change* is essentially independent of the temperature at which the change occurs, but it is pressure dependent for reactions involving gases. Therefore Equation 12.11 is restricted to determining ΔS at 1 atmosphere, the pressure at which $S°$ values are tabulated.

The application of Equation 12.11 to calculate $\Delta S°$ for a reaction is identical to the application of Equation 6.13 in finding ΔH. Because the entropy of a substance is based on an absolute scale, *$S°$ for elements is not zero,* as is the $\Delta H_f°$ of an element. Be sure to take this into account in solving the following example.

Example 12.6. Calculate $\Delta S°$ for the formation of one mole of ammonia from its elements. The equation for the reaction is $N_2(g) + 3\ H_2(g) \rightarrow 2\ NH_3(g)$.

Using absolute entropy values from Table III, substitute into Equation 12.11 to find $\Delta S°$.

12.6a. $\Delta S° = -47.38$ cal/°K.

$$\begin{aligned}\Delta S° &= \sum S° \text{ (products)} - \sum S° \text{ (reactants)}\\ &= 2\ S°_{NH_3} - S°_{N_2} - 3\ S°_{H_2}\\ &= 2(46.01) - 45.77 - 3(31.21) = -47.38 \text{ cal/°K}\end{aligned}$$

Certain changes can be described as occurring "reversibly." A **reversible change** has two identifying characteristics: First, a system undergoing reversible change is always essentially in a state of equilibrium with its surroundings; system and surrounding temperatures and pressures are closely balanced throughout the change, never differing by more than an infinitesimal amount.

Second, the direction of a reversible change may be reversed by an infinitesimal change in one of the conditions, e.g., temperature or pressure, under which the change occurs. *At constant temperature* ΔS is related to the reversible heat flow by the equation

$$\Delta S = \frac{Q_{rev}}{T} \qquad (12.12)$$

where T represents the absolute temperature at which the change occurs.

Reversible changes can be approached, but never quite realized in actual practice. One process that closely approaches reversibility is a change of state occurring at constant temperature and pressure, as in the boiling or freezing of a liquid. Heat flow for such a change is essentially equal to Q_{rev}. Furthermore, at constant pressure $\Delta H = Q_P$ (Equation 12.7). Consequently $\Delta H_{change\ of\ state}$ corresponds to Q_{rev}, and Equation 12.12 becomes

$$\Delta S_{change\ of\ state} = \frac{\Delta H_{change\ of\ state}}{T} \qquad (12.13)$$

at constant pressure and temperature.

Latent heats are readily determined in the laboratory. They may then be used to calculate ΔS for the change of state by means of Equation 12.13.

Example 12.7. Calculate ΔS in calories/mole−°K for the conversion of one mole of liquid water at 100°C to a vapor in an open beaker at 1 atmosphere. $\Delta H_{vap} = 9.72$ kcal/mole.

12.7a. $\Delta S = 26.1$ cal/mole−°K.

$$\Delta S = \frac{\Delta H_{vap}}{T} = \frac{9720 \text{ cal/mole}}{(273+100)°K} = 26.1 \text{ cal/mole-°K}$$

Notice that entropy increases as a substance changes from a liquid to a gas. This is compatible with our qualitative view of entropy as a measure of randomness or disorder; the particle movement of a gas is much more free, more random, than that of a liquid. By contrast, ΔS for the freezing of water is a negative quantity. A decrease in entropy on freezing reflects the more ordered arrangement of molecules in a solid state than in a liquid.

12.6 FREE ENERGY

At this point we have identified two basic driving forces that contribute to spontaneous change: minimization of energy and maximization of entropy. Sometimes these driving forces support each other, both tending to produce change or both opposing change. More often they are contrary, one tending toward change, the other opposing. Whether or not change is spontaneous under these circumstances depends on which of the two exerts the greater influence.

The net effect of these two tendencies is expressed in another thermodynamic state function called **free energy**,* symbolized by the letter G.

Free energy is defined by a mathematical equation:

$$G = H - TS. \tag{12.14}$$

It follows that

$$\Delta G = \Delta H - \Delta(TS).$$

If temperature is constant this becomes

$$\Delta G = \Delta H - T\Delta S, \tag{12.15}$$

in which form it is known as the Gibbs-Helmholtz equation. Free energy is a state function; ΔG is independent of the path between initial and final states. Free energy is dependent upon both temperature and pressure.

Energy is minimized when ΔH is negative. If ΔH is negative in Equation 12.15, it contributes to making ΔG negative. Entropy is maximized when ΔS is positive. Because the TΔS term is *subtracted* in finding ΔG, a positive ΔS also tends to make ΔG negative. *If ΔH and TΔS have values such that ΔG is negative, the reaction will be spontaneous as written.* In other words, ΔG is the thermodynamic criterion for spontaneity. If one calculates a positive value for ΔG the reaction is nonspontaneous; the reverse reaction takes place. If ΔG = 0 the system is in a state of equilibrium, with no *net* change taking place in either direction.

Like its energy counterparts E and H, free energy cannot be measured directly. As with standard enthalpies of formation, ΔH_f°, standard free energies of formation, ΔG_f°, of all elements are given a value of zero, and values of ΔG_f° of compounds are calculated accordingly. Values of ΔG_f° at 25°C are listed in Table III, page 285. ΔG° for a reaction may be calculated from these values in the same manner as ΔH°. The equation, which has the same form as Equation 6.13, is

$$\Delta G^\circ = \sum \Delta G_f^\circ \text{ (products)} - \sum \Delta G_f^\circ \text{ (reactants)}. \tag{12.16}$$

Example 12.8. Calculate ΔG° at 25°C for the burning of propane by the reaction $C_3H_8(g) + 5\ O_2(g) \rightarrow 3\ CO_2(g) - 4\ H_2O(l)$, using ΔG_f° values from Table III, page 285.

---

*Free energy is frequently referred to as Gibbs free energy after J. Willard Gibbs, whose contributions to our understanding of chemical thermodynamics are perhaps greater than those of any other individual.

12.8a. $\Delta G^\circ = -503.93$ kcal.

$$\Delta G^\circ = \sum \Delta G^\circ \text{ (products)} - \sum \Delta G^\circ \text{ (reactants)}$$
$$= 3(-94.26) + 4(-56.69) - (-5.61) = -503.93 \text{ kcal.}$$

Equation 12.15 suggests there is another way to find ΔG° for the reaction in the foregoing example. . . .

Example 12.9. Calculate ΔG° for the reaction in Example 12.8 by the equation $\Delta G^\circ = \Delta H^\circ - T\Delta S^\circ$.

As a first step, you must determine ΔH° as in Example 12.5, page 169.

12.9a. $\Delta H^\circ = -530.61$ kcal.

$$\Delta H^\circ = 3(-94.05) + 4(-68.32) - (-24.82) = -530.61 \text{ kcal}$$

ΔS° is also necessary, as in Example 12.6. . . .

12.9b. $\Delta S^\circ = -89.48$ cal/°K.

$$\Delta S^\circ = 3(51.05) + 4(16.72) - 64.51 - 5(49.00) = -89.48 \text{ cal/°K}$$

Recalling that the temperature at which standard state values have been tabulated is 25°C, or 298°K, you now have everything required for the Gibbs-Helmholtz equation. Notice that ΔS° is in calories per degree-Kelvin, whereas all other energy units are in kcal. Make the necessary conversion, and solve for ΔG° by the equation $\Delta G^\circ = \Delta H^\circ - T\Delta S^\circ$.

12.9c. $\Delta G^\circ = -503.94$ kcal.

$$\Delta G^\circ = -530.61 - 298(-0.08948) = -503.94 \text{ kcal}$$

In solving this problem, the ΔS° value has been changed from -89.48 cal/°K to -0.08948 kcal/°K. When multiplied by temperature, the °K units cancel, leaving only kcal. Within the limits of experimental error the values for ΔG° are the same by both methods of calculation.

Our next step is to use the free energy as a criterion for reaction spontaneity.

Example 12.10. Predict whether or not the following reaction will be spontaneous at 298°K and 1 atmosphere:

$$SO_3(g) + NO(g) \rightarrow SO_2(g) + NO_2(g).$$

For this reaction $\Delta H^\circ = 9.98$ kcal and $\Delta S^\circ = 13.69$ cal/°K $= 0.01369$ kcal/°K. Calculate ΔG° from the Gibbs-Helmholtz equation and make your prediction.

12.10a. $\Delta G^\circ = 5.90$ kcal. Because ΔG° is positive, the reaction is nonspontaneous as written, but would be spontaneous in the opposite direction.

$$\Delta G^\circ = 9.98 - 298(0.01369) = +5.90 \text{ kcal}$$

Inasmuch as $\Delta H°$ and $\Delta S°$ are nearly independent of temperature, the Gibbs-Helmholtz equation can be used to predict the value of $\Delta G°$ at temperatures other than 298°K. The reaction of Example 12.10 may be used to illustrate the point. . . .

Example 12.11. Assuming the values of $\Delta H°$ and $\Delta S°$ given in Example 12.10 for the reaction $SO_3(g) + NO(g) \rightarrow SO_2(g) + NO_2(g)$ are independent of temperature, determine $\Delta G°$ at 800°K, and predict whether or not the reaction would be spontaneous at that temperature.

The method is the same as in the foregoing example. . . .

12.11a. $\Delta G° = -0.972$ kcal; the reaction is spontaneous.

$$\Delta G° = 9.98 - 800(0.01369) = -0.972 \text{ kcal}$$

The negative value of $\Delta G°$ leads to the prediction that the reaction is spontaneous at 800°K. This is an example of how a chemical change that does not occur spontaneously at one temperature may be made spontaneous by adjusting the temperature.

12.7 SUMMARY

Your introduction to chemical thermodynamics is now complete. Before attempting a single problem that includes all the concepts presented in this chapter, you may find this expanded, full-chapter summary of important equations helpful. Remember that in this brief introduction these equations are for constant pressure changes which involve only pressure-volume work.

$$\Delta E = Q - W \qquad (12.1)$$

$$Q_{\text{change of state}} = n \times \Delta H_{\text{change of state}} \qquad (5.2)$$

$$W = P\Delta V \qquad (12.2)$$

$$\Delta H = \Delta E + P\Delta V \qquad (12.6)$$

$$\Delta H = Q_P \qquad (12.7)$$

$$\Delta H = \Delta E + RT\Delta n_{gas} \text{ (constant temperature)} \qquad (12.8)$$

$$W = RT\Delta n_{gas} \text{ (constant temperature)} \qquad (12.9)$$

$$\Delta H = \Delta H° = \sum \Delta H_f° \text{ (products)} - \Delta H_f° \text{ (reactants)} \qquad (6.13)$$

$$\Delta S° = \sum S° \text{ (products)} - \sum S° \text{ (reactants)} \qquad (12.11)$$

$$\Delta G = \Delta H - T\Delta S \text{ (constant temperature)} \qquad (12.15)$$

$$\Delta G° = \sum \Delta G° \text{ (products)} - \sum \Delta G° \text{ (reactants)} \qquad (12.16)$$

It is surprising what one can learn from no more than the equation for a chemical reaction. . . .

Example 12.12. Calculate Q, W, ΔE, ΔH°, ΔG° and ΔS° for the reaction $2\ CO(g) + O_2(g) \rightarrow 2\ CO_2(g)$ at 25°C and 1.00 atmosphere. Also predict whether or not the reaction will be spontaneous.

There are several places to begin. Perhaps the most direct is to find ΔH°, ΔG° and ΔS° from the tables of thermodynamic properties, using the sum of the product properties minus the sum of the reactant properties. It is convenient in approaching an overall problem of this sort to tabulate data from the tables, using reaction species as column headings:

	$2\ CO(g)$	+	$O_2(g)$	→	$2\ CO_2(g)$
ΔH_f°					
ΔG_f°					
ΔS°					

Enter the appropriate values into the blanks in this table.

12.12a.

	$2\ CO(g)$	+	$O_2(g)$	→	$2\ CO_2(g)$
ΔH_f°	−26.42		0		−94.05
ΔG_f°	−32.81		0		−94.26
S°	47.30		49.00		51.06

From this tabulation, calculate ΔH°, using the $\sum \Delta H^\circ$ process.

12.12b. $\Delta H^\circ = -135.26$ kcal.

$$\Delta H^\circ = 2(-94.05) - 2(-26.42) = -135.26 \text{ kcal}$$

ΔG° and ΔS° are found in the same way. Solve for both.

12.12c. $\Delta G^\circ = -122.90$ kcal; $\Delta S^\circ = -41.48$ cal/°K.

$$\Delta G^\circ = 2(-94.26) - 2(-32.81) = -122.90 \text{ kcal}$$
$$\Delta S^\circ = 2(51.06) - 2(47.30) - 49.00 = -41.48 \text{ cal/°K}$$

If you've done everything correctly, you should be able to confirm your result by substituting into the Gibbs-Helmholtz equation, $\Delta G° = \Delta H° - T\Delta S°$, and come out with an equality. Try it and see if it works. . . .

12.12d. $-122.90 = -135.26 - 298(-0.04148) = -122.90.$

The value of Q for the constant pressure process should be readily apparent. . . .

12.12e. $Q = -135.26$ kcal.

At constant pressure, $Q = \Delta H° = -135.26$ kcal.

Finding W will take a bit more work. Express it in kcal.

12.12f. $W = -0.593$ kcal.

$$W = RT\Delta n_{gas} = \frac{1.99 \text{ cal}}{\text{mole}-°\text{K}} \times 298°\text{K} \times (2-3) \text{ moles} \times \frac{1 \text{ kcal}}{1000 \text{ cal}} = -0.593 \text{ kcal}$$

Normally we would not double calculate any value, but this being a problem designed to illustrate all methods we have used in the chapter, let's solve for work as $P\Delta V$. Finding ΔV has a catch to it—see if you can calculate it correctly.

12.12g. $\Delta V = -24.5$ liters.

The change in the number of moles of gas is two product moles minus three reactant moles, a net of -1 mole. Molar volume at 25°C and one atmosphere is 24.5 liters/mole; it is not 22.4 liters/mole, the molar volume at STP:

$$V = \frac{nRT}{p} = \frac{1 \text{ mole}}{1 \text{ atm}} \times \frac{0.0821 \text{ liter}-\text{atm}}{\text{mole}-°\text{K}} \times (273 + 25)°\text{K} = 24.5 \text{ liters}$$

Now solve for work as $P\Delta V$, expressing the answer in kilocalories.

12.12h. $W = -0.593$ kcal.

$$W = P\Delta V = 1 \text{ atm} \times -24.5 \text{ liters} \times \frac{0.0242 \text{ kcal}}{1 \text{ liter}-\text{atm}} = -0.593 \text{ kcal}$$

The only thing left is ΔE from the first law. . . .

12.12i. $\Delta E = -134.67$ kcal.

$$\Delta E = Q - W = -135.26 - (-0.593) = -134.67 \text{ kcal}$$

And at last, will the reaction be spontaneous at 25°C and 1 atmosphere? Remember the criterion. . . .

12.12j. The reaction will be spontaneous, as predicted by the negative value for $\Delta G°$ (answer 12.11c).

Hopefully you are impressed and have gained confidence in seeing how much information can be derived from nothing more than the equation for a chemical reaction and a table of thermodynamic properties.

PROBLEMS

Section 12.2

12.1. In a certain endothermic change, 8.34 kcal of heat energy pass between the system and its surroundings, while the surroundings do 0.71 kcal of work on the system. Find ΔE.

12.2. Calculate W in kcal when 5.20 moles of ammonia are vaporized at the normal boiling point, −33°C, and 1.00 atmosphere.

12.3. 84.8 grams of steam are condensed at 1.00 atmosphere and 100°C. Assuming the gas laws apply to steam at its boiling point, calculate ΔE. ΔH_{vap} = 9.72 kcal/mole.

12.4. For the reaction $CaCO_3(s) \rightarrow CaO(s) + CO_2(g)$, Q = 42.5 kcal at 25°C and 1.00 atmosphere. Calculate ΔE for the decomposition of one mole of calcium carbonate.

12.23. $\Delta E = -4.65$ kcal in a chemical change where expanding gases result in a work exchange of 0.39 kcal between the system and surroundings. Calculate Q.

12.24. Carbon dioxide sublimes at −79°C. Calculate W in kcal as 34.6 grams of CO_2 sublime against a constant external pressure of 1.00 atmosphere. Assume gas laws are valid for the vapor.

12.25. Calculate ΔE when 86.6 grams of ether, $(C_2H_5)_2O$, vaporize at its normal boiling point of 35°C, and 1.00 atmosphere. ΔH_{vap} = 6.67 kcal/mole. Assume gas laws are valid for ether vapor.

12.26. Oxygen was discovered with the thermal decomposition of mercury(II) oxide: $HgO(s) \rightarrow Hg(l) + \frac{1}{2} O_2(g)$. Q = 21.7 kcal for this reaction at 25°C and 1.00 atmosphere. Calculate ΔE.

Section 12.4

12.5. Calculate Q, W, ΔE and ΔH for the vaporization of one mole of ethyl alcohol, C_2H_5OH, at one atmosphere and its normal boiling point of 78°C. At those conditions ΔH_{vap} = 9.40 kcal/mole. Assume the gas laws are valid for the vapor at its boiling point.

12.6. Elemental sodium melts at 98°C, at which ΔH_{fus} = 630 cal/mole. Find Q, W, ΔH and ΔE when 40.0 grams of sodium are melted at 98°C and 1 atmosphere. Assume solid and liquid densities equal at the melting point.

12.7. Using the table of Standard Heats of Formation, calculate the heat of combustion (see page 169) of methane, $CH_4(g)$, by the enthalpy summation method (Equation 6.13).

12.8. Calculate Q, W, and ΔE for the reaction in Question 12.7, with all values adjusted to the reference temperature of 25°C.

12.27. Bromine, Br_2, is one of the few elements that is a liquid at room conditions. It boils at 58°C, at which ΔH_{vap} = 7.42 kcal/mole. Calculate Q, W, ΔH and ΔE for the vaporization of 1.25 moles of Br_2 at 58°C and 1 atmosphere.

12.28. When mercury freezes at −39°C there is a negligible change in volume. If ΔH_{fus} = 557 cal/mole, find Q, W, ΔH and ΔE for the freezing of one pound (454 grams) of mercury at its freezing point and 1 atmosphere.

12.29. Calculate ΔH for the combustion of ethane by the reaction $2\ C_2H_6(g) + 7\ O_2(g) \rightarrow 4\ CO_2(g) + 6\ H_2O(l)$, by means of Equation 6.13 and $\Delta H_f°$ values from Table III (page 285).

12.30. Calculate Q, W and ΔE for the reaction in Problem 12.29, with all values adjusted to the reference temperature of 25°C.

12.9. Calculate ΔH and ΔE for the reaction $2\ MgO(s) + C(s) \rightarrow CO_2(g) + 2\ Mg(s)$ at 25°C and 1 atmosphere.

12.10. $\Delta E = 5.2$ kcal for the reaction $H_2C_2O_4(s) \rightarrow HCOOH(l) + CO_2(g)$ at 25°C and 1 atmosphere. Calculate ΔH without reference to Table III.

12.31. For the reaction $CuO(s) + H_2(g) \rightarrow Cu(s) + H_2O(l)$ at 1 atmosphere and 25°C, find ΔH and ΔE.

12.32. $2\ C_8H_{18}(l) + 25\ O_2(g) \rightarrow 16\ CO_2(g) + 18\ H_2O(l)$ is the equation for the burning of *n*-octane at 25°C. Calculate ΔH if $\Delta E = -2629.58$ kcal.

Section 12.5

12.11. Calculate $\Delta S°$ for the combustion of one mole of methane, $CH_4(g)$, from the absolute entropies listed in Table III. (See Problem 12.7.)

12.12. Find $\Delta S°$ at 1 atmosphere and 25°C for the reaction $2\ H_2(g) + O_2(g) \rightarrow 2\ H_2O(l)$.

12.13. If $\Delta H_{fus} = 1.44$ kcal/mole for water at one atmosphere and 0°C, calculate $\Delta S°$ when 100 grams of liquid water freezes.

12.33. Calculate $\Delta S°$ for the combustion of ethane by the equation given in Problem 12.29, using absolute entropies listed in Table III.

12.34. Calculate $\Delta S°$ at 1 atmosphere and 298°K for the formation of one mole of ammonia from its elements.

12.35. $\Delta H_{vap} = 6.67$ kcal/mole for ethyl ether, $(C_2H_5)_2O$, at its normal boiling point of 35°C. Calculate $\Delta S°$ for the vaporization of 50.0 grams of ether at its boiling point.

Section 12.6

12.14. Calculate $\Delta G°$ for the combustion of one mole of methane, $CH_4(g)$, using the Gibbs-Helmholtz equation (Equation 12.15) and the results of Problems 12.7 and 12.11.

12.15. Calculate $\Delta G°$ for the combustion of one mole of methane, $CH_4(g)$, using the free energies of formation listed in Table III (page 285). Compare this result with the answer to Problem 12.14.

12.16. $\Delta S° = 53.18$ cal/°K for the reaction $H_2C_2O_4(s) \rightarrow HCOOH(l) + CO_2(g)$. Calculate $\Delta G°$ for the reaction, recalling that you determined $\Delta H°$ to be 5.8 kcal in Problem 12.10.

12.17. A common laboratory method for preparing oxygen is to heat potassium chlorate: $2\ KClO_3(s) \rightarrow 2\ KCl(s) + 3\ O_2(g)$. We know the reaction is spontaneous at elevated temperatures—it must be, if it is used—but is it theoretically spontaneous at 25°C and 1 atmosphere? $\Delta H° = -21.36$ kcal, and $\Delta S° = 118.18$ cal/°K. Calculate $\Delta G°$ at 25°C. Also assume $\Delta H°$ and $\Delta S°$ are independent of temperature and find $\Delta G°$ at −20°K.

12.36. Calculate $\Delta G°$ for the combustion of ethane by the equation given in Problem 12.29, using the Gibbs-Helmholtz equation (Equation 12.15) and the results of Problems 12.29 and 12.33.

12.37. Calculate $\Delta G°$ for the combustion of ethane by the equation given in Problem 12.29, using free energies of formation listed in Table III (page 285). Compare your result with the answer to Problem 12.36.

12.38. In Problem 12.32 you determined that $\Delta H = -2634.92$ kcal for $2\ C_8H_{18}(l) + 25\ O_2(g) \rightarrow 16\ CO_2(g) + 18\ H_2O(l)$. Find $\Delta S°$ from the absolute entropies in Table III (page 285), and calculate $\Delta G°$ by the Gibbs-Helmholtz equation. Compare the result with $\Delta G°$ determined from ΔG_f° values from Table III.

12.39. Hydrogen fluoride, dissolved in water to make hydrofluoric acid, is extremely corrosive, dissolving such usually inert substances as lead and glass. But how about its reactivity as a gas? Predict the spontaneity of the reaction $Br_2(l) + 2\ HF(g) \rightarrow 2\ HBr(g) + F_2(g)$ at 25°C and 1 atmosphere, given that $\Delta H° = 111.08$ kcal and $\Delta S° = 24.14$ cal/°K. Calculate $\Delta G°$ for the reaction, both at 25°K and 300°K.

Section 12.7

12.18. In the vicinity of a large electrical discharge, as in a flash of lightning, diatomic oxygen molecules are converted to triatomic molecules, in which form the gas is called ozone: $3\ O_2(g) \rightarrow 2\ O_3(g)$. For this reaction at 1 atmosphere and 25°C, calculate $\Delta H°$, $\Delta G°$, $\Delta S°$, Q, W and ΔE.

12.19. An oxyacetylene cutting torch found in welding shops burns acetylene by the equation $C_2H_2(g) + 5\ O_2(g) \rightarrow 4\ CO_2(g) + 2\ H_2O(l)$. For this reaction at 1 atmosphere and 298°K, find $\Delta H°$, $\Delta G°$, $\Delta S°$, Q, W and ΔE.

12.20. When sulfur is burned in air, a deep blue flame is produced, along with a suffocating gas, sulfur dioxide: $S(s) + O_2(g) \rightarrow SO_2(g)$. Determine $\Delta G°$ for the reaction.

12.21. Sulfur dioxide is used in the manufacturing of sulfuric acid by converting it to sulfur trioxide in a reaction that reaches equilibrium. The equation for the reaction is $2\ SO_2(g) + O_2(g) \rightarrow 2\ SO_3(g)$. Calculate $\Delta G°$.

12.22. The reactions of problems 12.20 and 12.21 may be combined to represent the conversion of sulfur to sulfur trioxide in the equation $2\ S(s) + 3\ O_2(g) \rightarrow 2\ SO_3(g)$. Calculate $\Delta G°$ for this reaction. In solving problems 12.20 and 12.21, you calculated the corresponding $\Delta G°$ values for the individual steps. How does $\Delta G°$ for the combined reaction compare with the sum of the $\Delta G°$ values for the steps? Can you explain the relationship?

12.40. In the same manner that a diatomic molecule is formed by two atoms of the same halogen, a diatomic molecule can be formed by one atom of each of two different halogens. Iodine chloride, ICl, is an example. Determine $\Delta H°$, $\Delta G°$, $\Delta S°$, Q, W and ΔE for the reaction $I_2(s) + Cl_2(g) \rightarrow 2\ ICl(g)$.

12.41. One might suspect that formic acid could be produced by bubbling carbon monoxide through water: $CO(g) + H_2O(l) \rightarrow HCOOH(l)$. Determine whether or not this reaction would be spontaneous at 25°C and 1 atmosphere. Calculate also W and ΔE for the reaction under these conditions.

12.42. The initial chemical change in obtaining copper from an ore containing $Cu_2S(s)$ is to blast air through the molten ore. The equation for the reaction is $2\ Cu_2S(s) + 3\ O_2(g) \rightarrow 2\ Cu_2O(s) + 2\ SO_2(g)$. Find $\Delta G°$ for the reaction.

12.43. The copper(I) oxide produced in the reaction of problem 12.42 is converted into what is known as blister copper by further reaction with Cu_2S: $2\ Cu_2O(s) + Cu_2S(s) \rightarrow 6\ Cu(s) + SO_2(g)$. Find $\Delta G°$ for this reaction.

12.44. The preparation of blister copper by the steps shown in problems 12.42 and 12.43 can be combined in a single equation:

$$Cu_2S(s) + O_2(g) \rightarrow 2\ Cu(s) + SO_2(g).$$

Determine $\Delta G°$ for the reaction, and compare it to the sum of the $\Delta G°$ values from problems 12.42 and 12.43.

CHAPTER 13

CHEMICAL EQUILIBRIUM—GASEOUS REACTIONS

When a reversible reaction is confined to a closed system the potential for equilibrium exists. In the hypothetical reaction $A + B \rightleftharpoons 2\,C$, for example, if the rate of the forward reaction is greater than the rate of the reverse reaction, the concentrations of A and B will decline over a period of time and the concentration of C will increase. Inasmuch as reaction rate is directly related to the concentrations of the reactants, the rate of the forward reaction will decrease, whereas the rate of the reverse reaction will increase. Ultimately they will become equal, after which there will be no further net change in any concentrations. The system is then at *equilibrium*: forward and reverse reaction rates are equal in a closed system, producing no measurable changes in any property of the system.

13.1 THE EQUILIBRIUM CONSTANT

For any reaction system at equilibrium, the ratio of the product of the concentrations of the species on the right side of the equation, each raised to the power of its coefficient in the equation, divided by the corresponding product of the concentrations of the species on the left is a constant. The **equilibrium constant expression,** K, for the general reaction, $aA + bB \rightleftharpoons xX + yY$ is

$$K = \frac{[X]^x\,[Y]^y}{[A]^a\,[B]^b}, \qquad (13.1)$$

where enclosure of a chemical symbol in square brackets designates concentration in moles per liter.

It must be emphasized that the equilibrium constant expression is always related to a specific equation. The concentrations in the numerator are of those species on the *right* side of the equation, whereas the concentrations in the denominator are of those species on the *left* side. If the equilibrium equation is written in reverse, interchanging right and left sides, the new equilibrium constant is the reciprocal of what it first was. Thus, for A + B = 2 C

$$K = \frac{[C]^2}{[A]\,[B]};$$

but for 2 C = A + B

$$K = \frac{[A]\,[B]}{[C]^2}.$$

Clearly the K expressions are reciprocals.

13.2 LE CHATELIER'S PRINCIPLE

Any change in the concentration of any species in an equilibrium alters a reaction rate. If this alteration affects the forward and reverse rates differently, the equilibrium is disturbed: the rate in one direction will be greater than in the other. This leads to further changes in concentrations of the species on both sides of the equation.

Le Chatelier's Principle makes possible a qualitative prediction of the direction of these changes, which, in turn, helps in solving equilibrium problems. This principle may be stated as follows: **If an equilibrium system is subjected to a stress, processes occur that tend to counteract partially the imposed change, thereby bringing the system to a new position of equilibrium.** These processes take the form of the equilibrium "shifting" in either the forward or reverse direction.

In our hypothetical equilibrium, $A + B \rightleftharpoons 2\,C$, for example, suppose the concentration of B were increased. The forward reaction rate would be increased, becoming greater than the reverse rate. As a consequence, the concentration of C would gradually increase while the concentrations of A and B decreased until the new equilibrium was reached. The net change between the new and old equilibria would be a decrease in the concentration of A and an increase in that of C, both caused by the Le Chatelier shift, and a net increase in the concentration of B. Note that only a portion of the B that was added would react to form C. The statement of Le Chatelier's Principle says that the reaction counteracts only *partially* the imposed change.

Concentration decreases are also covered by Le Chatelier's Principle. Again referring to the $A + B \rightleftharpoons 2\,C$ equilibrium, what would happen if the concentration of C were reduced? The rate of the reverse reaction would drop, becoming less than that of the forward reaction. This would cause the equilibrium to shift to the right. As a result, the concentrations of A and B would drop, while that of C would increase, but not up to its original value. Finally a new equilibrium would be established.

Numerous other qualitative predictions may be made on the basis of Le

Chatelier's Principle, but they are not germane to the consideration of equilibrium problems, and therefore will not be covered here.

13.3 DETERMINATION OF EQUILIBRIUM CONSTANTS

If an equilibrium problem is concerned with only one reversible reaction, the solution of the problem usually falls into two steps:

1. Find the equilibrium concentration of each species, numerically if possible, and algebraically otherwise.
2. Insert these values into the equilibrium constant expression and solve for the desired quantity.

It is usually helpful to follow the thought sequence in these problems by preparing a table of moles per liter of each substance in the equilibrium equation, showing the number of moles per liter initially present, I; then those reacting, R; and those present when equilibrium is reached, E. Using a hypothetical equilibrium, A + 3 B ⇌ 2 C, the table would be as follows:

	A + 3 B ⇌ 2 C
Initial moles/liter — *I*	
Reacting moles/liter — *R*	
Equilibrium moles/liter — *E*	

The student is urged to prepare the table for each problem, and then follow the stepwise procedure by which it is filled in.

Example 13.1. Four moles of A and 8 moles of B are introduced into an empty one-liter container. When equilibrium is reached in accord with the equation A + 3 B ⇌ 2 C, the vessel contains four moles of C. Find the numerical value of K.

Read the problem carefully, and place in the table all moles per liter values possible *from the problem's statement.* There are four.

13.1a.

	A	+	3 B	⇌	2 C
I	4		8		0
R					
E					4

The initial concentration of C must be zero since the container was empty before A and B were introduced.

There is now sufficient information to determine the number of moles per liter of C involved in the reaction. Any time the initial and equilibrium moles per liter are known for any species, the number of reacting moles per liter can be found. In this case, you begin with zero moles per liter of C, and at equilibrium you have four. How many must have been produced in reaching equilibrium?

13.1b. Four. If you began with none, and ended with four, you must have gained four. A *gain* should be designated with a plus sign. Insert this number into the table in the C column.

	A	+	3 B	⇌	2 C
I	4		8		0
R					+4
E					4

Once the number of moles per liter of one species involved in the reaction is determined, the number of moles per liter of all others may be found by stoichiometry. The equation states that for every two moles of C formed, one mole of A and three moles of B must have been consumed. In this case, four moles of C were produced. How many moles of A and B were consumed in the process? Place these figures into the table, with minus signs to indicate that this material was consumed.

13.1c. Two moles of A and six moles of C.
The reacting moles of A and B respectively, according to a dimensional analysis setup, would be:

$$4 \text{ \sout{moles C}} \times \frac{1 \text{ mole A}}{2 \text{ \sout{moles C}}} = 2 \text{ moles A} \quad ; \quad 4 \text{ \sout{moles C}} \times \frac{3 \text{ moles B}}{2 \text{ \sout{moles C}}} = 6 \text{ moles B}$$

The table to this point follows:

	A	+	3 B	⇌	2 C
I	4		8		0
R	−2		−6		+4
E					4

The signs of both A and B are negative: A and B are consumed in the reaction.

The equilibrium concentrations of A and B are now found from the initial and reacting values. If you start with four moles per liter of A and use two of them, how many will be left at equilibrium? Also, if you begin with eight moles per liter of B and use six, how many will be left at equilibrium? Complete the tabulation.

13.1d. Two moles per liter of both A and B.

	A	+	3 B	⇌	2 C
I	4		8		0
R	−2		−6		+4
E	2		2		4

As you now see, the result is obtained by algebraic addition of the initial and reacting concentrations: $I + R = E$.

The second part of the problem involves inserting the equilibrium concentrations of all species into the equilibrium constant expression and solving for the numerical value of K. Write the expression for K, substitute the required numbers, and calculate.

13.1e. K = 1.

$$K = \frac{[C]^2}{[A]\,[B]^3} = \frac{4^2}{2 \times 2^3} = 1$$

Example 13.2. Suppose 12 moles of C are introduced into the empty one-liter flask at a different temperature. The quantity of C decomposes to reach equilibrium with A and B thus: 2 C → A + 3 B. Calculate the value of K for the reaction, A + 3 B ⇌ 2 C, if the vessel contains six moles of B when equilibrium is reached.

To see your goal clearly, write the equilibrium constant expression for the K value you are to find.

13.2a.

$$K = \frac{[C]^2}{[A]\,[B]^3}.$$

This is the same expression as in the last problem. K expressions are always associated with a *specific equation*, and they are not concerned with the *initial* direction of the reaction.

Set up the table, and insert the four given values.

13.2b.

	A	+	3 B	⇌	2 C
I	0		0		12
R					
E			6		

Completing the table requires precisely the same thought process required for parts b, c, and d of Example 13.1. Watch the signs on the reacting quantities. Complete the table.

13.2c.

	A	+	3 B	⇌	2 C
I	0		0		12
R	+2		+6		− 4
E	2		6		8

To outline the thought processes: You begin with zero moles of B, and reach equilibrium with six moles. Therefore the reacting quantity is +6. The reaction stoichiometry indicates that production of six moles of B will be accompanied by two moles of A (3:1 ratio of coefficients), and will consume four moles of C (3:2 ratio of coefficients). Starting with no A and producing two moles yields two moles at equilibrium; starting with 12 moles of C and using four leaves eight moles at equilibrium.

Now substitute into the K expression and solve for a numerical answer.

13.2d. K = 4/27 = 0.148.

$$K = \frac{[C]^2}{[A]\,[B]^3} = \frac{8^2}{2 \times 6^3} = \frac{4}{27} = 0.148$$

The answer may be expressed in decimal or rational form, whichever is more convenient.

Example 13.3. 14.0 moles of X and 20.0 moles of Y are introduced into an empty two-liter container. They react, eventually reaching equilibrium according to the equation $X + 2\,Y \rightleftharpoons 2\,Z$. The equilibrium concentration of Z is 6.0 moles per liter. Calculate K.

Begin as before, setting up the table and entering all the concentrations that may be derived directly from the problem. Caution! This one's different.

13.3a.

	X	+	2 Y	⇌	2 Z
I	7.0		10.0		0
R					
E					6.0

$$I_x = \frac{14.0 \text{ moles}}{2 \text{ liters}} = 7.0 \text{ moles/liter}; \qquad I_Y = \frac{20.0 \text{ moles}}{2 \text{ liters}} = 10.0 \text{ moles/liter}$$

Careful reading of the problem shows that the initial *quantities* of X and Y are given, and the final *concentration* of Z. Either concentrations or quantities may be used in reasoning through the problem to find conditions at equilibrium, but the ultimate use of the equilibrium constant *must involve concentrations.* Immediate changing of initial state quantities to a mole-per-liter basis is recommended in this book primarily because it is consistent with the practice followed almost without exception with ionic equilibria. Furthermore, it reduces the possibility of later forgetting to convert from quantity to concentration units.

The balance of this example is handled in the same manner as the two previous examples. Complete the problem.

13.3b. K = 9/16 = 0.563.

	X	+	2 Y	⇌	2 Z
I	7.0		10.0		0
R	−3.0		−6.0		+6.0
E	4.0		4.0		6.0

$$K = \frac{[Z]^2}{[X]\,[Y]^2} = \frac{(6.0)^2}{(4.0)\,(4.0)^2} = \frac{36}{(4.0)\,(16)} = \frac{9}{16} = 0.563$$

As before, the number of reacting moles per liter of Z was found from the initial and final values. The numbers of reacting moles per liter of X and Y follow from the stoichiometry of the reaction, and equilibrium concentrations are the algebraic summation, *I* + *R*.

Let's apply these techniques to a problem involving real chemicals. . . .

Example 13.4. 30.0 moles of NO and 18.0 moles of O_2 are placed in an empty 3.00 liter reaction vessel at a certain temperature. They react until equilibrium is established according to the equation $2\,NO(g) + O_2(g) = 2\,NO_2(g)$. At equilibrium the vessel contains 26.4 moles of NO_2. Determine the value of K at that temperature.

Begin by computing the concentrations of NO and O_2 initially and NO_2 at equilibrium, and then enter them into the usual table.

13.4a.

	2 NO	+	O_2	=	2 NO_2
I	10.0		6.0		0.0
R					
E					8.8

Because the reaction vessel has a volume of 3.00 liters, all molar quantities must be divided by 3.00 to obtain concentration in moles per liter.

From the initial and final concentrations of NO_2 the *R* entry may be determined. This leads to completion of the table. . . .

13.4b.

	2 NO	+	O_2	=	2 NO_2
I	10.0		6.0		0.0
R	−8.8		−4.4		+8.8
E	1.2		1.6		8.8

The R value for oxygen is half that of the other species because its coefficient in the equilibrium equation is half the coefficients of the other species.

Now write the equilibrium constant expression, substitute the equilibrium concentrations, and compute K.

13.4c. K = 34

$$K = \frac{[NO_2]^2}{[NO]^2[O_2]} = \frac{8.8^2}{(1.2)^2(1.6)} = 34$$

13.4 DETERMINATION OF EQUILIBRIUM CONCENTRATIONS

Most work with chemical equilibria involves reactions for which the equilibrium constant is known or may be calculated from thermodynamic data. This enables us to make predictions about equilibrium concentrations on the basis of the initial concentrations of the reactants.

Example 13.5. In a hypothetical reaction, six moles of M and ten moles of N are placed into an empty one-liter cylinder. They reach equilibrium according to the equation M + N = 2 P. If K = 0.50 at the temperature of the equilibrium, calculate the equilibrium concentrations of all species.

Starting the table with the given information is straightforward.

13.5a.

	M	+	N	$\rightleftharpoons$	2 P
I	6		10		0

This time no R or E values are given. The assignment of an algebraic symbol to the concentration of some species must be used, and then the equilibrium concentrations of all species must be determined in terms of this symbol. In this instance, let y equal the number of moles per liter of M reacting.* The stoichiometry of the reaction should now enable you to indicate the numbers of moles per liter of N and P in the reaction. Complete the *R* line of the table.

13.5b.

	M	+	N	⇌	2 P
I	6		10		0
R	$-y$		$-y$		$+2y$

From the equation coefficients, the molar quantities of M and N reacting are the same (y), and the moles of P formed (2y) are twice as great as the moles of M and N consumed.

Completion of the table on the basis of $E = I + R$ now follows logically.

13.5c.

	M	+	N	⇌	2 P
I	6		10		0
R	$-y$		$-y$		$+2y$
E	$6-y$		$10-y$		$2y$

Write the K expression for the equation.

13.5d.

$$K = \frac{[P]^2}{[M]\,[N]}.$$

Now substitute the equilibrium concentrations from the table into the K expression and equate to the given value of K, which is 0.50. This will yield a quadratic equation, which can be solved for y.

13.5e.

$$K = \frac{(2y)^2}{(6-y)\,(10-y)} = 0.50$$

Rearranging, to put the equation into the form $ay^2 + by + c = 0$,

$$4y^2 = 0.50(60 - 16y + y^2)$$

$$4y^2 = 30 - 8y + 0.5y^2$$

$$3.5y^2 + 8y - 30 = 0$$

*In many equilibrium problems exponential notation is used. This involves the symbol "×" to indicate multiplication. To avoid possible confusion, therefore, the symbol "x" will not be used for an unknown in this text.

There are no apparent short cuts for solving this problem, so you may as well dust off the old quadratic formula, which occasionally finds use in equilibrium problems. For a quadratic equation in the form $ay^2 + by + c = 0$,

$$y = \frac{-b \pm \sqrt{b^2 - 4ac}}{2a}.$$

Substitute into the formula and solve for y.

13.5f. y = 2.

$$y = \frac{-b \pm \sqrt{b^2 - 4ac}}{2a} = \frac{-8 \pm \sqrt{8^2 - (4)(3.5)(-30)}}{2(3.5)}$$
$$= \frac{-8 \pm \sqrt{64 + 420}}{7} = \frac{-8 \pm 22}{7}$$

As usual, a quadratic equation yields two mathematical solutions. Practical chemical considerations should eliminate one. A negative concentration cannot exist, so $-30/7$ is obviously *not* the answer and may be discarded. Consequently,

$$y = \frac{-8 + 22}{7} = \frac{14}{7} = 2.$$

The value of y is not the equilibrium concentration of any species; what you are asked to find are the three equilibrium concentrations. Your table gives these in terms of y. Substitute and solve for each.

13.5g.

$$[M] = 6 - y = 6 - 2 = 4$$
$$[N] = 10 - y = 10 - 2 = 8$$
$$[P] = 2y = 2 \times 2 = 4$$

The correctness of these results may be confirmed by substituting into the equilibrium constant expression:

$$K = \frac{[P]^2}{[M]\,[N]} = \frac{4^2}{4 \times 8} = 0.5.$$

The next two examples concern the same equilibrium system, but they are different in a manner that makes an important point about equilibria.

Example 13.6. At a certain temperature, K = 51.5 for the equilibrium $H_2(g) + I_2(g) \rightleftharpoons 2\ HI(g)$. If 1.000 mole of hydrogen and 1.000 mole of iodine are introduced into an empty 1.000-liter reaction vessel at that temperature, find $[H_2]$, $[I_2]$, and $[HI]$ when equilibrium is reached.

This problem is similar to Example 13.5. Begin the tabulation by entering initial concentrations of all species.

13.6a.

	$H_2(g)$	+	$I_2(g)$	$\rightleftharpoons$	$2\ HI(g)$
I	1.000		1.000		0

Again letting y equal the reacting moles per liter of either of the reactants, find all values of R in terms of y, and then all values of E on the basis that E = I + R. Complete the table.

13.6b.

	$H_2(g)$	+	$I_2(g)$	$\rightleftharpoons$	$2\ HI(g)$
I	1.000		1.000		0
R	$-y$		$-y$		$+2y$
E	$1.000-y$		$1.000-y$		$2y$

Now write the equilibrium constant expression, insert the equilibrium values from the table, equate to the known value of K, and solve for y.

13.6c. y = 0.782.

$$K = \frac{[HI]^2}{[H_2]\,[I_2]} = \frac{(2y)(2y)}{(1.000-y)(1.000-y)} = 51.5.$$

If we wished, we could solve this equation by using the quadratic formula. However, there is a much easier way to solve it. Note that the left side of the equation is a perfect square. That is,

$$\frac{(2y)(2y)}{(1.000-y)(1.000-y)} = \frac{(2y)^2}{(1.000-y)^2} = 51.5.$$

Taking the square root of both sides, we obtain

$$\frac{2y}{1.000-y} = \sqrt{51.5} = 7.18.$$

Solving:

$$2y = 7.18 - 7.18y; \quad 9.18y = 7.18; \quad y = 0.782$$

Finally, use the value of y to find the equilibrium concentrations of all three species.

13.6d. $[HI] = 2y = 2(0.782) = 1.564$ moles/liter

$[H_2] = [I_2] = 1.000 - y = 1.000 - 0.782 = 0.218$ moles/liter

Now let's approach the same equilibrium from the other side. . . .

Example 13.7. Suppose 2.000 moles of HI are introduced to the same empty 1.000-liter flask, and this decomposes, forming hydrogen and iodine. K remains 51.5 for the equilibrium $H_2(g) + I_2(g) \rightleftharpoons 2\ HI(g)$. Again find the equilibrium concentrations of all species.

This problem combines the features of Examples 13.2 and 13.5. Again letting y equal the reacting moles per liter of hydrogen or iodine, assemble the whole table.

13.7a.

	$H_2(g)$	+	$I_2(g)$	$\rightleftharpoons$	2 HI(g)
I	0		0		2.000
R	+y		+y		−2y
E	y		y		2.000−2y

Again write the equilibrium constant expression, insert the equilibrium values from the table, equate to the known value of K, and solve for y. Also find $[H_2]$, $[I_2]$, and [HI].

13.7b. $y = 0.218 = [H_2] = [I_2]; [HI] = 1.564.$

$$K = \frac{[HI]^2}{[H_2][I_2]} = \frac{(2.000 - 2y)^2}{y^2} = 51.5$$

Solving by extracting the square root of both sides,

$$y = 0.218$$

$$[HI] = 2.000 - 2y = 2.000 - 2(0.218) = 1.564$$

Observe that the answers are identical for Examples 13.6 and 13.7. This illustrates an important feature of equilibria, that the same position of equilibrium is reached regardless of where the reaction starts.

13.5 FREE ENERGY AND THE EQUILIBRIUM CONSTANT

In an all gas phase equilibrium the partial pressure of any component is an expression of the concentration of that component. For the general equilibrium $aA + bB = xX + yY$, this leads to an alternative expression for the equilibrium constant in which partial pressures are substituted for concentrations. This expression is given the symbol K_p:

$$K_p = \frac{(p_X)^x (p_Y)^y}{(p_A)^a (p_B)^b} \qquad (13.2)$$

It is customary, when working with both forms of the equilibrium constant, to identify that based on mole/liter concentration units as K_c, rather than just K, as we have done thus far. To avoid uncertainty, we shall adopt K_c for the concentration constant *in this section only*. Our use of K_p in this book is so very brief, however, we shall continue to use K for the concentration constant in all other sections. It can be shown that the two equilibrium constants are related by the equation

$$K_p = K_c (RT)^{\Delta n_g}. \qquad (13.3)$$

where R is the universal gas constant in liter-atm/mole–°K, T is temperature in degrees Kelvin, and Δn_g is the change in number of moles of gaseous species, $\Sigma n_{product\ gases} - \Sigma n_{reactant\ gases}$, indicated by the coefficients in the equilibrium equation.

K_p may be used in exactly the same way as K_c in solving equilibrium problems. We shall not explore this type of problem in this book, although a few sample problems appear at the end of the chapter. Our interest in K_p is in its relationship with the change in free energy, ΔG, for a reaction. This relationship comes from the equation

$$\Delta G = -RT \ln K_p + RT \ln Q = -2.303\ RT \log K_p + 2.303\ RT \log Q, \quad (13.4)$$

in which ΔG is the free energy change at absolute temperature T, R is the universal gas constant, ln indicates the natural (base e) logarithm, log represents the base 10 logarithm, and Q is a quotient in the form of the same ratio of partial pressures as K_p. The difference between K_p and Q is that K_p represents the partial pressure ratio when the system is in a state of equilibrium, which restricts the numerical value of the ratio to a single number, the equilibrium constant itself. Q represents the partial pressure ratio at any time, and most significantly when the system is *not* at equilibrium. As such Q may have any value, determined only by the instantaneous partial pressures that contribute to it. When the system is at standard state, i.e., when all gaseous partial pressures are 1 atmosphere, the value of Q becomes 1. Because the logarithm of 1 is zero, the second term in the two expressions of Equation 13.4 becomes zero, leading to

$$\Delta G^\circ = -RT \ln K_p = -2.303\ RT \log K_p. \quad (13.5)$$

Inasmuch as ΔG_f° values are most readily available at 25°C (298°K), Equation 13.5 is usually used to calculate equilibrium constants at that temperature. For instance,

Example 13.8. Calculate K_p and K_c for the conversion of oxygen into ozone at 298°K by the equation $3\ O_2(g) = 2\ O_3(g)$.

To apply Equation 13.5 we must know the value of ΔG° at the temperature of the equilibrium. Standard free energies of formation at this temperature are listed in Table III (page 285). Equation 12.16 (page 172) provides the basis for calculating ΔG°. Proceed that far.

13.8a. $\Delta G^\circ = 78.12$ kcal.

$$\Delta G^\circ = \Sigma\Delta G_f^\circ(\text{products}) - \Sigma\Delta G_f^\circ(\text{reactants})$$
$$= 2(39.06) - 3(0) = 78.12 \text{ kcal}$$

Using Equation 13.5, you may solve for log K_p, substitute known values and calculate the logarithm. Watch your value for R. You are working with ΔG° in kcal, so R must be in kcal/mole–°K.

13.8b. Log K_p = −57.2.

From $\Delta G° = -2.303\ RT \log K_p$,

$$\log K_p = \frac{\Delta G°}{-2.303\,RT} = \frac{78.12}{-2.303(0.00199)(298)} = -57.2$$

To express the value of K_p we must take the antilog of −57.2. Methods for handling this and other operations involving logarithms are discussed in Appendix II (page 279). Refer to this if necessary, and express K_p in exponential notation.

13.8c. $K_p = 6 \times 10^{-58}$.

$$\text{Log } K_p = -\ 57.2$$

$$K_p = 10^{-57.2} = 10^{0.8-58} = 10^{0.8} \times 10^{-58} = 6 \times 10^{-58}$$

Now that K_p has been evaluated, use Equation 13.3 to calculate K_c. First, however, what is the value of Δn_g, the change in total moles of gas represented by the equation $3\ O_2(g) = 2\ O_3(g)$?

13.8d. $\Delta n_g = 2 - 3 = -1$.

Now solve Equation 13.3, $K_p = K_c\,(RT)^{\Delta n_g}$, for K_c, substitute and calculate. Again watch the value used for R. The derivation of Equation 13.3 is based on the ideal gas law PV = nRT in which R is measured in liter-atmospheres/mole−°K, so the corresponding value must be used.

13.8e. $K_c = 1 \times 10^{-56}$.

$$K_c = \frac{K_p}{(RT)^{\Delta ng}} = \frac{6 \times 10^{-58}}{(0.0821 \times 298)^{-1}} = (6 \times 10^{-58})(0.0821 \times 298)$$

$$= 1 \times 10^{-56}$$

The conversion of oxygen to ozone at 25°C is not what would be called a spontaneous reaction. This is indicated twice in this example. First, the large positive value of ΔG, 78.12 kcal, does not suggest spontaneity. Second, the smallness of the equilibrium constants, 10^{-58} for K_p and 10^{-56} for K_c indicates that the equilibrium lies far to the reactant side, which is another way of saying a very, very small amount of O_2 is converted to O_3.

Example 13.9. If nitric oxide, NO, is released in air, there is a possibility it will react with oxygen to produce nitrogen dioxide, NO_2: $NO(g) + \frac{1}{2}\ O_2(g) = NO_2(g)$. Assuming such an equilibrium is reached,

(a) calculate K_p at 25°C;
(b) calculate K_c at 25°C;
(c) find the ratio $\frac{p_{NO_2}}{p_{NO}}$ at equilibrium;
(d) confirm that the values of ΔG°, K_p, K_c and the above ratio are consistent

in their qualitative prediction as to the direction favored when equilibrium is reached.

Assume the partial pressure of oxygen in air is constant at 0.20 atmosphere.

Calculation of K_p is as in the previous example. Complete part (a) of the problem, expressing K_p in exponential notation.

13.9a. $K_p = 1.3 \times 10^6$.
Drawing values from Table III for the reaction $NO(g) + \frac{1}{2}O_2(g) = NO_2(g)$.

$$\Delta G° = \Delta G_f°(NO_2) - \Delta G_f°(NO)$$
$$= 12.39 - 20.72 = -8.33 \text{ kcal}$$

Rearranging and substituting into Equation 13.5,

$$\log K_p = \frac{\Delta G°}{-2.303\ RT} = \frac{-8.33}{-2.303(0.00199)\ (298)} = 6.10$$

$$\text{By antilogs, } K_p = 10^{6.10} = 10^{0.10+6} = 10^{0.10} \times 10^6 = 1.3 \times 10^6$$

K_c is also found in the same manner as in the foregoing example. Watch your values of Δn_g and R. . . .

13.9b. $K_c = 6.4 \times 10^6$.
In finding Δn_g we have 1 mole of gas on the right side of the equation and 1-1/2, or 1.5, on the left. $\Delta n_g = 1 - 1.5 = -0.5$.

$$K_c = \frac{K_p}{(RT)^{\Delta n_g}} = \frac{1.3 \times 10^6}{(0.0821 \times 298)^{-0.5}} = 1.3 \times 10^6 \times 24.5^{0.5}$$
$$= 1.3 \times 10^6 \times \sqrt{24.5}$$
$$= 6.4 \times 10^6$$

Part (c) begins to take shape if you write the K_p expression for the reaction and solve it for the ratio $\frac{p_{NO_2}}{p_{NO}}$.

13.9c.

$$K_p = \frac{p_{NO_2}}{(p_{NO})(p_{O_2})^{0.5}}$$

Multiplying both sides by $(p_{O_2})^{0.5}$,

$$\frac{p_{NO_2}}{p_{NO}} = K_p(p_{O_2})^{0.5}$$

You have values for both K_p and p_{O_2}. Substitute and calculate the answer.

13.9d.

$$\frac{p_{NO_2}}{p_{NO}} = 5.8 \times 10^5$$

$$\frac{p_{NO_2}}{p_{NO}} = K_p(p_{O_2})^{0.5} = 1.3 \times 10^6 \, (0.20)^{0.5}$$

$$= 1.3 \times 10^6 \times \sqrt{0.20} = 5.3 \times 10^5$$

The final question involves an interpretation of the values of $\Delta G°$, K_p, K_c and the ratio calculated above. This is a bit beyond a problem book type of question, but see if you can tie together the significance of these three numerical answers.

13.9e. The negative value of $\Delta G°$ predicts that the reaction will occur spontaneously at 25°C and 1 atmosphere. This is supported by a large value of K_p, which indicates a greater "abundance" of product NO_2 than reactant NO. The calculated ratio confirms this relative concentration of product as being about 540,000 times as great as the concentration of reactant, indicating the reaction does indeed occur as written.

PROBLEMS

Section 13.3

13.1. 6.0 moles of SO_2 and 4.0 moles of O_2 are introduced into a 1.0-liter reaction vessel. At equilibrium the vessel contains 4.0 moles of SO_3. Calculate K for $2\,SO_2(g) + O_2(g) = 2\,SO_3(g)$.

13.2. 9.00 moles of CO and 15.0 moles of Cl_2 are placed in a 3.00-liter reaction chamber at a certain temperature. When equilibrium is reached according to the equation $CO(g) + Cl_2(g) = COCl_2(g)$, there are 6.30 moles of Cl_2 in the chamber. Calculate K.

13.3. 23 moles of hydrogen and 37 moles of iodine vapor are placed in an evacuated 5.0-liter chamber. Equilibrium is reached according to the equation $H_2(g) + I_2(g) = 2\,HI(g)$, at which time $[I_2] = 4.0$. Determine K at the temperature of the equilibrium.

13.14. The equilibrium described in Problem 13.1 is reached at a different temperature by introducing 10.0 moles of SO_3 into a 1.0 liter reaction chamber. At equilibrium, 7.0 moles of SO_3 remain. Find K for that temperature.

13.15. 7.5 moles of CO and 14.0 moles of steam are introduced into a 2.0-liter reaction vessel. Temperature is such that the reaction $CO(g) + H_2O(g) = CO_2(g) + H_2(g)$ reaches equilibrium with 6.0 moles of hydrogen present. Calculate K at that temperature.

13.16. 4.1 moles of oxygen and 4.0 moles of NO are introduced to an evacuated 0.50-liter reaction unit. At the temperature of the system, the equilibrium $2\,NO(g) + O_2(g) = 2\,NO_2(g)$ is reached when $[NO] = 1.6$. Calculate K.

Section 13.4

13.4. 2.4 moles of steam and 2.4 moles of CO are placed in an empty 1.0-liter reaction cylinder. Equilibrium is reached according to the equation $CO(g) + H_2O(g) = CO_2(g) + H_2(g)$. $K = 2.9$ at the temperature of the system. Calculate the concentration of each species at equilibrium.

13.5. 48 moles of PCl_5 are placed in a 4.0-liter reaction chamber at a temperature at which

13.17. 6.0 moles of H_2 and 3.0 moles of I_2 are introduced into an evacuated 1.0-liter reaction vessel at a temperature at which $K = 4.0$ for the equilibrium $H_2(g) + I_2(g) = 2\,HI(g)$. Find the equilibrium concentration of each species.

13.18. 4.0 moles of CO_2 and 8.0 moles of H_2 are placed in an empty 1.0-liter chamber. At

K = 0.050 for the equilibrium $PCl_5(g) = PCl_3(g) + Cl_2(g)$. Calculate the concentration of each species when equilibrium is reached.

13.6. Dinitrogen tetroxide dissociates to nitrogen dioxide, reaching the equilibrium $N_2O_4(g) = 2\ NO_2(g)$. K = 0.00690 at 27°C. If 12.0 moles of N_2O_4 are placed in a 3.00-liter reaction device, find $[N_2O_4]$ and $[NO_2]$ when equilibrium is reached at 27°C.

13.7. A certain quantity of HI is introduced to a 1.0-liter reaction vessel at a temperature at which K = 6.0 for $H_2(g) + I_2(g) = 2\ HI(g)$. When equilibrium is reached there are 1.8 moles of I_2 present. How many moles of HI were introduced originally?

13.8. The gaseous system $CO(g) + H_2O(g) = CO_2(g) + H_2(g)$ is at equilibrium with concentrations as follows: [CO] = 0.30; $[H_2O]$ = 0.10; $[CO_2]$ = 0.20; $[H_2]$ = 0.60. How many moles per liter of $H_2O(g)$ must be forced into the system to raise $[H_2]$ to 0.70? (Hint: Find K from the given concentrations.)

13.9. At a certain temperature HI decomposes to equilibrium concentrations of 0.80 for [HI], and 0.20 for $[H_2]$ and $[I_2]$. To this equilibrium is added 0.40 mole HI per liter. Find the concentrations of all species after a new equilibrium is established according to LeChatelier's Principle.

the temperature of the system K = 0.050 for the equilibrium $CO(g) + H_2O(g) = CO_2(g) + H_2(g)$. Determine the equilibrium concentrations of all species.

13.19. K = 150 for the equilibrium $CO(g) + Cl_2(g) = COCl_2(g)$ at certain temperature. If 15.0 moles of $COCl_2$ are introduced to a 5.00-liter container, it decomposes until reaching equilibrium at that temperature. Determine the final concentrations of all species.

13.20. How many moles of NH_3 must be placed in an evacuated 0.250-liter reaction device to yield an equilibrium concentration of 1.50 moles/liter for nitrogen if K = 8.00 for the reaction $N_2(g) + 3\ H_2(g) = 2\ NH_3(g)$?

13.21. Equilibrium concentrations are 0.31 mole/liter for CO; 0.14 mole/liter for Cl_2; and 0.78 mole/liter for $COCl_2$ in the system $CO(g) + Cl_2(g) = COCl_2(g)$. How many moles of Cl_2 must be added per liter to reduce [CO] to 0.25 mole/liter?

13.22. At a certain temperature, K = 2.0 for the reaction $N_2O_4(g) = 2\ NO_2(g)$. If 1.0 mole of N_2O_4 is introduced into a container at that temperature, what will be the equilibrium concentration of NO_2 if (a) the volume of the container is 1.0 liter; (b) the volume of the container is 10.0 liters?

Section 13.5

13.10. Calculate K_p and K_c for the reaction $H_2 + Br_2 = 2\ HBr$ at 298°K. Show that $\Delta G°$, K_p and K_c are consistent in their prediction as to the direction to which the equilibrium is favored at 298°K.

13.11. Determine K_p at 25°C for the formation of ammonia from its elements: $N_2(g) + 3\ H_2(g) = 2\ NH_3(g)$. Also find K_c.

13.23. Calculate K_p at 25°C for the equilibrium $SO_2(g) + NO_2(g) = SO_3(g) + NO(g)$. Also determine K_c, and predict which side of the equilibrium is favored at 25°C.

13.24. For the equilibrium $4\ NH_3(g) + 5\ O_2(g) = 4\ NO(g) + 6\ H_2O(g)$ at 298°K, calculate the values of K_p and K_c.

The following problems are added to illustrate equilibrium constant calculations using partial pressures and K_p.

13.12. A mixture of nitrogen at a partial pressure of 5.10 atmospheres and hydrogen at a partial pressure of 14.5 atmospheres reacts to form ammonia. When the system $N_2(g)$ +

13.25. CO(g) and $H_2O(g)$ are introduced to a reaction vessel under starting partial pressures of 1.60 atm and 3.50 atm respectively. Equilibrium is reached according to the equa-

$3\ H_2(g) = 2\ NH_3(g)$ reaches equilibrium, $p_{NH_3} = 8.80$ atmospheres. Calculate K_p.

13.13. Hydrogen iodide gas is introduced to a reaction device under an initial pressure of 6.40 atm. It decomposes, forming H_2 and I_2. If $K_p = 46.0$ for the reaction $H_2(g) + I_2(g) = 2\ HI(g)$, find the equilibrium partial pressures of all species.

tion $CO(g) + H_2O(g) = CO_2(g) + H_2(g)$. Calculate K_p if $p_{CO} = 0.25$ atm at equilibrium.

13.26. PCl_3 and Cl_2 are placed into a reaction cylinder at partial pressures of 3.80 atm and 4.60 atm respectively. The equilibrium $PCl_5(g) = PCl_3(g) + Cl_2(g)$ is reached at a temperature at which $K_p = 1.80$. Compute the equilibrium partial pressures of each species.

CHAPTER 14

THE WATER EQUILIBRIUM

14.1 THE IONIZATION OF WATER

Water is generally regarded as a nonelectrolyte. If, however, a sufficiently sensitive detector is used, it can be demonstrated that even pure water contains some ions. These are derived from the ionization of the water molecule, a reaction that may be represented as follows: $HOH \rightleftharpoons H^+ + OH^-$. Thus pure water is really a very dilute solution of hydrogen* and hydroxide ions. The equilibrium constant for this extremely important equilibrium is

$$K = \frac{[H^+]\,[OH^-]}{[HOH]}.$$

The molarity of pure water may be found in the same manner as any other molarity. At a density of 1 g/ml, the "concentration" of water may be said to be 1000 grams per liter. Thus

$$\frac{1000\ \cancel{\text{grams water}}}{1\ \text{liter}} \times \frac{1\ \text{mole water}}{18\ \cancel{\text{grams water}}} = 55.5\ \text{moles water/liter}.$$

This concentration does not change appreciably, at least in comparison to that of H^+ and OH^-. Being essentially constant, it is incorporated into a special

*Some chemists object to the use of the term hydrogen ion for the acidic species in aqueous solution. A hydrogen ion, which is actually a single proton, is quickly hydrated in aqueous solution, i.e., it associates with one or more molecules of water. This species is usually represented by the formula, H_3O^+, which is called the hydronium ion. While the H_3O^+ symbol is unquestionably closer to reality than simply H^+, it complicates calculations in equilibrium problems somewhat—not seriously, but enough to warrant avoiding those complications in a beginning chemistry course. Accordingly, in this book, the symbol H^+ will be used, and it will be called a hydrogen ion, with the unwritten understanding that, in aqueous solution, it is actually hydrated. On occasion $H^+(aq)$ will emphasize the H^+ ion in aqueous solution. The hydronium ion, H_3O^+, will be used only when there is special reason to call attention to the role of water in the reaction.

equilibrium constant, K_w, by multiplying both sides of the above equilibrium constant equation by [HOH]:

$$K[HOH] = K_w = [H^+][OH^-] = 1.0 \times 10^{-14}. \quad (14.1)$$

The value given for K_w is its value at 25°C.

If pure water ionizes according to the equation $HOH \rightleftharpoons H^+ + OH^-$, and the water is the only source of either ion, then $[H^+]$ must be equal to $[OH^-]$. Therefore

$$[H^+][OH^-] = [H^+][H^+] = [H^+]^2 = 1.0 \times 10^{-14}$$

$$[H^+] = \sqrt{1.0 \times 10^{-14}} = 1.0 \times 10^{-7}$$

Similarly, $[OH^-] = 1.0 \times 10^{-7}$ in pure water.

Water or solutions in which $[H^+] = [OH^-] = 1.0 \times 10^{-7}$ are said to be *neutral*, i.e., neither acidic nor basic. In acidic solution, $[H^+]$ is greater than 1.0×10^{-7} and $[OH^-]$ is less than 1.0×10^{-7}. In basic solutions, the concentration of OH^- is greater than 1.0×10^{-7} and that of H^+ less than 1.0×10^{-7}. In either case, the product of the concentrations remains 1.0×10^{-14}.

If either the hydrogen or hydroxide ion concentration of an aqueous solution is known, the other may be obtained readily by simple substitution into Equation 14.1. By appropriate division,

$$[H^+] = \frac{K_w}{[OH^-]} \quad (14.2)$$

and

$$[OH^-] = \frac{K_w}{[H^+]} \quad (14.3)$$

Example 14.1. Calculate $[OH^-]$ in a solution in which $[H^+]$ is 4.2×10^{-4}.

You should require no special guidance on this problem other than to recall that $K_w = 1.0 \times 10^{-14}$. Substitute into Equation 14.3 and solve.

14.1a. $[OH^-] = 2.4 \times 10^{-11}$.

$$[OH^-] = \frac{K_w}{[H^+]} = \frac{1.0 \times 10^{-14}}{4.2 \times 10^{-4}} = 0.24 \times 10^{-10} = 2.4 \times 10^{-11}$$

Did you happen to come up with 2.4×10^{-9} for this problem? This is a very common error, and one that can be avoided by a simple expedient that settles the decimal point and exponent questions at the beginning of the problem. 1.0×10^{-14} may also be expressed as 10×10^{-15}. If you use this form of K_w wherever it is to be divided by another number in decimal notation the answer emerges in proper form:

$$\frac{1.0 \times 10^{-14}}{4.2 \times 10^{-4}} = \frac{10 \times 10^{-15}}{4.2 \times 10^{-4}} = \frac{10}{4.2} \times \frac{10^{-15}}{10^{-4}} = 2.4 \times 10^{-11}$$

14.2 THE "p" CONCEPT: pH AND pOH

As may be seen from the first example, working with hydrogen and hydroxide ion concentrations in aqueous solutions leads to some rather small numbers and large ranges in orders of magnitude. So many chemical and biochemical phenomena occur in aqueous solutions and are so sensitive to the concentrations of these ions that it becomes expedient to express them in a manner less cumbersome than decimal notation. Such a method is the "p" concept in which **the p value of a number is defined as the negative base 10 logarithm of a number, or the logarithm of the reciprocal of the number.** To recall the nature of logarithms, if $X = 10^n$, then $\log X = n$. The pX, the *negative* logarithm of X, is $-n$. (See Appendix II, p. 279, for a more detailed review of logarithms.)

The pH of a solution is defined by the equation

$$pH = -\log[H^+] = \log\frac{1}{[H^+]}. \qquad (14.4)$$

It follows that

$$\log[H^+] = -pH. \qquad (14.5)$$

From the definition of logarithms,

$$[H^+] = 10^{-pH}. \qquad (14.6)$$

Similarly,

$$[OH^-] = 10^{-pOH}. \qquad (14.7)$$

The relationship between pH and pOH is significant. Substituting from Equations 14.6 and 14.7 into Equation 14.1,

$$[H^+][OH^-] = (10^{-pH})(10^{-pOH}) = 10^{-14}. \qquad (14.8)$$

From the principle that $(X^m)(X^n) = X^{(m+n)}$, Equation 14.8 becomes

$$10^{-pH + (-pOH)} = 10^{-14}. \qquad (14.9)$$

It follows that

$$-pH - pOH = -14 \qquad (14.10)$$

$$pH + pOH = 14. \qquad (14.11)$$

This is an opportune time to pause for a moment to develop a "feeling" for pH—a sense of what it's all about. It is a measure of acidity, at least in the sense that hydrogen ion concentration constitutes acidity. It is an inverse sort of measurement: the higher the pH, the lower the acidity, and vice versa. A few values should illustrate this picture:

$[H^+]$	$[H^+]$	pH	pOH	$[OH^-]$
1.0	10^0	0	14	10^{-14}
0.1	10^{-1}	1	13	10^{-13}
0.01	10^{-2}	2	12	10^{-12}
0.001	10^{-3}	3	11	10^{-11}
0.0000001	10^{-7}	7	7	10^{-7}
0.00000000000001	10^{-14}	14	0	10^0

As seen from this table, a change of 1 unit in pH represents a change in $[H^+]$ by a factor of 10. Thus a solution of pH 2 has 10 times the $[H^+]$ as a solution of pH 3: $\frac{10^{-2}}{10^{-3}} = 10$. If the difference in pH values is 2, the ratio of H^+ concentration is 10^2, or 100. If the pH difference is x, then this ratio is 10^x. Always remember that the solution with the higher pH has the lower $[H^+]$.

Example 14.2. Solution *A* has a pH of 4. Solution *B* has a pH of 9. Find the value of

$$\frac{[H^+] \text{ in solution } A}{[H^+] \text{ in solution } B}$$

to be expressed as a ratio of integers.

14.2a. The ratio is 100,000:1.

$$\frac{[H^+] \text{ in solution } A}{[H^+] \text{ in solution } B} = \frac{10^{-4}}{10^{-9}} = 10^{+5} = \frac{100{,}000}{1}$$

Now let's put some of this information to use:

Example 14.3. What is the pH of 0.001 M HCl, if the HCl is completely ionized according to the equation $HCl = H^+ + Cl^-$?

First, what is the hydrogen ion concentration of 0.001 M HCl?

14.3a. $[H^+] = 0.001$ mole/liter.
0.001 M HCl contains 0.001 mole of HCl in one liter of solution, and, on ionization, this produces 0.001 mole of H^+.

Now express this concentration in terms of 10 raised to a power.

14.3b. $0.001 = 1 \times 10^{-3} = [H^+]$.
Now, using Equation 14.6, what is the pH?

14.3c. pH = 3.0.

$$pH = -\log[H^+] = -\log(1 \times 10^{-3}) = 0 - (-3) = 3$$

Let's extend this problem:

Example 14.4. What is the pOH of the solution in Example 14.3, 0.001 M HCl?

Equation 14.11 makes this easy. Go ahead.

14.4a. pOH = 11.
Substituting into Equation 14.11,

$$3 + pOH = 14; \qquad pOH = 14 - 3 = 11.$$

Let's take the problem even further:

Example 14.5. What is the $[OH^-]$ of the solution in Example 14.3, in which pOH = 11?

This is essentially the reverse of Example 14.2, and the answer is readily seen from Equation 14.7.

14.5a. $[OH^-] = 10^{-11}$.

Example 14.6. On the basis of Example 14.5, which shows $[OH^-] = 10^{-11}$, calculate $[H^+]$.

The first section and Equation 14.2 will be handy here.

14.6a. $[H^+] = 10^{-3}$.

$$[H^+] = \frac{K_w}{[OH^-]} = \frac{10^{-14}}{10^{-11}} = 10^{-3}$$

It is no coincidence that $[H^+]$ has the same value here as in Example 14.3, inasmuch as Examples 14.3–14.6 all involve the same solution. These examples illustrate that if either $[H^+]$, $[OH^-]$, pH, or pOH is known, all the other quantities may be calculated. Moreover, starting at any one of these values, you should be able to "make the loop" and return to the same value, as was done in Examples 14.3–14.6.

14.3 INTERCONVERSION BETWEEN pH AND $[H^+]$

Only rarely are ionic concentrations simple exponentials as in the last four examples. More often they are expressed as the product of two numbers in decimal notation, as the $[H^+]$ of 4.2×10^{-4} in Example 14.1. To find the pH of such a solution you must find the logarithm of the hydrogen ion concentration, or $\log 4.2 \times 10^{-4}$. The pH is the negative of that logarithm. The method of such a calculation is explained in detail in Appendix II on page 279. The "line setup" for the conversion is:

$$[H^+] = 4.2 \times 10^{-4} = 10^{\log 4.2} \times 10^{-4} = 10^{0.62} \times 10^{-4} = 10^{0.62-4} = 10^{-3.38}.$$

$$pH = -\log [H^+] = -\log 10^{-3.38} = 3.38.$$

By examining the procedure it is seen that *the pH of a solution with hydrogen ion concentration expressed in exponential notation is found by subtracting the logarithm of the number factor* (4.2) *from the absolute value of the exponent of ten,* (−4). In the example above, $4 - \log 4.2 = 4 - 0.62 = 3.38$.

Example 14.7. What is the pH of a solution if $[H^+] = 2.9 \times 10^{-8}$?

First, find the logarithm of 2.9.

14.7a. $\log 2.9 = 0.46$.

If 0.46 is now subtracted from the absolute value of the exponent of 10 you will have the pH.

14.7b. $pH = 8 - 0.46 = 7.54$.

The reverse procedure, converting from pH to $[H^+]$, is also detailed in Appendix II, page 279. If the pH of a solution is 11.80, for example, $[H^+] = 10^{-11.80}$. The line setup for converting $10^{-11.80}$ to exponential notation is:

$$[H^+] = 10^{-11.80} = 10^{0.20-12} = 10^{0.20} \times 10^{-12} = (\text{antilog}\ 0.20) \times 10^{-12} = 1.6 \times 10^{-12}.$$

Example 14.8 Find $[H^+]$ in a solution having a pH of 2.43.

First express $[H^+]$ as an exponential of 10.

14.8a $[H^+] = 10^{-2.43}$.

Now change the exponential to standard exponential notation.

14.8b. $[H^+] = 3.7 \times 10^{-3}$.

$$[H^+] = 10^{-2.43} = 10^{0.57-3} = 10^{0.57} \times 10^{-3} = 3.7 \times 10^{-3}$$

The question of significant figures in conversions between $[H^+]$ and pH arises at this point. Both should express the same number of significant figures. Consider the number 2.42, which has three significant figures. Its logarithm is

0.384 to three significant figures. But what about the logarithm of 24.2, also a three-significant figure number? Is it simply 1.38—or is it 1.384? The answer is 1.384.

In logarithms only the numbers *after the decimal* are counted as significant figures. The number before the decimal in the logarithm simply gives the exponent. This becomes more apparent if you consider both 2.42 and 24.2 as 2.42×10^n. For 2.42, $n = 0$; for 24.2, $n = 1$. Taking logarithms,

$$\log 2.42 \times 10^n = \log 2.42 + \log^n$$
$$= 0.384 + n.$$

Here, n is an exponent. It tells us nothing about the significant figures in the number 2.42. Thus the logarithm of the number 2.42×10^{15} is, to the proper number of significant figures, 15.384.

In conversions between $[H^+]$ and pH the number of digits to the right of the decimal in pH should be the same as the number of significant figures in $[H^+]$. Both values will then be expressed in the same number of significant figures.

Example 14.9. As a final example, "make the loop" for a solution having $[OH^-] = 1.7 \times 10^{-5}$. Calculate in order pOH, pH and $[H^+]$, and then return to the starting $[OH^-]$.

14.9a. pOH = 4.77; pH = 9.23; $[H^+] = 5.9 \times 10^{-10}$

pOH: $\text{pOH} = 5 - \log 1.7 = 5 - 0.23 = 4.77$

pH: $\text{pH} = 14.00 - 4.77 = 9.23$

$[H^+]$: $[H^+] = 10^{-9.23} = 10^{0.77-10} = 10^{0.77} \times 10^{-10} = 5.9 \times 10^{-10}$

$[OH^-]$: $[OH^-] = \dfrac{K_w}{[H^+]} = \dfrac{10 \times 10^{-15}}{5.9 \times 10^{-10}} = 1.7 \times 10^{-5}$

PROBLEMS

14.1. Find the value of the ratio

$$\frac{[H^+] \text{ of Solution A}}{[H^+] \text{ of Solution B}}$$

if: (a) pH of A = 5 and pH of B = 8;
(b) pOH of A = 4 and pH of B = 7;
(c) pOH of A = 5 and pOH of B = 9

14.2. The pH of lemon juice is 2.3. Calculate $[H^+]$, $[OH^-]$ and pOH.

14.3. The pH of human gastric juices falls in the range 1.0–3.0. If the pOH happens to be 12.1, find the corresponding pH, $[H^+]$ and $[OH^-]$.

14.11. Find the value of the ratio

$$\frac{[H^+] \text{ of Solution X}}{[H^+] \text{ of Solution Y}}$$

if: (a) pH of X = 3 and pOH of Y = 6;
(b) pH of X = 8 and pH of Y = 6;
(c) pH of X = 10 and pOH of Y = 10

14.12. The pH of sea water is about 8.1. Calculate its pOH, $[OH^-]$ and $[H^+]$.

14.13. Household ammonia comes in a 1–5% range of concentrations. A pOH of 3.1 is typical. Calculate $[OH^-]$, $[H^+]$ and pH in such a solution.

14.4. Find the pH, pOH and $[OH^-]$ of milk if the hydrogen ion concentration is 4×10^{-7}.

14.5. $[OH^-] = 5 \times 10^{-12}$ is a particular carbonated soft drink. Determine $[H^+]$, pH and pOH.

14.6. Assuming the reaction $HCl(aq) \rightarrow H^+(aq) + Cl^-(aq)$ goes to completion, i.e., the HCl is totally ionized in solution, calculate $[H^+]$, pH, pOH and $[OH^-]$ in 0.00001 M HCl.

14.7. The CRC Handbook lists the solubility of magnesium hydroxide at 0.0009 gram per 100 ml. Find $[OH^-]$, $[H^+]$, pH and pOH in such a solution.

14.8. $K_w = 1.00 \times 10^{-14}$ at 25°C. But at 60°C, $K_w = 9.55 \times 10^{-14}$. Calculate the pH of water at 60°C.

14.9. Assuming the reaction $HCl(aq) + NaOH(aq) \rightarrow NaCl(aq) + HOH(l)$ will go to completion, i.e., until the limiting reagent is totally consumed, determine the pH of one liter of solution containing 0.015 mole of HCl and 0.005 mole of NaOH.

14.10. Complete the following table:

Solution	pH	pOH	$[OH^-]$	$[H^+]$
A	4.62			
B			1.8×10^{-9}	
C				8.6×10^{-3}
D		5.48		

14.14. $[H^+] = 2 \times 10^{-4}$ in orange juice. Find the $[OH^-]$, pOH and pH.

14.15. The hydroxide ion concentration in a particular brand of beer is 3×10^{-10}. Calculate the pOH, pH, and $[H^+]$.

14.16. Dilute solutions of sodium hydroxide ionize completely when dissolved in water: $NaOH(s) \rightarrow Na^+(aq) + OH^-(aq)$. If 0.4 gram of sodium hydroxide is dissolved in water and diluted to 100 liters, calculate $[OH^-]$, pOH, pH and $[H^+]$.

14.17. HF is about 8.4% ionized in 0.1 M HF. This means that only 8.4% of the 0.1 mole of HF in one liter of solution ionizes according to the equation $HF(g) \rightarrow H^+(aq) + F^-(aq)$. Determine $[H^+]$, pH, pOH and $[OH^-]$ in 0.1 M HF.

14.18. Calculate the pOH and pH of water at 10°C if $K_w = 2.95 \times 10^{-15}$ at that temperature.

14.19. Making the same assumption as in Problem 14.9, estimate the pH of the solution that results from the dissolving of 1.4 grams of NaOH in one liter of 0.012 M HCl.

14.20. Complete the following table:

Solution	pH	pOH	$[OH^-]$	$[H^+]$
W	10.84			
X			3.9×10^{-11}	
Y				6.6×10^{-8}
Z		0.38		

CHAPTER 15

IONIC EQUILIBRIA

15.1 PERCENT IONIZATION

When gaseous hydrogen chloride is bubbled into water it ionizes as follows:

$$HCl(g) \xrightarrow{H_2O} H^+(aq) + Cl^-(aq)$$

The ionization of hydrogen chloride in water is virtually complete: in most concentrations 100.0% of the solute is converted into H^+ and Cl^- ions, as indicated by the equation. This is not always the case. For example, structurally similar hydrogen fluoride is only 2.6% ionized in a 1.000 molar solution. This means that, in one liter of solution, equilibrium is reached after 0.026 mole of HF ionizes to produce 0.026 mole of H^+ and 0.026 mole of F^-. The balance of the HF, 0.974 mole, remains in solution as un-ionized HF molecules. This is indicated quantitatively in an *I–R–E* tabulation such as we used in Chapter 13:

	HF(aq)	=	H^+(aq)	+	F^-(aq)
I	1.000		0		0
R	−0.026		+0.026		+0.026
E	0.974		0.026		0.026

Percent ionization, as all other percentage concepts, is the ratio of the part to the whole, multiplied by 100:

$$\% \text{ ionization} = \frac{\text{amount of solute ionized}}{\text{total solute present}} \times 100. \quad (15.1)$$

A "strong" acid is one that is 100% ionized, or very close thereto; a "weak" acid is one having a low percentage ionization. Hydrochloric acid is therefore a strong acid, and hydrofluoric acid is a weak acid. The terms *strong* and *weak* should not be associated with the corrosive properties of the acid. Both HCl and HF are corrosive acids, but HF is the more corrosive of the two, dissolving even glass and lead, which are not affected by HCl.

In expressing solution concentrations, some chemists distinguish between *molarity* and *formality*. Formality is the number of moles of *solute* dissolved in a liter of solution, and molarity is the number of moles of each molecular or ionic species in one liter of resulting solution. If 1.0 mole of completely ionized HCl is dissolved in a liter of solution, its concentration is 1.0 F HCl. It has three molarities: 0.0 M HCl, 1.0 M H^+, and 1.0 M Cl^-. With weak acid HF, 1.000 F HF has concentrations 0.974 M HF, 0.026 M H^+ and 0.026 M F^-. In use, the concepts of formality and molarity are almost identical, and the context of the sentence removes any ambiguity. Therefore, the distinction between these closely related terms will not be made in this text.

15.2 DETERMINATION OF K_a FROM pH or PERCENT IONIZATION

A weak acid ionizes as follows: $HA \rightleftharpoons H^+ + A^-$. The equilibrium constant corresponding to the equation is called the acid constant, K_a. (Some texts refer to this as the ionization constant, K_i.)

$$K_a = \frac{[H^+][A^-]}{[HA]} .$$

If the pH of a solution of a weak acid is measured, its K_a may be found.

Example 15.1. If 1.00 mole of a weak acid, HW, is dissolved in 1.00 liter of solution, the pH is 1.30. Calculate (a) K_a, and (b) the percentage of ionization.

This acid has ionized and reached equilibrium by the equation $HW \rightleftharpoons H^+ + W^-$. From the pH, find $[H^+]$ at equilibrium.

15.1a. $[H^+] = 5.0 \times 10^{-2} = 0.050.$

$$[H^+] = 10^{-1.30} = 10^{0.70-2} = 10^{0.70} \times 10^{-2} = 5.0 \times 10^{-2}$$

The approach to this problem is the approach developed in Chapter 13. You will recall the process whereby the equilibrium concentrations of all species are found and entered into an *I–R–E* tabulation, and then substituted into the formula for the equilibrium constant. Review this material generally at this time if you are not at ease with it, as it will be used throughout this chapter.

From the information of the problem you can enter all *I* values, provided you regard the initial concentration of H^+ as zero. It is actually about 10^{-7} because of the ionization of water, but this is negligible compared to the equilibrium hydrogen ion concentration calculated above. Because all the W^- ion concentration comes from the same source as the H^+ ion—the ionization of HW—you should be able to state the equilibrium value of $[W^-]$. The balance of the table follows logically. Try for the entire table.

15.1b.

	HW	$\rightleftharpoons$	H^+	+	W^-
I	1.00		0		0
R	−0.050		+0.050		+0.050
E	0.95		0.050		0.050

Starting with zero H^+ and ending with 0.050 (10^{-7} certainly is negligible compared to 0.050!), the reaction must have produced 0.050 mole H^+ per liter. This fixes the R values for each species. Equilibrium concentrations follow.

Calculation of K_a follows by substitution of equilibrium concentrations into the K_a expression. Do this now.

15.1c. $K_a = 2.6 \times 10^{-3}$.

$$K_a = \frac{[H^+][W^-]}{[HW]} = \frac{(0.050)^2}{0.95} = 2.6 \times 10^{-3}$$

Percentage ionization is found by Equation 15.1. Initially there was 1.00 mole of HW present. Of this, 0.050 mole reacted. The percent that reacted is the part divided by the whole, converted to percent by multiplying by 100. Calculate the percent of ionization.

15.1d. 5.0%.

$$\% \text{ ionization} = \frac{HW_{reacted}}{HW_{initial}} \times 100 = \frac{0.050}{1.00} \times 100 = 5.0\%$$

In solving this and other ionic equilibrium problems, you should pay close attention to the rules of significant figures. The reason for this will become apparent in the next example. At this point, however, check your table again and see if you have recorded the values in the same number of significant figures as shown here, and if not, understand the necessary corrections before proceeding.

Example 15.2. A 0.10 M solution of another weak acid, HB, is 3.1% ionized. Calculate K_a for that acid and the pH of the solution.

Begin this example by setting up the *I–R–E* table and filling in the *I* line.

15.2a.

	HB	⇌	H^+	+	B^-
I	0.10		0		0

The concentration of the acid is given. Obviously there is no B^- ion present initially, because its sole source is the ionization of the acid. As before, we take the initial concentration of H^+ to be zero.

This problem is somewhat different from Example 15.1 in that the percentage of the initial acid that ionizes is given. The reacting quantity is 3.1% of the acid initially present, or 3.1% of 0.10. This amounts to 0.0031 mole/liter. It is possible, therefore, to complete the *R* line of the table—and the equilibrium line follows directly. Carry through to the calculation of K_a.

15.2b. $K_a = 9.6 \times 10^{-5}$.

	HB	⇌	H^+	+	B^-
I	0.10		0		0
R	−0.0031		+0.0031		+0.0031
E	0.10		0.0031		0.0031

$$K_a = \frac{[H^+][B^-]}{[HB]} = \frac{0.0031^2}{0.10} = 9.6 \times 10^{-5}$$

In calculating the equilibrium concentration of HB, the arithmetic would indicate the subtraction 0.10 − 0.0031. This difference is 0.0969. According to the rules of significant figures, however, the difference must be rounded off to the second decimal. In doing so it becomes simply 0.10. In essence, the 0.0031 is negligible compared to the 0.10 from which it is being subtracted, and may therefore be disregarded.

Now find the pH if $[H^+]$ is 0.0031, or 3.1×10^{-3}.

15.2c. $pH = 3 - \log 3.1 = 3 - 0.48 = 2.52$.

Problems in which reacting quantities are negligible compared to quantities initially present are common. If the number of moles reacting is very much smaller than the initial number of moles, then the number of moles at equilibrium may be taken as equal to the number initially present. In other words, if $A + B = C$, and $B \ll A$, then $A \approx C$. The following four generalizations may be stated about this procedure:

1. The fraction of acid ionized will usually be negligibly small when K_a has a large negative exponent.
2. The rules of significant figures as applied to addition and subtraction determine whether or not the reacting moles are negligible.
3. Very small numbers are never ignored as factors in multiplication or division.
4. Assumptions of negligibility must be checked and proved valid for each problem.

15.3 FINDING PERCENT IONIZATION AND pH FROM KNOWN MOLARITY AND K_a

If the molarity of a weak acid and its K_a value are known, the percent ionization can be determined. Though differing in detail, the procedure is essentially as outlined in Example 13.5 (page 186). No equilibrium concentrations are known, and an algebraic symbol is used to express one such concentration. When equilibrium concentrations are found in terms of that variable they are inserted into the K_a expression, equated to K_a, and the equation solved.

Example 15.3. Find the percent ionization of 1.0 M HB—the same acid as in Example 15.2, with $K_a = 9.6 \times 10^{-5}$.

Begin by setting up the table and entering the values that may be found in the statement of the problem.

15.3a.

	HB	⇌	H^+	+	B^-
I	1.0		0		0
R					
E					

To fill in the bottom portion of the table it is usually best to assign an algebraic variable to represent $[H^+]$ at equilibrium. Letting y equal $[H^+]$, complete the table.

15.3b.

	HB	⇌	H^+	+	B^-
I	1.0		0		0
R	−y		+y		+y
E	1.0−y		y		y

Now set up the K_a expression, substitute the equilibrium values from the table, and equate to K_a. Do not solve the equation.

15.3c.

$$K_a = \frac{[H^+][B^-]}{[HB]} = \frac{y^2}{1.0 - y} = 9.6 \times 10^{-5}.$$

Rearranging this equation gives

$$y^2 + 9.6 \times 10^{-5}y - 9.6 \times 10^{-5} = 0.$$

In principle, solving this equation for y requires the quadratic equation. Perhaps it may be avoided. . . .

Suppose, as is frequently the case, y is very small compared to 1.0—negligible, in fact, in terms of significant figures, so that $1.0 - y = 1.0$. (It was in the 0.10 M solution of Example 15.2.) If this is the case, note the simplification in the denominator of the equation: it becomes 1.0. Let us assume $1.0 - y = 1.0$, and add an E_a line to the table to show the assumption. The complete table now becomes

	HB	⇌	H^+	+	B^-
I	1.0		0		0
R	−y		+y		+y
E	1.0−y		y		y
E_a	1.0		y		y

Using this assumption, set up and solve for y.

15.3d. $y = 9.8 \times 10^{-3}$.

If $y^2/1.0 = 9.6 \times 10^{-5}$, then $y^2 = 9.6 \times 10^{-5} = 96 \times 10^{-6}$. The change in decimal and exponent is made to simplify extracting the square root in the next step. In taking the square root of an exponential you divide the exponent by 2: $\sqrt{X^{2n}} = x^n$. It is expedient, therefore, to rewrite the exponential so it has an even number for the exponent of 10. This adjustment should always be such as to yield a number factor between 1 and 100, so its square root will be between 1 and 10. Thus

$$y = \sqrt{96 \times 10^{-6}} = 9.8 \times 10^{-3}$$

In this simpler solution, no quadratic formula was necessary. But now, can you justify the simpler procedure? Checking is simple: *Does* $1.0 - y = 1.0$ if $y = 9.8 \times 10^{-3}$, or 0.0098?

According to the rules of significant figures it does: 1.0 − 0.0098 = 0.9902, which must be rounded off to the first decimal, or 1.0. The assumption checks. (If you doubt this procedure, solve the equation $y^2/1.0 - y = 9.6 \times 10^{-5}$ by the quadratic formula and see if you do not find $y = 9.8 \times 10^{-3}$, to two significant figures.)

This technique of making an assumption for purposes of simplifying calculations is a very important one in equilibrium problems. It should be thoroughly understood at this point. Further, it must be realized that any assumption must be checked before the result may be accepted.

Example 15.3 asks for the percent ionization, which is yet to be determined. Initially 1.0 mole of HB was present in a liter of solution. The reacting quantity was 9.8×10^{-3} moles. What percentage of 1.0 is this?

15.3e. 0.98%.

$$\frac{9.8 \times 10^{-3}}{1.0} \times 100 = 0.98\%$$

Example 15.4. Find the pH and the percent ionization in 0.0010 M HB—the same acid as in Example 15.2 with $K_a = 9.6 \times 10^{-5}$.

Except for the molarity of the acid, this problem is identical to Example 15.3. Make a corresponding assumption, add an E_a line to the table, set up the problem, solve as far as $[H^+]$, and check the assumption.

15.4a. $[H^+] = 3.1 \times 10^{-4}$. However, the assumption is NOT valid.

	HB	⇌	H^+	+	B^-
I	0.0010		0		0
R	−y		+y		+y
E	0.0010−y		y		y
E_a	0.0010		y		y

$$K_a = \frac{[H^+][B^-]}{[HB]} = \frac{y^2}{1.0 \times 10^{-3}} = 9.6 \times 10^{-5}$$

$$y^2 = 9.6 \times 10^{-8}; \quad y = 3.1 \times 10^{-4}$$

$$\begin{array}{r} 0.0010 \\ 0.00031 \\ \hline 0.0007 \end{array}$$

$$0.0007 \neq 0.0010$$

Now what? The quadratic formula? It always works on a quadratic equation, and this one is no exception. But it is tedious to solve with coefficients such as 9.6×10^{-5} and constant terms such as 9.6×10^{-8}! Fortunately the method of successive approximations is available as an alternative.

In the solution above it was *assumed* that $[HB] = 1.0 \times 10^{-3}$. But the result so obtained indicated that [HB] was closer to 7×10^{-4}. Let's take that as a second assumption, and solve the problem again, based on that assumption. In other words, using 7×10^{-4} as [HB] and y as $[H^+]$ and $[B^-]$, insert these values into the K_a expression, equate to 9.6×10^{-5}, and solve for y.

15.4b. $y = 2.6 \times 10^{-4}$.

$$K_a = \frac{[H^+][B^-]}{[HB]} = \frac{y^2}{7 \times 10^{-4}} = 9.6 \times 10^{-5}$$

$$y^2 = 6.7 \times 10^{-8}; \qquad y = 2.6 \times 10^{-4}$$

This second assumption must also be tested for validity:

$$\begin{array}{l} 0.0010 \\ \underline{0.00026} \\ 0.00074 = 0.0007 \end{array}$$

Assumption valid.

Through this example the student should become familiar with the method of successive approximations and use it when a first assumption fails.

Next you must find the pH of the solution, based on $[H^+] = 2.6 \times 10^{-4}$.

15.4c. $pH = 4 - \log 2.6 = 4 - 0.41 = 3.59$.

Now calculate the percentage of ionization.

15.4d. 26%.

$$\frac{2.6 \times 10^{-4}}{1.0 \times 10^{-3}} \times 100 = 26\%$$

Comparing molarities and percentage ionizations in Examples 15.2–15.4 illustrates a significant point:

Molarity	*% Ionization*
1.0	0.98
0.10	3.1
0.0010	26

Note that the solutions ionize more completely when dilute.

Let's try these techniques on a real chemistry problem.

Example 15.5. Find the pH of 0.60 M acetic acid, $HC_2H_3O_2$, or, as the "formula" is often written, HOAc. $K_a = 1.8 \times 10^{-5}$.

This example parallels Example 15.3. Set up completely and solve for pH.

15.5a. pH = 2.48.

	HOAc	⇌	H^+	+	OAc^-
I	0.60		0		0
R	−y		+y		+y
E	0.60−y		y		y
E_a	0.60		y		y

$$K_a = \frac{[H^+][OAc^-]}{[HOAc]} = \frac{y^2}{0.60} = 1.8 \times 10^{-5}$$

$$y^2 = 1.08 \times 10^{-5} = 10.8 \times 10^{-6}$$

$$y = 3.3 \times 10^{-3}$$

Assumption check: 0.60 − 0.0033 = 0.60. Valid.

$$pH = 3 - \log 3.3 = 3 - 0.52 = 2.48$$

The same principles may be applied to solutions of weak bases. When ammonia dissolves in water, for example, the following equilibrium is established: $NH_3 + HOH \rightleftharpoons NH_4^+ + OH^-$. The equilibrium constant for this reaction, designated K_b, is

$$K_b = \frac{[NH_4^+][OH^-]}{[NH_3]},$$

and has a value of 1.8×10^{-5}. Problems involving K_b are handled in the same manner as problems involving K_a.

Example 15.6. Dimethylamine comes to the following equilibrium in aqueous solutions:

$$(CH_3)_2NH + HOH = (CH_3)_2NH_2^+ + OH^-. \qquad K_b = 7.5 \times 10^{-4}.$$

Find the pH of 0.25 M $(CH_3)_2NH$.

Using the techniques developed in this chapter, prepare a table of *I*–*R*–*E* values. Include an assumption line.

15.6a.

	$(CH_3)_2NH$	+	HOH	⇌	$(CH_3)_2NH_2^+$	+	OH^-
I	0.25				0		0
R	−y				+y		+y
E	0.25−y				y		y
E_a	0.25				y		y

Now write the K_b expression, substitute the equilibrium values from the table, equate to K_b and solve for y. Check the assumption.

15.6b. y = 1.4×10^{-2}. Assumption not valid.

$$K_b = \frac{[(CH_3)_2NH_2^+][OH^-]}{[(CH_3)_2NH]} = \frac{y^2}{0.25} = 7.5 \times 10^{-4}$$

$$y^2 = 1.9 \times 10^{-4}; \qquad y = 1.4 \times 10^{-2}$$

$$0.25 - 0.014 = 0.24 \neq 0.25$$

The first approximation indicates the equilibrium concentration of $(CH_3)_2NH$ is probably closer to 0.24 than 0.25. Making this the second assumption, solve again for y and check the result against the assumption.

15.6c. $y = 1.3 \times 10^{-2}$.

$$K_b = \frac{[(CH_3)_2NH_2^+][OH^-]}{[(CH_3)_2NH]} = \frac{y^2}{0.24} = 7.5 \times 10^{-4}$$

$$y^2 = 1.8 \times 10^{-4}; \qquad y = 1.3 \times 10^{-2}$$

Assumption check: $0.25 - 0.013 = 0.24$. Valid.

You now have $[OH^-] = 1.3 \times 10^{-2}$. Compute pH.

15.6d. pH = 12.11.

One Method

$$pOH = 2 - \log 1.3 = 1.89$$

$$pH = 14.00 - 1.89 = 12.11$$

Second Method

$$[H^+] = \frac{K_w}{[OH^-]} = \frac{10 \times 10^{-15}}{1.3 \times 10^{-2}} = 7.7 \times 10^{-13}$$

$$pH = 13 - \log 7.7 = 12.11$$

15.4 HYDROLYSIS

In its broad definition, hydrolysis is the chemical reaction of any species with water. It is most frequently applied to the reaction between water and the conjugate bases of weak acids, or the conjugate acids of weak bases. When HOAc is added to water it functions as an acid, donating a proton to the water molecule:

$$HOAc + H_2O \rightleftharpoons H_3O^+ + OAc^- \qquad (15.2)$$

The acetate ion so produced is called the *conjugate base* of acetic acid. If it is introduced to water, along with a spectator ion, as sodium acetate, for example, it acts as a base by receiving a proton from the water molecule:

$$OAc^- + HOH = HOAc + OH^- \qquad (15.3)$$

This is a hydrolysis reaction. It is also the reaction of a base. They are the same.

To see that this is the case, compare Equation 15.3 with the reaction of ammonia as a base with water:

$$NH_3 + HOH \rightleftharpoons NH_4^+ + OH^- \tag{15.4}$$

That the base is a neutral molecule in one instance and an ion in the other does not alter the character of the reaction.

The equilibrium constant for Equation 15.3 could be called K_b, but because it is a hydrolysis reaction it is called the hydrolysis constant and designated K_h. Values of K_h are not usually tabulated in reference books, as are values of K_a and K_b. They are readily found from acid and base constants, however. From Equation 15.2,

$$K_a = \frac{[H_3O^+]\,[OAc^-]}{[HOAc]}\,. \tag{15.5}$$

From Equation 15.3,

$$K_b = \frac{[HOAc]\,[OH^-]}{[OAc^-]} = K_h \tag{15.6}$$

Multiplying Equation 15.6 by Equation 15.5 yields an interesting result:

$$\begin{aligned} K_a K_b &= \frac{[H_3O^+]\,[\cancel{OAc^-}]}{[\cancel{HOAc}]} \times \frac{[\cancel{HOAc}]\,[OH^-]}{[\cancel{OAc^-}]} \\ &= [H_3O^+]\,[OH^-] = K_w \end{aligned} \tag{15.7}$$

Dividing by K_a,

$$K_b = \frac{K_w}{K_a} = K_h\,. \tag{15.8}$$

What happens if the ammonium ion, NH_4^+, which is the *conjugate acid* of ammonia, is introduced to water along with a spectator chloride ion? Again hydrolysis occurs this time with the ammonium ion functioning as an acid:

$$NH_4^+ + H_2O \rightleftharpoons H_3O^+ + NH_3 \tag{15.9}$$

Reactions 15.2 and 15.9 are analogous: both represent the ionization of a weak acid.

As before, the product of the equilibrium constants from Equations 15.4 and 15.9 is K_w. This leads to

$$K_h = \frac{K_w}{K_b}\,. \tag{15.10}$$

Hydrolysis problems are straightforward applications of the methods developed already in this chapter, once the K_h value is determined.

Example 15.7. Find the pH of 0.10 M NaOAc. $K_a = 1.8 \times 10^{-5}$ for HOAc.

The first step is to write the hydrolysis equation. This has been done in the preceding discussion: it is Equation 15.3. The second step is to compute the value of K_h, using Equation 15.8.

15.7a.

$$K_h = \frac{K_w}{K_a} = \frac{10 \times 10^{-15}}{1.8 \times 10^{-5}} = 5.6 \times 10^{-10}$$

Next set up an *I–R–E* tabulation for the equilibrium equation. The initial concentration of the acetate ion is stated in the problem. HOAc and OH^- are "produced" by the reaction. Regard the initial $[OH^-]$ from the water as negligible, as in earlier problems. Complete the table, with the usual assumption.

15.7b.

	OAc^-	+	HOH	$\rightleftharpoons$	HOAc	+	OH^-
I	0.10				0		0
R	−y				+y		+y
E	0.10−y				y		y
E_a	0.10				y		y

Substitute the concentrations from the table into the K_h expression, equate it to the numerical value of the constant, and solve for y, the OH^- ion concentration.

15.7c. $[OH^-] = 7.5 \times 10^{-6}$.

$$K_h = \frac{[HOAc][OH^-]}{[OAc^-]} = \frac{y^2}{0.10} = 5.6 \times 10^{-10}; \qquad y = 7.5 \times 10^{-6}$$

Assumption check: $0.10 - 7.5 \times 10^{-6} = 0.10$. Valid.

The problem asks for the pH. Find this from the now known $[OH^-]$.

15.7d. pH = 8.87.

One Method

$$pOH = 6 - \log 7.5 = 5.13$$
$$pH = 14.00 - 5.13 = 8.87$$

Second Method

$$[H^+] = \frac{K_w}{[OH^-]} = \frac{10 \times 10^{-15}}{7.5 \times 10^{-6}}$$
$$= 1.33 \times 10^{-9}$$
$$pH = 9 - \log 1.33 = 8.88$$

Example 15.8. What is the pH of a 1.0 M Na_2CO_3 solution?

The hydrolysis this time will be of the CO_3^{2-} ion. It is the conjugate base of what acid?

15.8a. HCO_3^-.

$$HCO_3^- \rightleftharpoons H^+ + CO_3^{2-}$$

Now write the hydrolysis equation, which will be analogous to Equation 15.3.

15.8b. $CO_3^{2-} + HOH \rightleftharpoons HCO_3^- + OH^-$.

Using Equation 15.8, write the hydrolysis constant expression and calculate its value. $K_a = 4.8 \times 10^{-11}$ for HCO_3^-.

15.8c.

$$K_h = \frac{[HCO_3^-][OH^-]}{[CO_3^{2-}]} = \frac{K_w}{K_a} = \frac{10 \times 10^{-15}}{4.8 \times 10^{-11}} = 2.1 \times 10^{-4}.$$

Next determine the equilibrium concentrations of all species in the usual manner—by an *I–R–E* table.

15.8d.

	CO_3^{2-}	+	HOH	$\rightleftharpoons$	HCO_3^-	+	OH^-
I	1.0				0		0
R	−y				+y		+y
E	1.0−y				y		y
E_a	1.0				y		y

Substitute the assumed equilibrium concentrations into the K_h expression, equate to K_h and solve for y.

15.8e. $y = [OH^-] = 1.4 \times 10^{-2}$.

$$K_h = \frac{[HCO_3^-][OH^-]}{[CO_3^{2-}]} = \frac{y^2}{1.0} = 2.1 \times 10^{-4}; \qquad y = 1.4 \times 10^{-2}$$

Assumption check: $1.0 - 0.014 = 1.0$. Valid.

Now convert $[OH^-] = 1.4 \times 10^{-2}$ to pH.

15.8f. pH = 12.15.

One Method

$$pOH = 2 - \log 1.4 = 1.85$$
$$pH = 14.00 - 1.85 = 12.15$$

Second Method

$$[H^+] = \frac{K_w}{[OH^-]} = \frac{10 \times 10^{-15}}{1.4 \times 10^{-2}}$$
$$= 7.1 \times 10^{-13}$$
$$pH = 13 - \log 7.1 = 12.15$$

15.5 SIMULTANEOUS EQUILIBRIA

If Example 15.8 leaves you feeling that you haven't been told all of a story, good for you! We have an unanswered—and so far unasked—question: What happens to the HCO_3^- ion produced in the hydrolysis of CO_3^{2-}? It is the conjugate base of a weak acid, H_2CO_3. Doesn't it hydrolyze too?

In fact, it does. The hydrolysis of CO_3^{2-} becomes an example of a two-step reaction process. Having already gone through the first step, let's carry it to completion.

Example 15.9. In Example 15.8 it was found that $[OH^-] = [HCO_3^-] = 1.4 \times 10^{-2}$ as the result of the hydrolysis of 1.0 M CO_3^{2-}. How will these be changed by the resulting hydrolysis of HCO_3^-?

First, write the hydrolysis equation for HCO_3^-.

15.9a. $HCO_3^- + HOH \rightleftharpoons H_2CO_3 + OH^-$.

Now compute the value of K_h for this hydrolysis if $K_a = 4.4 \times 10^{-7}$ for H_2CO_3.

15.9b.

$$K_h = \frac{K_w}{K_a} = \frac{10 \times 10^{-15}}{4.4 \times 10^{-7}} = 2.3 \times 10^{-8}$$

Next set up the *I–R–E* table for the hydrolysis of HCO_3^- *beginning with the results of Example 15.8:* $[HCO_3^-] = [OH^-] = 1.4 \times 10^{-2}$. Let y equal $[H_2CO_3]$ at equilibrium. Carry it all the way to assumed equilibrium concentrations.

15.9c.

	HCO_3^-	+	HOH	$\rightleftharpoons$	H_2CO_3	+	OH^-
I	0.014				0		0.014
R	−y				+y		+y
E	0.014−y				y		0.014+y
E_a	0.014				y		0.014

It appears that, if the assumptions are valid, neither the $[HCO_3^-]$ nor the $[OH^-]$ change appreciably as a result of the second step hydrolysis. Substitute into the K_a expression, solve for y, and check the validity of the assumptions.

15.9d. $y = 2.3 \times 10^{-8}$.

$$K_h = \frac{[H_2CO_3]\,[OH^-]}{[HCO_3^-]} = \frac{(1.4 \times 10^{-2})y}{1.4 \times 10^{-2}} = 2.3 \times 10^{-8}$$

$y = 2.3 \times 10^{-8}$

Assumption check: $1.4 \times 10^{-2} - 2.3 \times 10^{-8} = 1.4 \times 10^{-2}$. Valid.

Three important points emerge from Examples 15.8 and 15.9.

First, they illustrate simultaneous equilibria, where two or more equilibrium systems exist at the same time. Here there are at least three: the hydrolysis of the carbonate ion is the first. Hydrolysis of the HCO_3^- ion is another. The third equilibrium is the water equilibrium, $HOH \rightleftharpoons H^+ + OH^-$. This equilibrium is always present in aqueous systems. It was used in Example 15.8; the interconversion between OH^- and H^+, as well as pH and pOH, is based on this equilibrium, as developed in Chapter 14.

Simultaneous equilibrium problems can become quite complex, requiring numerous simultaneous equations for the determination of several unknowns. In essence, you have done this in Examples 15.8 and 15.9—and this is as far as we shall attempt to go with these problems in this text.

A second major point from Examples 15.8 and 15.9 is that any species that appears in two or more simultaneous equilibria can have but one concentration for all equilibria. This is reasonable: a given species cannot have two concentrations in the same solution at the same time. In the examples, $[HCO_3^-]$ was the same for both equilibria in which it appeared, and $[OH^-]$ was the same in all three.

The third major point is that in simultaneous equilibria in which a second step is the direct result of a first, as the HCO_3^- hydrolysis was the direct result of the CO_3^{2-} hydrolysis, the first step alone *frequently* governs the concentrations of the species involved in it. This was true in Examples 15.8 and 15.9: the first hydrolysis alone fixed the hydrogen carbonate and hydroxide ion concentrations. How can you tell when this first-step control will be in effect? In sequential equilibria such as these examples, each step has its own equilibrium constant. Differentiation between the constants is usually made by calling the constant for the first step K_1, and that for the second K_2. In these examples K_1 was about 10^{-4} and K_2 was about 10^{-8}. When there is a difference of several orders of magnitude between K_1 and K_2, as in this case, the first step governs the concentration of all species in it.

Example 15.10 Find the following for 0.200 M H_2S: $[H^+]$; $[H_2S]$; $[HS^-]$; $[S^{2-}]$; For $H_2S_1 = 1 \times 10^{-7}$ and $K_2' = 1 \times 10^{-15}$.

Here, where K_1 is 10^8 times K_2, we can be quite confident that the first ionization will set all concentrations except $[S^{2-}]$. In fact, considering the first step only, you find everything but $[S^{2-}]$ exactly as in Example 15.3. Solve for the first three desired quantities now.

15.10a. $[H^+] = 4.5 \times 10^{-5}$; $[H_2S] = 0.020$; $[HS^-] = 4.5 \times 10^{-5}$.

	H_2S	$\rightleftharpoons$	H^+	+	HS^-
I	0.020		0		0
R	−y		+y		+y
E	0.020 − y		y		y
E_a	0.020		y		y

$$K_1 = \frac{[H^+][HS^-]}{[H_2S]} = \frac{y^2}{0.020} = 1 \times 10^{-7}; \qquad y = 4.5 \times 10^{-5}$$

Assumption check: 0.020 − 0.000045 = 0.020. Valid.

Starting with $[HS^-]$ and $[H^+]$ from the first ionization, use K_2 to find $[S^{2-}]$ and confirm that neither $[HS^-]$ nor $[H^+]$ changes appreciably in the second step. The usual *I–R–E* approach for $HS^- = H^+ + S^{2-}$. . . .

15.10b. $[S^{2-}] = 1 \times 10^{-15}$.

	HS^-	$\rightleftharpoons$	H^+	+	S^{2-}
I	4.5×10^{-5}		4.5×10^{-5}		0
R	$-y$		$+y$		$+y$
E	$4.5 \times 10^{-5} - y$		$4.5 \times 10^{-5} + y$		y
E_a	4.5×10^{-5}		4.5×10^{-5}		y

$$K_2 = \frac{[H^+][S^{2-}]}{[HS^-]} = \frac{(4.5 \times 10^{-5})y}{4.5 \times 10^{-5}} = 1 \times 10^{-15} \qquad y = 1 \times 10^{-15}$$

Assumption check: $4.5 \times 10^{-5} - 1 \times 10^{-15} = 4.5 \times 10^{-5}$. Valid.

15.6 BUFFER SOLUTIONS

It doesn't take much to change the pH of most solutions. In "pure" water, or a neutral aqueous solution, the pH = 7, indicating $[H^+] = 10^{-7} = 0.0000001$. Adding 0.001 mole of H^+ to a neutral solution raises $[H^+]$ to 0.0010001, or 10^{-3}, which represents a pH of 3. $[H^+]$ has increased by a factor of 10,000:

$$\frac{0.001}{0.0000001} = \frac{10^{-3}}{10^{-7}} = \frac{10^{+4}}{1} = \frac{10{,}000}{1} \; .$$

Addition of 0.001 mole of OH^- to a neutral solution changes the pH from 7 to 11, a reduction of $[H^+]$ to $\frac{1}{10{,}000}$ of what it was. Some solutions, however, have the capacity to "consume" H^+ and OH^- in quantities even larger than these without an appreciable change in pH. These solutions are called **buffer** solutions.

In Example 15.5 you calculated the pH of a solution of a weak acid, HOAc. In Example 15.7 you found the pH of a solution of a weak base, the OAc^- ion, introduced in the form of NaOAc. A buffer solution must have both solutes, a weak acid and a weak base. These may be conjugates of each other, but this is not essential. The weak acid reacts with and consumes any OH^- that may be added, while the role of the weak base is to consume the H^+ that may be introduced. This prevents significant disturbance of the $HOH = H^+ + OH^-$ equilibrium on which the pH of the solution depends.

How a buffer controls pH may be seen through an example. . . .

Example 15.11. $K_a = 1.8 \times 10^{-5}$ for HOAc. Calculate the pH of a buffer prepared by dissolving 0.25 mole of HOAc and 0.40 mole of NaOAc in water and diluting to one liter.

In an *I–R–E* approach to this problem the initial concentrations of *both* HOAc and OAc^- are given. Assuming as usual that initial $[H^+]$ is negligible, and using y to represent the equilibrium $[H^+]$, complete the table through the assumptions.

15.11a.

	$HOAc$	=	H^+	+	OAc^-
I	0.25		0		0.40
R	−y		+y		+y
E	0.25 −y		y		0.40 +y
E_a	0.25		y		0.40

The ionization of y moles of HOAc reduces [HOAc] by y and yields y moles per liter of both H^+ and OAc^-, as before. The OAc^- is added to the 0.40 mole per liter already present from the sodium acetate. If y is negligible compared to 0.25 and 0.40, both of these values are unchanged, as shown by the E_a entries.

Substitute the assumed equilibrium concentrations into the K_a expression, solve for $[H^+]$, confirm the assumptions and calculate the pH.

15.11b. pH = 4.96

$$K_a = \frac{[H^+]\,[OAc^-]}{[HOAc]} = \frac{[H^+]\,(0.40)}{0.25} = 1.8 \times 10^{-5}$$

$$[H^+] = 1.1 \times 10^{-5}$$

Assumption check: 0.25 − 0.000011 = 0.25; 0.40 − 0.000011 = 0.40. Valid.

$$pH = 5 - \log 1.1 = 4.96.$$

Stabilization of pH by a buffer is explained by solving the K_a expression for the general weak acid equilibrium, $HA = H^+ + A^-$, for $[H^+]$:

$$[H^+] = K_a \frac{[HA]}{[A^-]} \qquad (15.11)$$

Inasmuch as K_a is constant, it follows that *$[H^+]$ depends entirely upon—is proportional to—the ratio of the acid concentration to the base concentration.* By fixing this ratio, the pH of the solution is determined.

To illustrate buffer action quantitatively, consider the buffer of Example 15.11. In one liter of solution there are 0.25 mole of un-ionized HOAc and 0.40 mole of OAc^- in equilibrium with each other according to the equation $HOAc = H^+ + OAc^-$. Suppose to this solution there is added some H^+, say 0.001 mole. (This is the same quantity which, in the opening paragraph of this section, produced a pH change of four units, starting with a neutral solution.) The added H^+ would react with some of the 0.40 mole of OAc^- ion present, and produce more HOAc. If *all* of the H^+ reacted with OAc^-—and no more than part of it can react, according to LeChatelier's Principle—it would reduce $[OAc^-]$ by only 0.001 and increase [HOAc] by the same amount. The $[HOAc]/[OAc^-]$ ratio that determines $[H^+]$ would become $\frac{0.25 + 0.001}{0.40 + 0.001}$, which, to the two significant figures justified, is unchanged at $\frac{0.25}{0.40}$. Clearly these changes are meaningless, and pH would remain at 4.96.

Note that if the $\frac{[HOAc]}{[OAc^-]}$ ratio had been $\frac{0.025}{0.040}$, the solution would still have been buffered at the same pH—but not nearly so effectively. Addition of 0.001 mole of H^+ per liter would then have decreased $[OAc^-]$ to 0.039 and increased [HOAc] to 0.026. The new $[H^+]$ would then be

$$[H^+] = K_a \frac{[HOAc]}{[OAc^-]} = 1.8 \times 10^{-5} \times \frac{0.026}{0.039} = 1.2 \times 10^{-5}.$$

This is an increase of about nine percent over the original $[H^+]$ of the buffer (1.1×10^{-5}), yielding a pH of 4.92. This points up the necessity of being sure the concentrations of weak acid and weak base are sufficiently large to consume additions of H^+ and OH^- without being significantly altered themselves.

Additions of 0.001 mole of OH^- per liter to the solution of Example 15.11 would have no more effect than the comparable quantity of acid. This time the OH^- would react with the un-ionized HOAc, reducing its concentration negligibly, while producing a negligible amount of OAc^-. The ratio would remain intact, and the pH hold at 4.96.

To summarize: A buffer solution is one which contains a substantial concentration of a weak acid and a weak base. The acid and base are frequently a conjugate pair, i.e., HOAc and OAc^-, as in the example above. The pH of the solution will depend upon the ratio of the concentrations of the weak acid and base. The concentrations must be sufficient to consume significant additions of H^+ or OH^- without being changed appreciably themselves.

Buffers of conjugate acid-base pairs may be prepared by some "backdoor" methods too. . . .

Example 15.12. What would be the pH of a solution that is prepared by dissolving 2.46 grams of sodium acetate (MW = 82.0) in 100 ml 0.20 M HCl?

Chemically the acetate ion will combine with the hydrogen ion to produce un-ionized acetic acid, all three species achieving concentrations that will satisfy the value of K_a. The first step is to find the molarity of the sodium acetate.

15.12a. 0.300 M NaOAc.

$$\frac{2.46 \text{ \sout{g NaOAc}}}{0.100\ l} \times \frac{1 \text{ mole NaOAc}}{82.0 \text{ \sout{g NaOAc}}} = 0.300 \text{ mole NaOAc}/l$$

The reaction $H^+ + OAc^- \rightarrow HOAc$ goes nearly to completion, forming un-ionized HOAc. Accordingly the determination of [HOAc] and $[OAc^-]$ becomes an excess stoichiometry problem. Thinking on a "per liter" basis, the problem asks how many moles of HOAc will form from 0.20 mole of H^+ (from HCl) and 0.300 mole of OAc^- (from NaOAc). Also, how many moles of OAc^- will remain unreacted? Find the [HOAc] and $[OAc^-]$ that result from the reaction.

15.12b. [HOAc] = 0.20; $[OAc^-]$ = 0.10.

0.20 mole H^+ is the limiting species, being converted to 0.20 mole HOAc. This requires 0.20 mole OAc^- from the 0.300 mole OAc^- present, leaving 0.10 mole OAc^- in excess. These quantities are per liter of solution, so they represent concentrations.

$[H^+]$ may now be found by Equation 15.11, and pH follows. Complete the problem.

15.12c. pH = 4.44.

$$[H^+] = K_a \frac{[HOAc]}{[OAc^-]} = 1.8 \times 10^{-5} \times \frac{0.20}{0.10} = 3.6 \times 10^{-5}$$

$$pH = 5 - \log 3.6 = 5 - 0.56 = 4.44.$$

An acetic acid–acetate ion type buffer may also be made by the partial neutralization of excess weak acid by a strong base:

Example 15.13. Find the pH of a solution of 2.0 grams of NaOH in 3.0×10^2 ml 0.28 M HOAc.

Again, begin by finding the molarity of the NaOH. . . .

15.13a. 0.17 M NaOH.

$$\frac{2.0 \text{ g NaOH}}{0.300\ l} \times \frac{1 \text{ mole NaOH}}{40.0 \text{ g NaOH}} = 0.17 \text{ M NaOH}$$

As long as it lasts, the strong base, 0.17 mole OH^-/liter, will react with the weak acid, 0.28 mole HOAc/liter,

$$OH^- + HOAc \rightarrow HOH + OAc^-.$$

What are the resulting [HOAc] and $[OAc^-]$?

15.13b. [HOAc] = 0.11; $[OAc^-]$ = 0.17.

The OH^- is the limiting species, and will release OAc^- until it is completely consumed. The excess HOAc is the 0.28 mole/liter initially present minus 0.17 mole/liter reacted, leaving 0.11 mole/liter unreacted.

Proceed to $[H^+]$ and pH as in the last example.

15.13c. pH = 4.94.

$$[H^+] = K_a \frac{[HOAc]}{[OAc^-]} = 1.8 \times 10^{-5} \times \frac{0.11}{0.17} = 1.16 \times 10^{-5}$$

$$pH = 5 - \log 1.16 = 5 - 0.07 = 4.94.$$

The truly practical problem with buffers is to decide what quantities of weak acid and base are required to buffer a solution at a predetermined pH. As previously noted, it is the *ratio* of the constituents that fixes the pH. The problem thereby becomes one of determining the ratio that will establish the desired pH. This follows readily if K_a and the desired $[H^+]$ are known.

Example 15.14. What quantities of glacial acetic acid (17.5 M) and sodium acetate (MW = 82.0) must be dissolved and diluted to 1.00 liter to produce a solution buffered at pH = 5.12?

First, calculate the $[H^+]$ at which the solution is to be buffered.

15.14a. $[H^+] = 7.6 \times 10^{-6}$ mole/liter.

$$[H^+] = 10^{-pH} = 10^{-5.12} = 10^{0.88-6} = 10^{0.88} \times 10^{-6} = 7.6 \times 10^{-6}$$

Now start with the K_a expression, solve for the desired ratio, and substitute the known K_a and $[H^+]$ values to obtain the numerical value of the ratio.

15.14b.

$$\frac{[HOAc]}{[OAc^-]} = \frac{[H^+]}{K_a} = \frac{7.6 \times 10^{-6}}{1.8 \times 10^{-5}} = 0.42$$

or

$$\frac{[OAc^-]}{[HOAc]} = \frac{K_a}{[H^+]} = \frac{1.8 \times 10^{-5}}{7.6 \times 10^{-6}} = 2.4$$

You may have solved for either of these ratios, which are obviously reciprocals of each other.

From this point on, there are as many correct solutions as there are people to do the problem. The only requirement is that the ratio calculated be maintained in translating into measured quantities of acid and sodium acetate. From practical considerations, however, it is customary to keep molarities of acid and base between 0.1 and 1.0. The remaining arithmetic is therefore simplified if further consideration is directed to the *fractional* ratio. The easiest solution, then, is to adopt the ratio $[HOAc]/[OAc^-] = 0.42/1.00$, making the solution 0.42 M in HOAc and 1.00 M in NaOAc. Following this approach, what quantity of NaOAc would you use for the liter of final solution?

15.14c. 82.0 grams NaOAc.

If the solution is to be 1.00 M we would dissolve 1.00 mole in one liter, and the molar weight is given as 82.0 g/mole.

Now, how many ml of concentrated HOAc will you use to get 0.42 mole?

15.14d. Use 24 ml HOAc.

$$0.42 \cancel{\text{mole}} \times \frac{1000 \text{ ml}}{17.5 \cancel{\text{moles}}} = 24 \text{ ml}$$

15.7 COMPLEX ION EQUILIBRIA

Complex ions contain metal ions bonded chemically to molecules or anions called ligands. The $Cu(NH_3)_4^{2+}$ and $Cu(CN)_4^{2-}$ ions are examples. In the first, the Cu^{2+} ion is complexed with four molecules of ammonia as the ligands. In the second the ligand is the cyanide ion, CN^-. The charge on the complex is the algebraic sum of the charges on the metal ion and the ligands.

Complex ions have a measurable tendency to dissociate into their components, establishing an equilibrium. With our two examples,

$$Cu(NH_3)_4^{2+}(aq) = Cu^{2+}(aq) + 4\ NH_3(aq)$$
$$Cu(CN)_4^{2-}(aq) = Cu^{2+}(aq) + 4\ CN^-(aq)$$

The equilibrium constants are referred to as dissociation constants, K_d, or instability constants, K_{inst}; we shall employ the former term in this text. For the above complexes these constants and their values are

$$K_d \text{ for } Cu(NH_3)_4^{2+} = \frac{[Cu^{2+}]\,[NH_3]^4}{[Cu(NH_3)_4^{2+}]} = 2 \times 10^{-13}$$

$$K_d \text{ for } Cu(CN)_4^{2-} = \frac{[Cu^{2+}]\,[CN^-]^4}{[Cu(CN)_4^{2-}]} = 1 \times 10^{-25}.$$

The relative values of this constant indicates that there is much less tendency for the $Cu(CN)_4^{2-}$ ion to dissociate than for the $Cu(NH_3)_4^{2+}$ complex; the former is more stable than the latter.

A table of dissociation constants appears in the Appendix (Table VII, page 292). Be wary of the values in this table, or any other K_d table you encounter. Values of these constants vary considerably from different sources, depending, no doubt, on the experimental data from which they are derived. Unless other values are given in the statement of a problem or example, values from Table VII have been used for all problems in this book, and answers reflect these values. If you use K_d values from another source you are apt to come up with a different —but equally correct—numerical answer.

Problems dealing with complex ions can be solved by the techniques of this chapter, with some modification to the assignment of variables.

Example 15.15. 1.59 grams of anhydrous $CuSO_4$ are dissolved in 1.00 liter of 0.50 M NH_3. Find the concentrations of Cu^{2+}, $Cu(NH_3)_4^{2+}$, and NH_3 in the resulting complex ion equilibrium.

To begin, find the initial molarity of the copper(II) ion, assuming 100% dissociation of $CuSO_4$.

15.15a. 0.0100 M Cu^{2+}.

$$\frac{1.59\ \cancel{g\ CuSO_4}}{l} \times \frac{1\ \cancel{mole\ CuSO_4}}{159\ \cancel{g\ CuSO_4}} \times \frac{1\ mole\ Cu^{2+}}{1\ \cancel{mole\ CuSO_4}} = 0.0100\ mole\ Cu^{2+}/l$$

Now set up the *I–R–E* table, but hold back on the E_a line, which requires a bit of new thinking this time. Watch out for ammonia.

15.15b.

	$Cu(NH_3)_4^{2+}$	$\rightleftharpoons$	Cu^{2+}	+	$4\ NH_3$
I	0		0.0100		0.50
R	+y		−y		−4y
E	y		0.0100−y		0.50−4y

Before making assumptions this time, let's see what might be predicted chemically. $K_d = 2 \times 10^{-13}$ for the reaction. What does that predict as to which side of the reaction will be favored at equilibrium, right or left?

15.15c. Left.

A very small K indicates that the numerator factors, which are the concentrations of the species on the right side of the equation, will be very small relative to the denominator factors, which are the concentrations of the species on the left side of the equation. In other words, the reaction will go nearly to completion *to the left* as written, leaving a very small concentration of the limiting species on the right. The limiting species this time is obviously copper(II) ion. From this consideration, estimate the value of y, which is $[Cu(NH_3)_4^{2+}]$ at equilibrium.

15.15d. y is very close to 0.0100.

Only if y does approach 0.0100 can the value of $[Cu^{2+}]$ be small enough to produce a K value of 2×10^{-13}.

The assumption in this problem, then, is that y = 0.0100, and that $[Cu^{2+}]$ is very small at equilibrium—equal to 0.0000 to four decimal places. The *Ea* values are therefore 0.0100 for $[Cu(NH_3)_4^{2+}]$, and 0.50 − 4(0.0100) = 0.46 for $[NH_3]$. For $[Cu^{2+}]$ the concentration cannot be zero, but it is too small to be detected through a subtraction limited to four decimal places. Let's just leave it as $[Cu^{2+}]$.

Write the equilibrium constant expression, substitute the values found in the above paragraph, equate to 2×10^{-13}, and solve for $[Cu^{2+}]$.

15.15e. $[Cu^{2+}] = 4 \times 10^{-14}$.

$$K_d = \frac{[Cu^{2+}][NH_3]^4}{[Cu(NH_3)_4^{2+}]} = \frac{[Cu^{2+}]\,(0.46)^4}{0.01} = 2 \times 10^{-13}$$

$$[Cu^{2+}] = 4 \times 10^{-14}$$

Assumption check: $4 \times 10^{-14} = 0.0000$. Valid.

All concentrations are now known.

Example 15.16. Calculate $[Ag^+]$, $[Ag(NH_3)_2^+]$, and $[NH_3]$ at equilibrium when 1.0 gram of $AgNO_3$ is dissolved in 500 ml 1.0 M NH_3. For $Ag(NH_3)_2^+ \rightleftharpoons Ag^+ + 2\ NH_3$, $K_d = 4 \times 10^{-8}$.

The method is identical to the previous example. Solve completely.

15.16a. $[Ag(NH_3)_2^+] = 0.012$; $[Ag^+] = 5 \times 10^{-10}$; $[NH_3] = 1.0$.

$$\text{Initial } Ag^+ \text{ molarity: } \frac{1.0\ \cancel{g\ AgNO_3}}{0.500\ l} \times \frac{\text{mole } AgNO_3}{170\ \cancel{g\ AgNO_3}} = 0.012 \text{ mole } AgNO_3/l$$

	$Ag(NH_3)_2^+$	$\rightleftharpoons$	Ag^+	+	$2\ NH_3$
I	0		0.012		1.0
R	$+y$		$-y$		$-2y$
E	y		$0.012-y$		$1.0-2y$
E_a	0.012		$[Ag^+]$		1.0

$$K_d = \frac{[Ag^+][NH_3]^2}{[Ag(NH_3)_2^+]} = \frac{[Ag^+][1.0]^2}{0.012} = 4 \times 10^{-8}$$

$$[Ag^+] = 5 \times 10^{-10}$$

Example 15.17. Find the ammonia concentration at equilibrium that is required if $[Cu^{2+}]$ is equal to $[Cu(NH_3)_4^{2+}]$ in the equilibrium $Cu(NH_3)_4^{2+} \rightleftharpoons Cu^{2+} + 4\ NH_3$. $K_d = 2 \times 10^{-13}$.

This problem is solved by setting up the equilibrium constant expression and equating it to the known value of K_d. Cancellation of the equal values of $[Cu^{2+}]$ and $[Cu(NH_3)_4^{2+}]$ permits a straightforward calculation for $[NH_3]$.

15.17.a. You may have reached $[NH_3]^4 = 2 \times 10^{-13}$ and not known what to do next. From the beginning . . . .

$$K_d = \frac{[\cancel{Cu^{2+}}][NH_3]^4}{[\cancel{Cu(NH_3)_4^{2+}}]} = [NH_3]^4 = 2 \times 10^{-13}; \qquad [NH_3] = \sqrt[4]{2 \times 10^{-13}}$$

The cancellation is possible, of course, because the two ion concentrations are equal.

To find the fourth root of 2×10^{-13}, the best, most general method uses the mathematics of logarithms. See page 279 in the Appendix if necessary, and then complete the problem.

15.17b. $[NH_3] = 6.6 \times 10^{-4}$, or 7×10^{-4} to the one significant figure justified by the K_d value.

$$\log[NH_3] = \tfrac{1}{4}\log(2 \times 10^{-13}) = \tfrac{1}{4}(\log 2 + \log 10^{-13})$$
$$= \tfrac{1}{4}[0.30 + (-13)]$$
$$= \tfrac{1}{4}(-12.70) = -3.18$$
$$[NH_3] = 10^{-3.18} = 10^{0.82-4} = 10^{0.82} \times 10^{-4} = 6.6 \times 10^{-4}$$

Obviously a very small concentration of ammonia is sufficient to complex half the copper present.

Example 15.18. Find the CN^- concentration at equilibrium that will result in $[Cu^{2+}]$ being equal to $[Cu(CN)_4^{2-}]$ in $Cu(CN)_4^{2-} \rightleftharpoons Cu^{2+} + 4\,CN^-$. For $Cu(CN)_4^{2-}$, $K_d = 1 \times 10^{-25}$. On the basis of your answer and the answer to Example 15.17 determine which is the "stronger" complexing agent, CN^- or NH_3.

The numerical part of this problem follows the procedure of Example 15.18. Find the required cyanide ion concentration. Then compare the "strengths" of the complexing agents.

15.18a. $[CN^-] = 6 \times 10^{-7}$.

$$K_d = \frac{\cancel{[Cu^{2+}]}[CN^-]^4}{\cancel{[Cu(CN)_4^{2-}]}} = [CN^-]^4 = 1 \times 10^{-25}; \qquad [CN^-] = \sqrt[4]{1 \times 10^{-25}}$$

$$\log [CN^-] = \tfrac{1}{4} \log (1 \times 10^{-25}) = -6.25$$

$[CN^-] = 5.6 \times 10^{-7}$, or 6×10^{-7} to one significant figure.

It requires a lower concentration of CN^- than NH_3 to complex half the Cu^{2+} present. The CN^- is the stronger complexing agent. This is consistent with the statement concerning the greater stability of the $Cu(CN)_4^{2-}$ complex at the beginning of this section.

Example 15.19. What concentration of ammonia would be required to reduce $[Ni^{2+}]$ to 1×10^{-10} in a solution initially 0.10 *M* in Ni^{2+}? $K_d = 2 \times 10^{-9}$ for $Ni(NH_3)_6^{2+} \rightleftharpoons Ni^{2+} + 6\ NH_3$.

Though this problem is distinctly different from the others, you should be able to solve it with few suggestions. Remember it is the total ammonia added—the "concentration" before equilibrium is reached—that is sought, not the equilibrium molarity. Solve the problem.

15.19a. 1.72 M NH_3, or 2 M NH_3, to one significant figure.
The *I–R–E* table is based on the fact that both initial and final nickel(II) ion concentrations are given: 0.10 and 1×10^{-10}. *R* for Ni^{2+} is therefore found from $0.10 - R = 1 \times 10^{-10}$. In terms of significant figures, *R* equals 0.10. The table follows:

	$Ni(NH_3)_6^{2+}$	$\rightleftharpoons$	Ni^{2+}	+	$6\ NH_3$
I	0		0.10		y+0.60
R	+0.10		−0.10		−0.60
E	0.10		1×10^{-10}		y

$$K_d = \frac{[Ni^{2+}][NH_3]^6}{[Ni(NH_3)_6^{2+}]} = \frac{1 \times 10^{-10} y^6}{0.10} = 2 \times 10^{-9}$$

$y^6 = 2$

$y = 1.12 = [NH_3]$ at equilibrium

$0.60 = [NH_3]$ reacted

$1.72 = [NH_3]$ initially

In all examples and discussions to this point it has been assumed that the ligand has been present in significant excess over the metal ion as equilibrium is reached. Formation of complex ions is actually a stepwise process, somewhat similar to the ionization of a polyprotic acid, such as H_2S. Each step has its own K_d, and the overall K_d is the product of $K_1 \times K_2 \times K_3$, etc. For example, with silver ion and ammonia,

$$Ag(NH_3)_2^+ \rightleftharpoons Ag(NH_3)^+ + NH_3 \quad K_1$$
$$Ag(NH_3)^+ \rightleftharpoons Ag^+ + NH_3 \quad K_2$$
$$Ag(NH_3)_2^+ \rightleftharpoons Ag^+ + 2\ NH_3 \quad K_d$$

If ammonia is not in excess of that required to form $Ag(NH_3)_2^+$ with almost all of the Ag^+ present, *both* stepwise equilibria are significant in problems and must be taken into account. We shall not, however, consider such problems within the scope of this book.

PROBLEMS

Section 15.2

15.1. Calculate K_a for $HCOOH = H^+ + HCOO^-$ if 0.50 M HCOOH is 2.0% ionized. Also find the pH of the solution.

15.2. The pH of 0.20 M benzoic acid, C_6H_5COOH, is 2.44. Calculate K_a for benzoic acid and the percent ionization in a 0.20 molar solution.

15.3. The pH of 1.0 M lactic acid ($HOCH_2CH_2COOH$—call it HL for short if you wish) is 1.93. Calculate K_a for lactic acid and the percent ionization of a 1.0 molar solution.

15.21. 0.10 M C_2H_5COOH is 1.1% ionized according to the equation $C_2H_5COOH = H^+ + C_2H_5COO^-$. Find K_a and the pH of the solution.

15.22. The hydrogen sulfate ion is 11% ionized in 1.0 molar solution: $HSO_4^- = H^+ + SO_4^{2-}$. Find K_a for this reaction, and the pH of 1.0 M $NaHSO_4$.

15.23. Calculate K_a for the first ionization step in maleic acid,

$$HOOCCH_2CHOHCOOH = H^+ + HOOCCH_2CHOHCOO^-$$

(which you may write $H_2M = H^+ + HM^-$ if you wish), if the pH = 1.51 for a 0.10 M solution.

Section 15.3

15.4. Calculate the pH and percent ionization of 0.10 M HNO_2.

15.5. Determine the percent ionization and pH of 0.30 M HOAc.

15.6. $[CO_2(aq)]$ in soda water (carbonated water) is about 3.3×10^{-2}. Calculate the pH of soda water, assuming all H^+ comes from the dissociation of carbonic acid, which is essentially aqueous carbon dioxide.

15.24. Find the pH and percent ionization of 1.2 M HF.

15.25. Cream of tarter, or potassium hydrogen tartarate, whose formula may be written $KHC_4H_4O_6$, is used as the hydrogen ion source in some baking powders: $HC_4H_4O_6^- = H^+ + C_4H_4O_6^{2-}$. Find the pH of a 0.10 molar solution of this compound. $K_a = 3.0 \times 10^{-5}$.

15.26. Natural water in contact with air contains dissolved carbon dioxide in an equilibrium that may be represented by $CO_2(aq) + H_2O(l) = H^+(aq) + HCO_3^-(aq)$. This is, in essence, the first step of the carbonic acid equilibrium. Assuming all H^+ and HCO_3^- in the solution come from this equilibrium, calculate the pH of natural water in contact with air if $[CO_2(aq)] = 1 \times 10^{-5}$.

Section 15.4

15.7. Find the pH of 0.50 M $NaNO_2$.

15.27. What will be the pH of 0.10 M NaClO, a solution commonly used as a household bleach?

15.8. What will be the pH of 1.0 M NH_4Cl if $K_b = 1.8 \times 10^{-5}$ for $NH_3 + H^+ = NH_4^+ + OH^-$? This problem is to be solved without using Table VI.

15.28. Anilinium chloride, $C_6H_5NH_3^+Cl^-$, ionizes in water to free the anilinium ion, $C_6H_5NH_3^+$. This ion hydrolyzes as follows:

$$C_6H_5NH_3^+ + H_2O = C_6H_5NH_2 + H_3O^+.$$

If $K_b = 4.2 \times 10^{-10}$ for aniline ($C_6H_5NH_2$), find the pH of 0.15 M $C_6H_5NH_3Cl$.

NOTE: When a strong acid is titrated with a strong base, such as an HCl–NaOH titration, the equivalence point—that point in the titration when the moles of acid equal the moles of base—appears at a pH of 7.00. If a weak acid is used, quite a different result is observed. When the general titration reaction $HA + OH^- = A^- + HOH$ reaches the equivalence point, there is an equilibrium with equal concentrations of HA and OH^- remaining. The same equilibrium, viewed in the reverse direction, is simply the hydrolysis of the A^- ion. The pH of the equivalence point is therefore the pH of the hydrolysis reaction at a comparable concentration. The titration of a weak base with a strong acid is similar. Caution: Watch changes in solution volumes when determining concentration in titration problems.

15.9. Calculate the pH at the equivalence point when titrating 0.50 M benzoic acid with 0.50 M NaOH.

15.29. Find the equivalence point pH when titrating 0.20 M HCl with 0.40 M NH_3 if $K_b = 1.8 \times 10^{-5}$ for ammonia.

Section 15.5

15.10. Calculate the pH of 0.20 M H_2SO_3. Show that the ionization of HSO_3^- has no effect on the pH.

15.11. Compute the pH of 0.25 M Na_3PO_4. Does the second or third hydrolysis step contribute to the pH of the solution? Confirm your answer.

15.30. Tartaric acid, $H_2C_4H_4O_6$, as well as cream of tartar, is used in baking powders. Find the pH of 0.10 M tartaric acid. Prove that the second ionization step does not affect the pH. $K_1 = 1.0 \times 10^{-3}$; $K_2 = 4.6 \times 10^{-5}$.

15.31. Find the pH of 0.40 M Na_2SO_3. Show how the hydrolysis of HSO_3^- affects the pH—if it does.

Section 15.6

15.12. Find the pH of a solution of 0.25 M $NaNO_2$ in 0.75 M HNO_2.

15.13. 28.0 grams of sodium acetate, $NaC_2H_3O_2$, are dissolved in 500 ml 0.12 M HOAc. Find the pH at which the solution is buffered.

15.14. 20.0 ml 4.0 M HCl and 11.5 grams of sodium acetate are dissolved in 150 ml water. Find the pH of the buffer solution produced. Also determine the pH if the water volume had been 250 ml.

15.15. What concentration ratio of benzoic acid and sodium benzoate will produce a buffer having a pH of 4.80? How many grams of sodium benzoate, C_6H_5COONa, would be required with 16.0 grams of benzoic acid, C_6H_5COOH, to produce this buffer?

15.32. Find the pH of a buffer 0.050 molar in benzoic acid and 0.40 molar in sodium benzoate.

15.33. 5.30 grams of sodium carbonate and 4.20 grams of sodium hydrogen carbonate are dissolved in water in a beaker. Calculate the pH of the solution produced.

15.34. 50.0 ml 0.60 M NH_4Cl are combined with 30.0 ml 0.40 M NaOH. Calculate the pH of the buffer that results. $K_b = 1.8 \times 10^{-5}$ for $NH_3 + HOH = NH_4^+ + OH^-$.

15.35. A CO_2–HCO_3^- buffer is highly critical in maintaining the pH of blood at 7.40±0.01. Calculate the $[CO_2(aq)]/[HCO_3^-(aq)]$ ratio that yields this pH. Consider $CO_2(aq)$ as $H_2CO_3(aq)$.

15.16. Determine specific quantities of sodium acetate and concentrated acetic acid (17M) that will produce 1.0 liter of buffer with a pH = 4.15. Express the sodium acetate in grams and the acetic acid in milliliters.

15.36. Recommend specific quantities of concentrated HCl (12 M) and sodium nitrite that will produce 500 ml of buffer with pH = 3.80. Answer in grams for sodium nitrite and milliliters for HCl.

Section 15.7

15.17. Find the equilibrium concentrations for all species in $Ag(S_2O_3)_2^{3-} = Ag^+ + 2\,S_2O_3^{2-}$ in a solution originally 0.025 molar in Ag^+ and 0.50 molar in $S_2O_3^{2-}$. Assume this equilibrium is the only reaction of significance.

15.18. 4.0 grams of anhydrous $CuSO_4$ are dissolved in water and diluted to 960 milliliters. The volume is then brought to 1.00 liter with 40.0 ml concentrated ammonia (15 M). Calculate $[Cu^{2+}]$, $[NH_3]$ and $[Cu(NH_3)_4^{2+}]$ at equilibrium, assuming the complexing of Cu^{2+} with NH_3 is the only significant reaction.

15.19. 250 ml of a solution is 0.050 M in Co^{2+}. What volume of 15 M NH_3 would be required to reduce $[Co^{2+}]$ to 1×10^{-15}? Must the dilution of the solution with concentrated ammonia be accounted for in this calculation? Why, or why not? $Co(NH_3)_6^{2+} = Co^{2+} + 6\,NH_3$.

15.20. How many milliliters of 15 M NH_3 are required to complex as $Ni(NH_3)_6^{2+}$ half the Ni^{2+} in 50 ml 0.24 M Ni^{2+}?

15.37. A solution is prepared with initial concentrations of 0.10 molar in NH_3 and 0.0040 molar in Cu^{2+}. Find the equilibrium values for $[Cu^{2+}]$, $[NH_3]$ and $[Cu(NH_3)_4^{2+}]$. Assume no other reactions.

15.38. Find the equilibrium concentrations for all species when 1.89 grams of $Zn(NO_3)_2$ are dissolved in 100 ml 0.60 M NH_3, assuming no reaction other than the complex ion equilibrium.

15.39. How many milliliters of 0.50 M $S_2O_3^{2-}$ are required to reduce to 1.0×10^{-8} the silver ion concentration in 50 milliliters of 0.040 M Ag^+ by complexing the Ag^+ to $Ag(S_2O_3)_2^{3-}$? Disregard any dilution effect and other possible reactions.

15.40. How many drops (20 drops = 1 ml) of 0.2 M KCN are required to complex as $Cu(CN)_4^{2-}$ and reduce to 1×10^{-20} M the Cu^{2+} in 0.50 ml 0.010 M Cu^{2+}? Assume no other significant reactions.

CHAPTER 16

SOLUBILITY EQUILIBRIA

If NaCl is placed into water it dissolves to the limit of its solubility, at which point the solution is said to be saturated. Excess salt will settle, with all apparent activity ceasing. Actually, however, the dissolving process continues, but so does the reverse process of crystallization—both at the same rate. In other words, equilibrium is reached according to the equation $NaCl \rightleftharpoons Na^+ + Cl^-$.

The same process occurs regardless of the solubility of the salt in question. With a so-called "insoluble" salt, such as AgCl, so little solute will dissolve that it will appear totally insoluble. Such is not the case, however. The ionic concentrations will be very small, but in terms of chemical equilibrium, very significant. The study of low solubility ionic equilibria is the subject of this chapter.

16.1 THE SOLUBILITY PRODUCT CONSTANT

If a low solubility salt, M_aX_b, is placed in water it dissolves until it reaches an equilibrium with a saturated solution of its own ions:

$$M_aX_b(s) = aM^{b+}(aq) + bX^{a-}(aq) \quad (16.1)$$

The equilibrium constant for this expression is

$$K = \frac{[M^{b+}]^a[X^{a-}]^b}{[M_aX_b]}. \quad (16.2)$$

The "concentration" of a solid is effectively constant. Therefore, multiplication of both sides of the above equation by $[M_aX_b]$ yields a constant on the left:

$$K[M_aX_b] = K_{sp} = [M^{b+}]^a[X^{a-}]^b. \quad (16.3)$$

The special constant designated K_{sp} is called the solubility product constant, and is defined by Equation 16.3. Accordingly, for

$$AgCl \rightleftharpoons Ag^+ + Cl^-, \qquad K_{sp} = [Ag^+]\,[Cl^-];$$

and for

$$PbF_2 \rightleftharpoons Pb^{2+} + 2\,F^-, \qquad K_{sp} = [Pb^{2+}]\,[F^-]^2.$$

Recorded values of solubility product constants (See Table VIII, page 293.) are most useful for solids in which both ions are monovalent. For such solutions the "effective concentration" or activity of each ion is approximately equal to the molarity. If polyvalent and/or spectator ions are present in the solution, experimental results may deviate considerably from those predicted by K_{sp}. In this book, however, it will be assumed that equilibrium phenomena depend upon molarities of species only, leaving to more advanced courses the refinement of the principles involved.

As with all equilibrium constants, K_{sp} is temperature dependent. Tabulated values for K_{sp} are generally given at 25°C. Deviations from this temperature will cause discrepancies between theoretical and observed results.

There are two steps involved in solving any solubility product constant problem. They are:

1. Write the equilibrium equation for the reaction.
2. Write the solubility product constant expression.

16.2 DETERMINATION OF K_{sp} FROM SOLUBILITY

Example 16.1. The solubility of silver chloride is 1.25×10^{-5} mole per liter. Find K_{sp}.

First, the equation and K_{sp} expression:

16.1a. $AgCl \rightleftharpoons Ag^+ + Cl^-$; $K_{sp} = [Ag^+]\,[Cl^-]$.
The solubility in moles per liter is the molarity of the solution. The molarities of the ions should be readily apparent. Substitute these into the K_{sp} expression and calculate its value.

16.1b. $K_{sp} = 1.56 \times 10^{-10}$.

From the equation, the moles of AgCl dissolved equal both the moles of Ag^+ and the moles of Cl^-. Therefore $1.25 \times 10^{-5} = [Ag^+] = [Cl^-]$.

$$K_{sp} = (1.25 \times 10^{-5})(1.25 \times 10^{-5}) = 1.56 \times 10^{-10}.$$

From this problem there emerges a direct, two-step approach to finding K_{sp} from solubility:

1. Determine the molarities of the ions in solution.
2. Substitute into the K_{sp} expression and solve.

Example 16.2. Calculate K_{sp} for BaF_2 if its solubility is 1.32 grams per liter.

First, the equation and K_{sp} expression:

16.2a. $BaF_2 \rightleftharpoons Ba^{2+} + 2\,F^-$; $K_{sp} = [Ba^{2+}][F^-]^2$.

Now find the molarity of each ion (Step 1).

16.2b. Solution is 7.54×10^{-3} M in Ba^{2+}, and 15.1×10^{-3} M in F^-.

$$\frac{1.32\ \cancel{g\ BaF_2}}{l} \times \frac{1\ \cancel{mole\ BaF_2}}{175\ \cancel{g\ BaF_2}} \times \frac{1\ mole\ Ba^{2+}}{1\ \cancel{mole\ BaF_2}} = 7.54 \times 10^{-3}\ mole\ Ba^{2+}/l$$

Molarity of F^- = twice the molarity of Ba^{2+} (by equation)

$= 2 \times 7.54 \times 10^{-3} = 15.1 \times 10^{-3}$.

Now substitute these values into the K_{sp} expression and solve (Step 2).

16.2c. $K_{sp} = (7.54 \times 10^{-3})(15.1 \times 10^{-3})^2 = 1.72 \times 10^{-6}$.
Students sometimes rebel at doubling the molarity to get the fluoride ion concentration, and then squaring it. The two steps in the procedure are separate, and in no way dependent upon each other. That the concentration of F^- is twice that of Ba^{2+} is the result of the stoichiometry of the reaction and occurs when the two ions come only from the solute. Later there will be problems in which the ions will come from different sources. You will not double and square then, nor will the ions be present in the stoichiometric ratio. The logic of the two-step approach is evident in the first example, and is applicable in this. Keep the steps separate, and perform them faithfully.

16.3 DETERMINATION OF SOLUBILITY FROM K_{sp}

Inasmuch as K_{sp} values are tabulated in most handbooks, the more relevant problem is to determine the solubility of a substance from that value. This is simply the reverse of what has already been done.

Example 16.3. Find the solubility of AgBr if its $K_{sp} = 5.0 \times 10^{-13}$.

First, the equation and K_{sp} expression.

16.3a. $AgBr \rightleftharpoons Ag^+ + Br^-$; $K_{sp} = [Ag^+][Br^-] = 5.0 \times 10^{-13}$.

This time the goal is to find the molarity of the solution. The equation indicates that the number of moles of AgBr that dissolve is equal to the number of moles of Ag^+ ion. Therefore, if $[Ag^+]$ can be found, it must equal the moles of AgBr dissolved in one liter, or the molarity. Following an algebraic approach, let s = $[Ag^+]$ at equilibrium. Therefore, s = solubility of AgBr in moles per liter. Now, if s = $[Ag^+]$, what is $[Br^-]$?

16.3b. $[Br^-] = s$. According to the equation, one mole of Br^- is formed for every mole of Ag^+. AgBr being the only source of each, they must be present in equal concentrations.

If both concentrations are known, can they not be substituted into the K_{sp} expression, equated to the known value of K_{sp}, and the equation solved? This is the method; go ahead.

16.3c. $s = 7.1 \times 10^{-7}$ = solubility of AgBr in moles/liter.

$$[Ag^+][Br^-] = s^2 = 5.0 \times 10^{-13} = 50 \times 10^{-14}$$
$$s = \sqrt{50 \times 10^{-14}} = 7.1 \times 10^{-7}$$

This time a three-step procedure emerges:

1. Assign a variable to represent one of the ionic species in the equilibrium.
2. Determine the concentration of the other ionic species in terms of the same variable.
3. Substitute the concentrations from 1 and 2 into the K_{sp} expression, equate to the K_{sp} value, and solve.

Example 16.4. Find the solubility of PbF_2 in grams/100 ml if its K_{sp} is 2.7×10^{-8}.

As usual, first give the equation and the K_{sp} expression:

16.4a. $PbF_2 \rightleftharpoons Pb^{2+} + 2\,F^-$; $K_{sp} = [Pb^{2+}][F^-]^2$.

Now let the variable *s* be the concentration of one of the ionic species (Step 1). You have a choice. Use it so *s* will also be equal to the solubility (molarity) of lead fluoride. What is your choice?

16.4b. Let $s = [Pb^{2+}]$.

Because of the 1:1 relationship between Pb^{2+} and PbF_2, s will also be equal to the solubility of PbF_2. This would not be true if s were made equal to $[F^-]$.

Now if $s = [Pb^{2+}]$, what is $[F^-]$ (Step 2)?

16.4c. $[F^-] = 2s$. For each mole of Pb^{2+} formed there will be 2 moles of F^-, according to the equation. Therefore $[F^-]$ is twice $[Pb^{2+}]$, or 2s.

Now substitute into the K_{sp} equation and solve for s (Step 3). Careful!

16.4d. $s = 1.89 \times 10^{-3}$ mole per liter.

$$[Pb^{2+}][F^-]^2 = (s)(2s)^2 = 4s^3 = 2.7 \times 10^{-8} = 27 \times 10^{-9}$$
$$s^3 = 6.75 \times 10^{-9}$$
$$s = 1.89 \times 10^{-3}$$

A common error here is to fail to square the 2 in 2s: $(2s)^2 = 4s^2$, not $2s^2$. Note this problem involves doubling and squaring, but again they are two separate and independent steps in the solution of the problem.

You have the solubility in moles of PbF_2 per liter. The problem asked for grams per 100 ml. Complete the problem.

16.4e. Solubility = 4.6×10^{-2} g/100 ml.

$$\frac{1.89 \times 10^{-3} \text{ mole}}{\text{l}} \times \frac{245 \text{ g}}{1 \text{ mole}} \times \frac{0.100 \text{ l}}{100 \text{ ml}} = 4.63 \times 10^{-2} \text{ g}/100 \text{ ml}$$

The reported answer is rounded off to two significant figures, based on the K_{sp}.

16.4 THE SIGNIFICANCE OF K_{sp}

K_{sp} is an equilibrium constant. It says that *when* an equilibrium exists, the product of the ionic concentrations as they appear in the K_{sp} expression has a certain value. There is no requirement that the ionic concentrations be equal, nor must they be in the same stoichiometric ratio as the species in the equilibrium equation. Indeed, rarely will they be in that ratio, unless both ions come only from the solute itself. Such has been the case in the problems so far considered. If they come from different sources their value will be determined by the sources alone. Their product, determined according to the K_{sp} expression, may not equal the K_{sp} value. This product, incidentally, is called the **ion product:** we shall have frequent occasion to use this term, and we shall abbreviate it **I.P.**

Suppose AgCl is placed in water and it dissolves until equilibrium is reached, $AgCl \rightleftharpoons Ag^+ + Cl^-$. At that time I.P. = $[Ag^+][Cl^-] = K_{sp}$. Now suppose chloride ion is added without increasing volume, thereby increasing $[Cl^-]$. What happens? Le Chatelier's Principle predicts a shift to the left, precipitating some AgCl. Until when? And what would I.P. be during and at the end of the process?

When $[Cl^-]$ increases through the addition of Cl^-, I.P. temporarily exceeds K_{sp}. Precipitation occurs. Both $[Ag^+]$ and $[Cl^-]$ decrease until the I.P. again equals K_{sp}. At this point a new equilibrium will have been established, one in which $[Cl^-]$ is a little more than it was in the original equilibrium, and $[Ag^+]$ a little less. But their product—the I.P.—will equal the K_{sp}, as must be the case at equilibrium.

From this it is clear that K_{sp} is the *maximum* I.P. a solution will tolerate without precipitation. If I.P. is greater than K_{sp}, precipitation should occur, and continue until the I.P. is reduced to the value of K_{sp} once again.

What if I.P. is less than K_{sp}? In this case, we have an unsaturated solution. If there is solute still in contact with the solution, the rate of dissolving will exceed the rate of crystallization, so the concentration of the resulting ions will increase with time, thereby increasing I.P. If solute still remains when I.P. reaches K_{sp}, saturation will have been reached and an equilibrium established. If the solute runs out before I.P. equals K_{sp}, the solution will simply remain unsaturated.

In summary, the I.P. of a solution may be less than or equal to K_{sp} without precipitating; but if I.P. exceeds K_{sp}, precipitation will occur and continue until they are again equal.

16.5 PRECIPITATION PREDICTION

If two potentially precipitating ions are combined in a solution in such concentrations that their ion product is greater than their K_{sp}, precipitation will occur. If their I.P. is equal to or less than K_{sp} there will be no precipitation. Using this criterion one can predict whether or not a precipitate will form when two solutions are mixed.

Example 16.5. Will precipitation occur if a solution 0.001 M in $BaCl_2$ is made 1×10^{-4} M in SO_4^{2-}? $K_{sp} = 1.1 \times 10^{-10}$ for $BaSO_4$.

To begin, write the equation and K_{sp} expression for the possible precipitation of barium sulfate.

16.5a. $BaSO_4 \rightleftharpoons Ba^{2+} + SO_4^{2-}$

$$K_{sp} = [Ba^{2+}]\,[SO_4^{2-}] = 1.1 \times 10^{-10}.$$

The molarity of the SO_4^{2-} ion is given. What is the molarity of Ba^{2+}?

16.5b. $[Ba^{2+}] = 0.001$. 1 mole $BaCl_2$ yields one mole of barium ion in dissolution. Now calculate the I.P. of the barium and sulfate ions.

16.5c. I.P. $= [Ba^{2+}]\,[SO_4^{2-}] = (1 \times 10^{-3})(1 \times 10^{-4}) = 1 \times 10^{-7}$.
Now determine whether or not precipitation will occur by comparing I.P. with K_{sp}. Do this by writing an inequality between the two values, and then stating your conclusion.

16.5d. $1 \times 10^{-7} > 1.1 \times 10^{-10}$
I.P. $> K_{sp}$; therefore, precipitation will occur.

Example 16.6. Equal volumes of 2.0×10^{-4} M Tl_2SO_4 and 6.0×10^{-2} M NaCl are combined. Will TlCl precipitate? $K_{sp} = 1.9 \times 10^{-4}$ for TlCl.

First, the equation and K_{sp} expression *for the potential precipitate:*

16.6a. $TlCl \rightleftharpoons Tl^{+} + Cl^{-}$; $K_{sp} = [Tl^{+}]\,[Cl^{-}] = 1.9 \times 10^{-4}$.

Find the molarity of the chloride ion in the *total solution.* Note that the volume of the combined solutions must be considered. Each solution effectively dilutes the other. Since no volumes are specified, let's assume that 1.0 liter of each solution is used. On that basis, what is the number of moles of Cl^{-} available in 1.0 liter of 6.0×10^{-2} M NaCl?

16.6b. 6.0×10^{-2} mole Cl^{-}.

$$1.0\,\not{l} \times \frac{6.0 \times 10^{-2}\text{ mole Cl}^{-}}{1\,\not{l}} = 6.0 \times 10^{-2}\text{ mole Cl}^{-}$$

This is the number of moles of chloride ions. What is the volume of the combined solutions?

16.6c. 2.0 liters.
If two equal volumes of 1.0 liter each are combined, the total volume will be 2.0 liters.

You now have the number of moles of chloride ions and the volume of solution in which they are present. What is the concentration of Cl^-?

16.6d. $[Cl^-] = 3.0 \times 10^{-2}$ mole/liter.

$$\frac{6.0 \times 10^{-2} \text{ mole } Cl^-}{2.0 l} = 3.0 \times 10^{-2} \text{ mole } Cl^-/l$$

What we have done here is to calculate the concentration of Cl^- by first obtaining the number of moles and then dividing by the total volume in liters. There is, however, a method by which the diluted concentration of a species may be quickly and readily calculated. For convenience's sake, it is offered here.

Let's set up steps 16.6b and 16.6d together:

$$1.0 \cancel{l \text{ NaCl solution}} \times \frac{6.0 \times 10^{-2} \text{ mole } Cl^-}{1 \cancel{l \text{ NaCl solution}}} \times \frac{1}{2.0 l \text{ dilute solution}}$$

Rearranging,

$$\frac{6.0 \times 10^{-2} \text{ mole } Cl^-}{l} \times \frac{1.0 l}{2.0 l} = 3.0 \times 10^{-2} \text{ mole } Cl^-/l$$

The last equation shows that the initial concentration of the chloride ion was multiplied by a volume ratio, initial volume/diluted volume. Whenever a solution of a given solute is diluted by water or another solution that does not contain the same solute, nor react with it in the first solution, the diluted concentration of that solute may always be obtained by multiplying its initial concentration by the ratio, initial volume/diluted volume. This ratio is called a **dilution factor.**

Notice that the dilution factor for this problem would be one-half regardless of the volume we chose to consider. Final volume is twice initial volume when equal volumes are combined.

Now find the molarity of the Tl^+ ion in the diluted solution.

16.6e. $[Tl^+] = 2.0 \times 10^{-4}$ mole/liter.

Could it be that you came up with 1.0×10^{-4}? If so, you overlooked the fact that 1 mole Tl_2SO_4 yields 2 moles Tl^+. Therefore,

$$\frac{2.0 \times 10^{-4} \cancel{\text{mole } Tl_2SO_4}}{\cancel{l}} \times \frac{2 \text{ moles } Tl^+}{1 \cancel{\text{mole } Tl_2SO_4}} \times \frac{1.0 \cancel{l}}{2.0 l} = 2.0 \times 10^{-4} \text{ mole } Tl^+/l$$

Now that the concentrations of both Tl^+ and Cl^- are known, their ion product may be calculated. Do this, compare with K_{sp}, and then decide whether or not TlCl will precipitate.

16.6f. No precipitate.

$$\text{I.P.} = [Tl^+][Cl^-] = (2.0 \times 10^{-4})(3.0 \times 10^{-2}) = 6.0 \times 10^{-6}$$

$6.0 \times 10^{-6} < 1.9 \times 10^{-4}$ (I.P. $< K_{sp}$), so no precipitate forms.

It is recommended that the I.P.– K_{sp} inequality be stated at the conclusion of each problem in which precipitation is to be predicted, followed by a statement as to whether or not precipitation will occur.

Looking back, Examples 16.5 and 16.6 fell into a three-step pattern:

1. Calculate the molarities of the potentially precipitating ions in the final solution.
2. Determine the ion product.
3. Compare the ion product with K_{sp}. If I.P. $>$ K_{sp}, a precipitate will form; if I.P. $<$ K_{sp}, no precipitate will form.

Example 16.7. 25 ml 0.10 M LiBr and 75 ml 1.0 M Na_2CO_3 are combined. Will Li_2CO_3 precipitate? $K_{sp} = 1.7 \times 10^{-5}$ for Li_2CO_3.

As usual, begin with the equilibrium equation and the K_{sp} expression.

16.7a. $Li_2CO_3 \rightleftharpoons 2\ Li^+ + CO_3^{2-}$;

$$K_{sp} = [Li^+]^2[CO_3^{2-}] = 1.7 \times 10^{-5}.$$

Now calculate the diluted molarities of the lithium and carbonate ions (Step 1).

16.7b.

$$\frac{0.10 \text{ mole } Li^+}{l} \times \frac{25 \text{ ml}}{100 \text{ ml}} = 0.025 \text{ mole } Li^+/l$$

$$\frac{1.0 \text{ mole } CO_3^{2-}}{l} \times \frac{75 \text{ ml}}{100 \text{ ml}} = 0.75 \text{ mole } CO_3^{2-}/l$$

With the lithium ion, 25 ml initial solution become 100 ml diluted solution, so the dilution factor is 25/100; with CO_3^{2-}, 75 ml initial volume becomes 100 ml, so the dilution factor is 75/100.

Now calculate I.P., compare with K_{sp}, and decide whether or not precipitation will occur (Steps 2 and 3).

16.7c. Precipitate forms.

$$\text{I.P.} = [Li^+]^2\ [CO_3^{2-}] = (0.025)^2(0.75) = 4.7 \times 10^{-4}$$

$4.7 \times 10^{-4} > 1.7 \times 10^{-5}$. Therefore precipitation should occur.

16.6 COMMON ION EFFECTS ON SOLUBILITY

When silver bromide is placed in water the following process occurs: $AgBr \rightleftharpoons Ag^+ + Br^-$. The number of moles of AgBr that dissolve equals the number of moles of Ag^+ ion produced, and the number of moles of Br^- ion produced. As long as all the Ag^+ or Br^- ion comes exclusively from AgBr, the concentration of either ion in moles per liter is equal to the solubility of AgBr in moles per liter. Ion concentration, then, expresses solubility. But note the restriction: *all* of the ions must come from the salt.

If bromide ion is introduced to the equilibrium in the form of a soluble bromide salt, Le Chatelier's Principle predicts a shift to the left, with a corresponding reduction in $[Ag^+]$. In other words, the solubility of silver bromide in a solution containing bromide from another source is less than its solubility in water. This is called the **common ion effect:** the solubility of a low solubility substance is reduced when a common ion—one already present in the solute—is introduced from another source. Let's explore this relationship quantitatively.

Example 16.8. Calculate the solubility of silver bromide in 0.0030 M NaBr. K_{sp} for AgBr is 5.0×10^{-13}.

And what is the usual starting point?

16.8a. $AgBr \rightleftharpoons Ag^+ + Br^-$; $K_{sp} = [Ag^+][Br^-] = 5.0 \times 10^{-13}$.

A bit of reasoning is in order now. Initially $[Br^-] = 0.0030$. To this will be added Br^- from the AgBr, which is only slightly soluble. Could it be that the Br^- from AgBr will be negligible compared to the 0.0030 mole/liter already present? This seems reasonable. Perhaps an *I–R–E* table will show this more clearly:

AgBr	⇌	Ag^+	+	Br^-
I		0		0.0030
R		+y		+y
E		y		0.0030 + y
E_a		y		0.0030

Let's solve the problem on the basis of the assumption, and then confirm its validity. Substitute into the K_{sp} expression, equate to K_{sp}, and solve for the silver ion concentration, which is also the solubility of AgBr in 0.0030 M NaBr.

16.8b. Solubility $= 1.7 \times 10^{-10}$ mole/liter

$$[Ag^+] = \frac{K_{sp}}{[Br^-]} = \frac{5.0 \times 10^{-13}}{0.0030} = 1.7 \times 10^{-10}$$

$$0.0030 + 1.7 \times 10^{-10} = 0.0030 \qquad \text{Assumption valid.}$$

In Example 16.3 it was found that the solubility of AgBr in water is about 10^{-6} mole per liter. In Example 16.8 it is about 10^{-10} mole per liter. It is evident that the solubility of a low solubility solute is repressed even more by the presence of a common ion, as predicted by Le Chatelier's Principle.

Example 16.9. How many grams of PbI_2 will dissolve in 250 ml 0.025 M $Pb(NO_3)_2$? $K_{sp} = 7.1 \times 10^{-9}$ for PbI_2.

As always, the equation and K_{sp} expression:

16.9a. $PbI_2 \rightleftharpoons Pb^{2+} + 2I^-$; $K_{sp} = [Pb^{2+}][I^-]^2 = 7.1 \times 10^{-9}$.
Can you now estimate the value of $[Pb^{2+}]$ in the saturated solution?

16.9b. $[Pb^{2+}] = 0.025$. This assumes the lead ion coming from the solution of PbI_2 is negligible compared to that from the lead nitrate, an assumption that will have to be confirmed.
Now if $[Pb^{2+}] = 0.025$, you should be able to find $[I^-]$ from the K_{sp} expression.

16.9c. $[I^-] = 5.33 \times 10^{-4}$. (An extra significant figure will be carried through the calculations, and the final result will be rounded off to the two significant figures justified by the K_{sp} value.)

$$[I^-]^2 = \frac{K_{sp}}{[Pb^{2+}]} = \frac{7.1 \times 10^{-9}}{2.5 \times 10^{-2}} = 2.84 \times 10^{-7} = 28.4 \times 10^{-8}$$

$$[I^-] = \sqrt{28.4 \times 10^{-8}} = 5.33 \times 10^{-4}$$

Assumption check: If $[I^-]$ from PbI_2 is 5.33×10^{-4}, what is $[Pb^{2+}]$ from PbI_2? Hint: the equation. . .

16.9d. $[Pb^{2+}] = 2.66 \times 10^{-4}$ from PbI_2. The equation shows that the number of moles of Pb^{2+} from PbI_2 is one-half the moles of I^-, or 2.66×10^{-4} mole/liter.
Is the assumption valid?

16.9e. Yes. $0.025 + 0.000266 = 0.025$.

The solubility in moles/liter of PbI_2 in 0.025 M $Pb(NO_3)_2$ should be readily apparent.

16.9f. Solubility = 2.66×10^{-4} mole per liter.
As with Pb^{2+}, the equation shows that the moles of PbI_2 that dissolve in one liter is one-half the moles of I^- produced.
The problem asks the number of grams of PbI_2 that will dissolve in 250 ml of 0.025 M $Pb(NO_3)_2$. Complete the calculations.

16.9g. 0.0307 gram, or, to two significant figures, 0.031 gram of PbI_2 will dissolve.

$$0.250\,\not{l} \times \frac{2.66 \times 10^{-4}\,\text{\sout{mole}}\,PbI_2}{1\,\not{l}} \times \frac{461\text{ g}}{1\,\text{\sout{mole}}} = 0.031\text{ g }PbI_2$$

Example 16.10. A solution is 1×10^{-3} M in both Cl^- and Br^-. Ag^+ ion is added slowly, and with negligible dilution of the original halide ion concentrations. Which halide ion will begin to precipitate first, and at what $[Ag^+]$ will this occur? For AgBr, $K_{sp} = 5.3 \times 10^{-13}$; for AgCl, $K_{sp} = 1.6 \times 10^{-10}$.

The first question can be answered by reason alone, without recourse to calculation. Either precipitate will begin when the I.P. for that compound exceeds K_{sp}. The factors that are to be multiplied, i.e., $[Br^-]$ for one precipitate and $[Cl^-]$ for the other, and $[Ag^+]$ for both, are identical in magnitude. In other words, the I.P. values are the same for both compounds prior to precipitation. As the I.P. increases for both salts, which K_{sp} will it exceed first, the larger K_{sp} or the smaller?

16.10a. The smaller.

If a six-foot-tall father and a five-foot, six-inch-tall mother (two constant K_{sp} values) have a growing son (an increasing I.P. value), the son will be taller than his mother before he is taller than his father!

Now, which silver halide will precipitate first, the chloride, or the bromide?

16.10b. The bromide.

10^{-13} is smaller than 10^{-10}. Therefore, the increasing I.P. will exceed 10^{-13} first, thereby producing a precipitate.

Now, based on the known K_{sp} for AgBr and the known $[Br^-]$, what $[Ag^+]$ will be required to initiate precipitation?

16.10c. $[Ag^+] > 5.3 \times 10^{-10}$ when AgBr begins to precipitate. Precipitation presumably begins when I.P. just exceeds K_{sp}. When I.P. $= K_{sp}$,

$$1 \times 10^{-3}[Ag^+] = 5.3 \times 10^{-13}$$
$$[Ag^+] = 5.3 \times 10^{-10}$$

Example 16.11. Continuing to introduce Ag^+ to the solution of Example 16.10, (a) at what $[Ag^+]$ will AgCl begin to precipitate? (b) what will $[Br^-]$ be when AgCl begins to precipitate? (c) what percentage of the original bromide ion will still be in solution when AgCl begins to precipitate? As in Example 16.10, disregard any dilution effect.

When the silver chloride begins to precipitate, simultaneous equilibria will prevail. Solid silver bromide will be present at equilibrium with silver and bromide ions in the solution. For silver bromide, K_{sp} will equal I.P. When silver chloride first begins to precipitate, solid silver chloride will be in equilibrium with silver and chloride ions. K_{sp} for silver chloride will equal I.P. for silver chloride. The silver ion concentration can have but one value: $[Ag^+]$ must be the same in *both* equilibria.

Part (a) is handled in exactly the same way as the last part of Example 16.10. Answer, please . . .

16.11a. $[Ag^+] = 1.6 \times 10^{-7}$.

When I.P. $= K_{sp}$,

$$[Ag^+] = \frac{K_{sp}}{[Cl^-]} = \frac{1.6 \times 10^{-10}}{1 \times 10^{-3}} = 1.6 \times 10^{-7}.$$

Comparing answers 16.10c and 16.11a, $[Ag^+]$ must be roughly 1000 times as concentrated before it will begin to precipitate chloride ion.

At the time AgCl begins to precipitate, solid AgBr will be in equilibrium with a solution containing both Ag^+ and Br^- ions. Under these circumstances I.P. = K_{sp}. Both K_{sp} and $[Ag^+]$ are known; solve for $[Br^-]$.

16.11b. $[Br^-] = 3.3 \times 10^{-6}$.

$$[Br^-] = \frac{K_{sp}}{[Ag^+]} = \frac{5.3 \times 10^{-13}}{1.6 \times 10^{-7}} = 3.3 \times 10^{-6}$$

Now let us calculate the per cent of Br^- remaining in solution when AgCl starts to precipitate. The number just calculated, 3.3×10^{-6}, is the concentration of Br^- remaining in a liter of solution when AgCl begins to precipitate. Initially the concentration of Br^- was 1×10^{-3} M. Take it from there. . . .

16.11c. 0.3% of the original bromide ion remains in solution when AgCl beings to precipitate.

$$\frac{3.3 \times 10^{-6}}{1 \times 10^{-3}} \times 100 = 0.3\%$$

16.7 COMPLEX IONS AND SOLUBILITY

Consider the equilibrium $AgCl \rightleftharpoons Ag^+ + Cl^-$. What would happen if ammonia were added? Ammonia complexes the silver ion: $Ag^+ + 2\ NH_3 \rightleftharpoons Ag(NH_3)_2^+$. Le Chatelier's Principle therefore predicts a shift to the right in the AgCl equilibrium, dissolving some of the AgCl. In other words, the solubility of AgCl in NH_3 solution is greater than its solubility in water.

The equation for the reaction between NH_3 and AgCl can be regarded as a two-step process,

$$\begin{array}{rcll} AgCl & \rightleftharpoons & Ag^+ + Cl^- & \quad K = K_{sp} \\ 2\ NH_3 + Ag^+ & \rightleftharpoons & Ag(NH_3)_2^+ & \quad K = 1/K_d \\ \hline AgCl + 2\ NH_3 & = & Ag(NH_3)_2^+ + Cl^- & \end{array}$$

As pointed out at the close of the last chapter, the overall equilibrium constant for sequential equilibria is the product of the stepwise equilibrium constants, a relationship known as the **multiple equilibrium rule.** Therefore, K for the final equation may be determined by multiplying K_{sp} by $1/K_d$:

$$\begin{array}{ccccc} \cancel{[Ag^+]}\,[Cl^-] & \times & \dfrac{[Ag(NH_3)_2^+]}{[NH_3]^2\cancel{[Ag^+]}} & = & \dfrac{[Ag(NH_3)_2^+]\,[Cl^-]}{[NH_3]^2} \\ (K_{sp}) & \times & (1/K_d) & = & K \end{array}$$

Rearranging,

$$K = \frac{K_{sp} \text{ of precipitate}}{K_d \text{ of complex}}. \qquad (16.4)$$

As usual, the solid is not included in the equilibrium constant expression.

Example 16.12. What is the solubility of AgCl in (a) water, and (b) 1.0 M NH_3? $K_{sp} = 1.6 \times 10^{-10}$ for $AgCl \rightleftharpoons Ag^+ + Cl^-$; $K_d = 4 \times 10^{-8}$ for $Ag(NH_3)_2^+ \rightleftharpoons Ag^+ + 2\ NH_3$.

Part (a) is the same as Example 16.3. Complete this step.

16.12a. Solubility = 1.3×10^{-5} mole per liter.

$$K_{sp} = [Ag^+]\,[Cl^-] = s^2 = 1.6 \times 10^{-10}$$
$$s = \sqrt{1.6 \times 10^{-10}} = 1.27 \times 10^{-5}$$

This is quite a low solubility—about 2 milligrams per liter.

For part (b), write the K expression for the equilibrium $AgCl + 2NH_3 \rightleftharpoons Ag(NH_3)_2^+ + Cl^-$ and calculate its value by Equation 16.4.

16.12b. $K = 4.0 \times 10^{-3}$

$$K = \frac{[Ag(NH_3)_2^+]\,[Cl^-]}{[NH_3]^2} = \frac{K_{sp}}{K_d} = \frac{1.6 \times 10^{-10}}{4 \times 10^{-8}} = 4.0 \times 10^{-3}$$

Now develop an I–R–E–E_a table for the problem.

16.12c.

	AgCl(s) +	2 NH_3 ⇌	$Ag(NH_3)_2^+$ +	Cl^-
I		1.0	0	0
R		−2y	+y	+y
E		1.0−2y	y	y
E_a		1.0	y	y

Substitute values from E_a in the table into the K expression, equate to the known value of K and solve for y. Check the assumption.

16.12d. The assumption is not valid.

$$K = \frac{[Ag(NH_3)_2^+]\,[Cl^-]}{[NH_3]^2} = \frac{y^2}{1.0} = 4.0 \times 10^{-3} \qquad y = 0.063$$

Assumption check: $1.0-2y = 1.0-0.126 \neq 1.0$. Not valid.

Return to the table and substitute values from the E line into the K expression and solve again. (You could work with a second assumption, but the algebraic solution is direct and does not require the quadratic formula.)

16.12e. $y = [Cl^-] = 5.9 \times 10^{-2}$.

$$K = \frac{[Ag(NH_3)_2^+]\,[Cl^-]}{[NH_3]^2} = \frac{y^2}{(1.0-2y)^2} = 40 \times 10^{-4}$$

Taking the square root of both sides,

$$\frac{y}{1.0 - 2y} = 6.3 \times 10^{-2}; \qquad y = 5.6 \times 10^{-2} \text{ mole/liter (about 8.4 grams/liter)}$$

Comparing this with the answer to part (a) it is evident that the solubility of AgCl has been increased more than a thousandfold by adding NH_3. Ammonia is used, therefore, to dissolve AgCl precipitates in certain laboratory procedures.

The greater solubility of a substance in the presence of a complexing ligand is useful in preventing unwanted precipitations. For example, based on a K_{sp} of 1.4×10^{-11}, the solubility of zinc carbonate is 3.6×10^{-6} mole/liter. Suppose you have a solution 0.0030 M in Zn^{2+} in which you wish to dissolve sodium carbonate until the carbonate ion concentration is 2.0×10^{-4} M without precipitating $ZnCO_3$. The ion product of such a solution would be $(3.0 \times 10^{-3})(2.0 \times 10^{-4}) = 6.0 \times 10^{-7}$, which is several orders of magnitude larger than the K_{sp} that would cause precipitation. By adding cyanide ion, CN^-, the zinc ion could be tied up as $Zn(CN)_4^{2-}$, reducing $[Zn^{2+}]$ to the point the ion product would be smaller than K_{sp}, thereby preventing precipitation. Let's look at the problem in detail. . . .

Example 16.13. How many milliliters of 0.20 M KCN are required to prevent precipitation of $ZnCO_3$ in one liter of 3.0×10^{-3} M Zn^{2+} as soluble carbonate is added until $[CO_3^{2-}] = 2.0 \times 10^{-4}$?
$K_{sp} = 1.4 \times 10^{-11}$ for $ZnCO_3$; $K_d = 1 \times 10^{-17}$ for $Zn(CN)_4^{2-}$.

Let's begin this problem by establishing a plan of attack so you can see where you're going:

1. First determine from K_{sp} the $[Zn^{2+}]$ that can be tolerated without precipitation when $[CO_3^{2-}] = 2.0 \times 10^{-4}$.
2. Subtracting the $[Zn^{2+}]$ that can be tolerated from that initially present tells how much must be "removed" by complexing as a $Zn(CN)_4^{2-}$ complex—in other words, $[Zn(CN)_4^{2-}]$ at equilibrium.
3. From K_d, $[Zn^{2+}]$ and $[Zn(CN)_4^{2-}]$, calculate $[CN^-]$ at equilibrium. This is part of the CN^- that must be added.
4. Calculate the amount of cyanide ion tied up in the $Zn(CN)_4^{2-}$ complex. This is the rest of the CN^- that must be added.
5. Determine the total cyanide to be added as the sum of steps 3 and 4.

Now that you see the pattern, let's begin. What maximum concentration of Zn^{2+} ion can exist in a solution in which $[CO_3^{2-}] = 2.0 \times 10^{-4}$ without $ZnCO_3$ precipitating?

16.13a. $[Zn^{2+}] = 7.0 \times 10^{-8}$.

$$[Zn^{2+}] = \frac{K_{sp}}{[CO_3^{2-}]} = \frac{1.4 \times 10^{-11}}{2.0 \times 10^{-4}} = 7.0 \times 10^{-8}$$

According to this calculation and the statement of the problem, the initial $[Zn^{2+}]$ of 3.0×10^{-3} must be reduced to 7.0×10^{-8} by the reaction $Zn^{2+} + 4\,CN^- \rightarrow Zn(CN)_4^{2-}$. How many moles of $Zn(CN)_4^{2-}$ will be formed per liter of solution? What will $[Zn(CN)_4^{2-}]$ become when this reduction of $[Zn^{2+}]$ is completed?

16.13b. $[Zn(CN)_4^{2-}] = 3.0 \times 10^{-3}$.

$[Zn^{2+}] = 0.0030$ initially, and 0.00000007 finally. In other words, virtually *all* of the 0.0030 mole of Zn^{2+} present is complexed. According to the 1:1 ratio between Zn^{2+} and the complex in the equation, an equal number of $Zn(CN)_4^{2-}$ ions will form; the final $[Zn(CN)_4^{2-}]$ = initial $[Zn^{2+}] = 3.0 \times 10^{-3}$.

Now that you know $[Zn^{2+}]$ and $[Zn(CN)_4^{2-}]$ at equilibrium, substitution into the K_d expression leads to $[CN^-]$ at equilibrium. . . .

16.13c. $[CN^-] = 8.1 \times 10^{-4}$.

$$K_d = \frac{[Zn^{2+}][CN^-]^4}{[Zn(CN)_4^{2-}]} = 1 \times 10^{-17} = \frac{(7.0 \times 10^{-8})[CN^-]^4}{3.0 \times 10^{-3}}$$

$$[CN^-]^4 = 4.29 \times 10^{-13}$$

$$\log [CN^-]^4 = 4 \log [CN^-] = -13 + \log 4.29 = -12.368$$

$$\log [CN^-] = -3.092$$

$$[CN^-] = 8.1 \times 10^{-4}$$

This accounts for the CN^- *still in solution.* But some of the added CN^- is tied up in the complex. If there are 3.0×10^{-3} mole of $Zn(CN)_4^{2-}$ in one liter, how many moles of CN^- are in the complex?

16.13d. 12×10^{-3} mole CN^-.

$$3.0 \times 10^{-3} \cancel{\text{mole } Zn(CN)_4^{2-}} \times \frac{4 \text{ moles } CN^-}{1 \cancel{\text{mole } Zn(CN)_4^{2-}}} = 12 \times 10^{-3} \text{ moles } CN^-$$

To the liter of solution you have added sufficient cyanide ion to produce 12×10^{-3} mole in the complex (step 16.13d) and 8.1×10^{-4} mole still in solution (step 16.13c). What is the total number of moles of CN^-, and how many milliliters of 0.20 M KCN were required to obtain them?

16.13e. 0.013 = total moles CN^-; 65 ml 0.20 M KCN.

$$\text{Total moles} = 12 \times 10^{-3} + 0.81 \times 10^{-3} = 12.8 \times 10^{-3} = 0.013 \text{ mole } CN^-$$

$$0.013 \cancel{\text{mole } CN^-} \times \frac{1 \cancel{\text{mole KCN}}}{1 \cancel{\text{mole } CN^-}} \times \frac{1000 \text{ ml}}{0.20 \cancel{\text{mole KCN}}} = 65 \text{ ml solution.}$$

This problem might be approached—and indeed solved—by an *I–R–E* table based on the equilibrium

$$ZnCO_3(s) + 4\ CN^-(aq) = Zn(CN)_4^{2-}(aq) + CO_3^{2-}(aq)$$

There is no solid zinc carbonate present, of course—but in the saturated solution $[Zn^{2+}]$ and $[CO_3^{2-}]$ are such that their I.P. just equals K_{sp}. The K for the reaction is, according to Equation 16.4,

$$K = \frac{K_{sp} \text{ of } ZnCO_3}{K_d \text{ of } Zn(CN)_4^{2-}} = \frac{1.4 \times 10^{-11}}{1 \times 10^{-17}} = 1.4 \times 10^{+6}$$

The I–R–E tabulation is

	$ZnCO_3(s)$	+	$4\ CN^-(aq)$	=	$Zn(CN)_4^{2-}(aq)$	+	$CO_3^{2-}(aq)$
I			$12 \times 10^{-3}+y$		0		0
R			12×10^{-3}		3.0×10^{-3}		–
E			y		3.0×10^{-3}		2.0×10^{-4}

This tabulation is developed from the initial and final values of $[Zn(CN)_4^{2-}]$. The initial value is zero. We conclude the final value is equal to the given initial value of $[Zn^{2+}]$ if virtually all of the Zn^{2+} initially present is successfully complexed by CN^-—which was the reason for adding the CN^-. (This same conclusion was reached quantitatively in 16.13b above.) The I and E values of $[Zn(CN)_4^{2-}]$ fix the R value at 3.0×10^{-3} for $Zn(CN)_4^{2-}$, and $4 \times 3.0 \times 10^{-3}$ or 12×10^{-3} for CN^-. This latter figure is the moles of CN^- in the complex, 16.13d above. We are not concerned with the R value of CO_3^{2-}, as its equilibrium concentration is given as 2.0×10^{-4} in the statement of the problem.

Using y for the symbol of $[CN^-]$ and substituting into the K expression for the equilibrium yields

$$K = \frac{[Zn(CN)_4^{2-}]\,[CO_3^{2-}]}{[CN^-]^4} = 1.4 \times 10^{+6} = \frac{(3.0 \times 10^{-3})\,(2.0 \times 10^{-4})}{y^4}$$

Solving for y yields $[CN^-] = 8.1 \times 10^{-4}$, the same as in 16.13c above. Total moles of CN^- are $12 \times 10^{-3} + 0.81 \times 10^{-3}$, or 0.013, as in 16.13e above.

You might well inquire about the dilution effect of adding 65 milliliters to one liter of solution, a 6.5% increase in volume. Sometimes additions in this type of problem are mathematically negligible, sometimes not. From a practical standpoint they are usually of little importance. In most problems we work with order-of-magnitude changes and concentration *ratios* in which dilution effects partially cancel. Also it is customary to introduce a substantial safety factor by administering an overdose of a calculated addition, unless concentrations are so critical that other harmful reactions result. Finally, the aforementioned variations in equilibrium constant values (page 224) from different sources cloud all results with some uncertainty.

The handling of significant figures also warrants comment. Because K_d has only one significant figure, the final answer should presumably be rounded off to 70 ml in Example 16.13e. The common practice of carrying more significant figures than are justified has been used in this example. To do otherwise in a problem involving logarithms and fourth roots can easily lead to 10% variations in answers, depending on how intermediate results are rounded off. You should be aware of this when comparing your answers with those of solved problems in the back of the book. If your answer is fairly close to, but not the same as, the answer given, compare methods. You may be able to account for differences in some rounding off operation.

PROBLEMS

Section 16.2

16.1. The solubility of cadmium sulfide, CdS, is 8.8×10^{-14} mole per liter. Calculate K_{sp}.

16.2. 1.0×10^{-3} gram of CuBr dissolve in 100 milliliters of water. Calculate its K_{sp}.

16.16. $Co(OH)_2$ dissolves in water to the extent of 3.7×10^{-6} mole per liter. Find its K_{sp}.

16.17. 250 milliliters of water will dissolve 8.7 milligrams of silver carbonate, Ag_2CO_3. What is the K_{sp} of silver carbonate?

Section 16.3

16.3. From the K_{sp} values given, calculate the solubilities of the following compounds in moles per liter: (a) $CaCO_3$, $K_{sp} = 8.7 \times 10^{-9}$; (b) NiS, $K_{sp} = 3 \times 10^{-19}$; (c) $Cu(IO_3)_2$, $K_{sp} = 7.4 \times 10^{-8}$.

16.4. How many grams of $PbSO_4$ will dissolve in 400 ml water if $K_{sp} = 1.6 \times 10^{-8}$?

16.5. So-called "carbonate hardness" in water is the result of dissolved calcium and magnesium bicarbonates. It may be removed by treatment with slaked lime, $Ca(OH)_2$:

$$Ca(HCO_3)_2(aq) + Ca(OH)_2(aq) = 2\ CaCO_3(s) + 2\ H_2O(l);$$
$$Mg(HCO_3)_2(aq) + 2\ Ca(OH)_2(aq) = Mg(OH)_2(s) + 2\ CaCO_3(s) + 2\ H_2O(l)$$

Calculate the maximum $[Ca^{2+}]$ that will remain in solution after such treatment, assuming equilibrium is reached between solid calcium carbonate and dissolved calcium ion. Also express the residual hardness in the conventional terms of "ppm," meaning the equivalent grams $CaCO_3$ per million grams of water.

16.18. Find the moles/liter solubilities of each of the following compounds, based on the K_{sp} values listed: (a) $CaSO_4$, $K_{sp} = 1.9 \times 10^{-4}$; (b) $Co(OH)_2$, $K_{sp} = 2 \times 10^{-16}$; (c) Ag_3PO_4, $K_{sp} = 1.3 \times 10^{-20}$.

16.19. $K_{sp} = 2.2 \times 10^{-12}$ for Ag_2CrO_4. How many grams of silver chromate will dissolve in 750 ml water?

16.20. Non-carbonate hardness in water involves Ca^{2+} and Mg^{2+} ions with anions other than HCO_3^-.The softening process this time is called the lime-soda process, using $Ca(OH)_2$ for the Mg^{2+} ions and sodium carbonate for the Ca^{2+} ions:

$$Mg^{2+}(aq) + 2\ OH^-(aq) = Mg(OH)_2(s)$$
$$Ca^{2+}(aq) + CO_3^{2-}(aq) = CaCO_3(s)$$

Assuming equilibrium is reached, calculate $[Mg^{2+}]$ in the softened water. Express this magnesium ion molarity in the equivalent grams of $CaCO_3$ per million grams of water (ppm) that would yield the same hardness. Also find the total residual hardness, both Mg^{2+} and Ca^{2+}, at equilibrium in ppm, using the answer to Problem 16.5.

Section 16.5

Problems 16.6–16.8 and 16.21–16.23: For each combination of solutions given, calculate the ion product of the potentially precipitating ions, compare it with the K_{sp}, and by means of an inequality predict whether or not a precipitate will form.

16.6. Equal volumes 1.5×10^{-5} M Ag^+ and 4.8×10^{-8} M Br^-. $K_{sp} = 5.0 \times 10^{-13}$ for AgBr.

16.7. 150 ml 8.2×10^{-4} M $Pb(NO_3)_2$ and 25 ml 4.2×10^{-6} M KI. $K_{sp} = 7.1 \times 10^{-9}$ for PbI_2.

16.8. 65 ml 0.28 M Ca $(NO_3)_2$ and 30 ml 0.42 M NaOH. $K_{sp} = 5.5 \times 10^{-5}$ for $Ca(OH)_2$.

16.21. 40 ml 1.2 M Ni^{2+} and 70 ml 6.8×10^{-5} M CO_3^{2-}. $K_{sp} = 6.6 \times 10^{-9}$ for $NiCO_3$.

16.22. 24 ml 1.5×10^{-4} M $AgNO_3$ and 59 ml 5.2×10^{-3} M $CaCl_2$. $K_{sp} = 1.6 \times 10^{-10}$ for AgCl.

16.23. 29 ml 6.5×10^{-7} M Na_3PO_4 and 41 ml 5.5×10^{-5} M $CaCl_2$. $K_{sp} = 2.0 \times 10^{-29}$ for $Ca_3(PO_4)_2$.

Section 16.6

16.9. Find the solubility of $BaCO_3$ in 0.010 M $Ba(NO_3)_2$. $K_{sp} = 8.1 \times 10^{-9}$ for $BaCO_3$.

16.10. How many grams of $PbSO_4$ will dissolve in 400 ml 0.020 M Na_2SO_4? $K_{sp} = 1.6 \times 10^{-8}$ for $PbSO_4$. Compare your answer with that of Problem 16.4.

16.11. A solution 1.0×10^{-4} molar both in Cl^- and I^- is subjected to the slow addition of silver ion, without effectively increasing the volume.
- (a) Which will first precipitate, AgCl or AgI?
- (b) At what $[Ag^+]$ will the first precipitate appear?
- (c) At what $[Ag^+]$ will the second precipitate begin to appear?
- (d) What will be the concentration of the halide ion first precipitated at the time the second precipitate begins to appear?
- (e) What percentage of the original halide ion that first precipitates remains in solution when the second precipitate begins to appear?

16.12. One method for the quantitative determination of chloride ion concentration is to titrate with standardized silver nitrate, using chromate ion as an indicator. The titration reaction precipitates white silver chloride: $Ag^+ + Cl^- \rightarrow AgCl$. When the chloride is virtually exhausted, the next silver ion added combines with chromate ion, yielding a red silver chromate precipitate that serves as a visible indicator: $2\ Ag^+ + CrO_4^{2-} \rightarrow Ag_2CrO_4$. $K_{sp} = 1.6 \times 10^{-10}$ for AgCl; $K_{sp} = 2.2 \times 10^{-11}$ for Ag_2CrO_4.
- (a) In a certain titration, a 100 ml sample contains 2.5×10^{-3} mole of chloride ion. How many drops (20 drops = 1 ml) of 0.10 M $AgNO_3$ will be added before AgCl precipitate first appears?
- (b) At what Ag^+ concentration will Ag_2CrO_4 begin to precipitate if $[CrO_4^{2-}] = 2.6 \times 10^{-5}$? Disregard any dilution effect.
- (c) What will be the $[Cl^-]$ when Ag_2CrO_4 begins to precipitate?
- (d) Again disregarding dilution factors, what percentage of the chloride ion originally present will still be in solution when Ag_2CrO_4 begins to precipitate?

16.24. Compute the moles/liter solubility of calcium hydroxide in 0.050 M NaOH. $K_{sp} = 5.5 \times 10^{-5}$ for calcium hydroxide.

16.25. Calculate the number of grams of $Cu(IO_3)_2$ that will dissolve in 800 ml 0.35 M KIO_3.

16.26. 1.0 M NaF is added slowly, with no significant increase in volume, to a solution 1.0×10^{-3} M in both Ba^{2+} and Sr^{2+}.
- (a) Which fluoride, BaF_2 or SrF_2, will precipitate first?
- (b) What will $[F^-]$ be when the first precipitate begins to appear?
- (c) At what $[F^-]$ will the second precipitate appear?
- (d) What will be the concentration of the first ion to precipitate with fluoride when the second precipitate first appears?
- (e) What percentage of the first cation to precipitate remains in solution when the second precipitate begins to form?

16.27. In making MgO for use in refractory brick, half the magnesium comes from the double oxide, $CaO \cdot MgO$, and half from the ocean in which $[Mg^{2+}] = 5.0 \times 10^{-2}$. First let's look at the *fresh* water reaction. When the double oxide is treated with fresh water and allowed to come to equilibrium, both low solubility hydroxides precipitate:
$CaO \cdot MgO + 2\ H_2O \rightarrow Ca(OH)_2 + Mg(OH)_2$.
Consequently both equilibria exist:
$Ca(OH)_2 = Ca^{2+} + 2\ OH^-$ and $Mg(OH)_2 = Mg^{2+} + 2\ OH^-$.
- (a) What would $[OH^-]$ be from $Ca(OH)_2$ alone; from $Mg(OH)_2$ alone; from the combination?
- (b) Determine $[Ca^{2+}]$, $[Mg^{2+}]$ and pH in the fresh water equilibrium mixture.
- (c) With excess sea water flowing steadily through the reaction vessel, the discharge back to the ocean is saturated in $Mg(OH)_2$ with $[Mg^{2+}] = 5.0 \times 10^{-2}$. What are the $[OH^-]$ and pH of the water returned to the ocean? Compare with (a) and (b) above.
- (d) At the $[OH^-]$ calculated in (c), what happens to the Ca^{2+} from the double oxide?

Section 16.7

16.13. Find the solubility of AgSCN ($K_{sp} = 1.2 \times 10^{-12}$) in 0.50 M NH_3 if $K_d = 4 \times 10^{-8}$ for $Ag(NH_3)_2^+$.

16.14. What is the solubility of $NiCO_3$ in (a) water, and (b) a solution in which $[NH_3] = 1.0$ at equilibrium? $K_{sp} = 6.6 \times 10^{-9}$ for $NiCO_3$ and $K_d = 2 \times 10^{-9}$ for $Ni(NH_3)_6^{2+}$.

16.15. You will use 0.25 M KCN to prevent precipitation of $Fe(OH)_3$ in 6.0×10^{-4} M Fe^{3+} as pH is adjusted to 10 (assume 10.00 for significant figures in calculations). What volume of KCN solution is required per liter of Fe^{3+} solution? Disregard dilution effects. $K_{sp} = 1.1 \times 10^{-36}$ for $Fe(OH)_3$, and $K_d = 1 \times 10^{-31}$ for $Fe(CN)_6^{3-}$.

16.28. Determine the solubility of CuS in 1.0 M NH_3 if $K_{sp} = 9 \times 10^{-36}$ for CuS and $K_d = 2 \times 10^{-13}$ for $Cu(NH_3)_4^{2+}$.

16.29. Calculate the solubility of AgBr in 1.0 M $S_2O_3^{2-}$. $K_{sp} = 5.0 \times 10^{-13}$ for AgBr, and $K_d = 1 \times 10^{-13}$ for $Ag(S_2O_3)_2^{3-}$.

16.30. How many milliliters of 15 M NH_3 must be added to one liter of solution to dissolve 0.010 mole of silver chloride—i.e., make the solubility of AgCl 0.010 mole/liter? $K_{sp} = 1.6 \times 10^{-10}$ for AgCl, and $K_d = 4.0 \times 10^{-8}$ for $Ag(NH_3)_2^+$.

CHAPTER 17

OXIDATION-REDUCTION; ELECTROCHEMISTRY

Oxidation-reduction reactions in chemistry, or "redox" reactions, as they are commonly called, involve the transfer of electrons. The terms **oxidation** and **reduction** are variously defined. For purposes of this text, **oxidation will be defined as an increase in oxidation number or oxidation state, and reduction will be defined as a reduction in oxidation number or oxidation state.** Oxidation "number" and "state" are synonymous terms, and will be used interchangeably.

Oxidation and reduction are simultaneous, mutually dependent, processes. In a reaction, all electrons lost by one species are gained by another: there are neither excesses nor shortages. As a result the net increase in oxidation number by the oxidized reactant is exactly balanced by the net decrease in oxidation number by the reduced reactant. No redox reaction is possible in which two species are oxidized and none reduced, or vice versa.

17.1 OXIDATION NUMBERS

The oxidation state of an element is fixed by a set of arbitrary rules which, when applied consistently, keep track of the electrons transferred in redox reactions. The system is artificial: it has no basis in experiment. It does, however, have great utility in introducing order to redox reactions. It provides a vocabulary, and simplifies the writing of equations describing some very complex reactions.

The rules for assignment of oxidation numbers to elements, be they in elemental, compound or ionic states, are as follows:

1. The oxidation number of elements in the elemental state is zero.
2. The oxidation number of a monatomic ion is the charge on the ion.
3. Combined hydrogen has an oxidation number of $+1$ *except* in binary hydrides, in which hydrogen is the more electronegative element, when it has an oxidation number of -1.

4. Combined oxygen has an oxidation number of −2, *except* in peroxides and superoxides, in which the oxidation number is fixed by rule 5.

5. The algebraic sum of the oxidation numbers of all elements in a species is equal to the charge on the species.

Application of the first four of these rules requires little explanation. Typical +1 oxidation states for hydrogen according to rule 3 include such species as NH_3, NH_4^+, H_2O, NaOH, OH^-, and H_2SO_4. Oxygen is in a −2 oxidation state wherever it appears in these species, as well as in such ions as MnO_4^-, ClO_4^-, ClO_3^-, ClO_2^-, ClO^- and $Cr_2O_7^{2-}$.

Rule 5 is a bit more complex. The peroxide exceptions to rule 4 are covered by rule 5. For example, hydrogen in hydrogen peroxide, H_2O_2, has its usual oxidation number of +1, in accord with rule 3. The charge on the species, or compound, is zero. With two atoms at +1 the total positive oxidation state is +2. This must be balanced by −2 from the two atoms of oxygen, or −1 for each atom. Oxygen in peroxides, then, is at a −1 oxidation state. In superoxides, such as KO_2, potassium contributes +1, which must be balanced by *two* oxygen atoms. Each oxygen, then, has an oxidation number of $-\frac{1}{2}$. Fractional oxidation numbers are not common, but they do exist. They illustrate the artificiality of the oxidation number concept.

Other examples of rule 5 appear in polyatomic ions, such as MnO_4^-. The total oxidation number is −1, the charge on the ion. Each oxygen contributes −2, for a total of −8. The oxidation number of Mn, added to −8, yields −1. Mn, then, must have an oxidation number of +7: $+7 - 8 = -1$. In $Cr_2O_7^{2-}$, the contribution of oxygen is seven times −2, or −14. The chromium must contribute +12, so that $+12 - 14 = -2$, the charge on the ion. The +12 of the chromium comes from two atoms, so the contribution of each atom is +6.

You should be able to verify the following oxidation numbers by applying similar reasoning:

S in SO_4^{2-}	+6	N in NH_4^+	−3
N in N_2O	+1	N in NO_3^-	+5
N in NO	+2	Cl in ClO_4^-	+7
N in N_2O_3	+3	Cl in ClO_3^-	+5
N in NO_2	+4	Cl in ClO_2^-	+3
N in N_2O_5	+5	Cl in ClO^-	+1

In this book, the oxidation number system will be used primarily to establish a vocabulary with which redox reactions may be considered. Its great value in balancing redox equations will be left to the major textbook.

17.2 ELECTROLYTIC CELLS

Figure 17.1 illustrates an **electrolytic cell.** Such cells consist generally of two pieces of metal called **electrodes** immersed in an ionic solution called an **electrolyte.** The electrodes are connected by wires to a battery or some other source of direct current. Electric charge flows through the completed circuit;

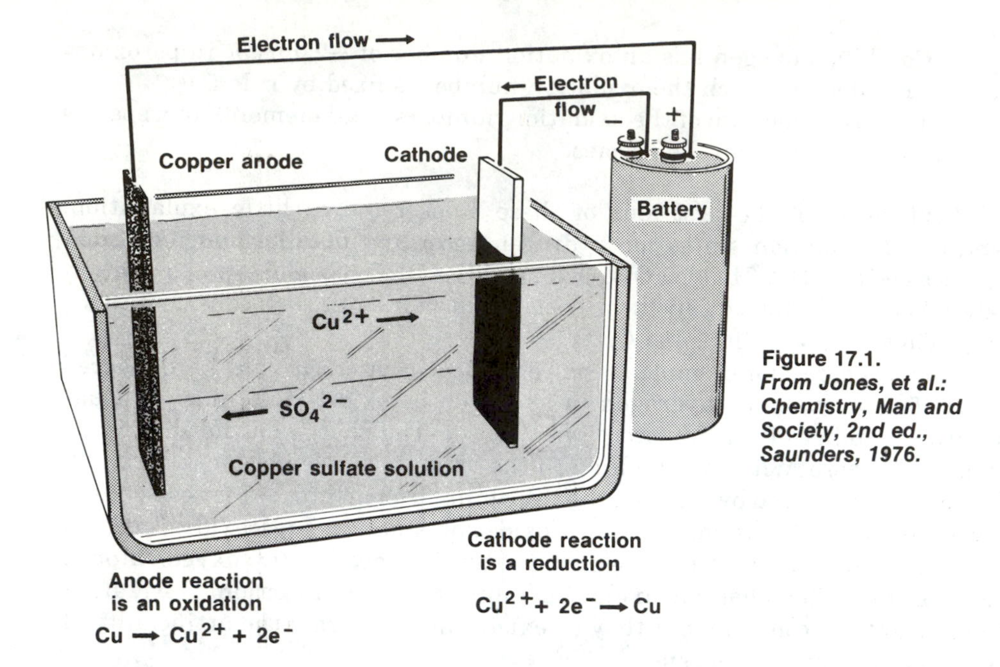

Figure 17.1. *From Jones, et al.: Chemistry, Man and Society, 2nd ed., Saunders, 1976.*

this constitutes electric current. Through the metallic parts of the circuit, the charge flow is carried by electrons; through the electrolyte the flow of charge is carried by ions. Oxidation or reduction reactions occur at the two electrodes: the electrode where oxidation occurs is called the **anode,** and the electrode at which reduction takes place is the **cathode.** The individual oxidation and reduction reactions are referred to as **half-reactions,** suggesting again that you must have both an oxidation half *and* a reduction half in a redox reaction.

The electrolytic cell shown in Figure 17.1 is a typical copper electroplating cell. The cathode is the object on which copper is plated. The reduction half-reaction is $Cu^{2+} + 2\,e^- \rightarrow Cu$. In this half-reaction the oxidation state of copper is *reduced* from +2 to 0, satisfying the definition of reduction. It also illustrates another widely used definition of reduction, namely *gain of electrons.* By *gaining* two electrons, the Cu^{2+} ion gains two units of negative charge to reduce its oxidation state from +2 to 0, in which form it deposits as a neutral copper atom on the cathode. The anode is a piece of copper that dissolves off into the solution, replenishing the Cu^{2+} ions removed at the cathode: $Cu \rightarrow Cu^{2+} + 2e^-$. The increase from 0 to 2+ in the oxidation state of copper identifies this as an oxidation half-reaction. Subtracting two electrons from both sides of the equation yields $Cu - 2e^- \rightarrow Cu^{2+}$, in which form it illustrates a common definition of oxidation as *loss of electrons.* As shown in the sketch, electrons move from the copper anode through the circuit to the cathode, replenishing those delivered to Cu^{2+} ions.

To study electrolytic cells it is necessary to establish a vocabulary of electrical units. The basic units of electricity were originally interdefined with respect to each other. These are the ampere, volt, ohm, watt, coulomb and joule. We have immediate interest in two of these units, the **ampere** and the **coulomb. The coulomb is a quantity of electric charge, symbolized by the letter C; the ampere is a rate of flow of charge, measured in coulombs per**

second, or C/s. If current in amperes, or coulombs/second, is multiplied by time in seconds, the product is quantity of charge in coulombs. Thus 1 ampere-second = 1 coulomb.

In chemistry the coulomb is not of itself a particularly useful unit of charge. We are more interested in the **quantity of charge carried by one mole of electrons,** called a **faraday,** and designated $\mathcal{F}$. One mole of electrons carries a quantity of charge equal to 96,500 coulombs, or 96,500 ampere-seconds. This relationship,

$$96{,}500 \text{ amp-sec} = \text{charge on 1 mole of electrons} = 1 \text{ faraday}, \quad (17.1)$$

is basic in electrochemical problems.

17.3 FARADAY'S LAWS OF ELECTROLYSIS

In his early work in electrochemistry, Michael Faraday discovered two important relationships which have become known as Faraday's Laws of Electrolysis. These are:

1. The quantity (mass) of an elemental substance released or deposited in electrolysis is proportional to the quantity of electric charge (coulombs, or ampere-seconds) that has passed through the system.
2. The masses of different substances released or deposited in electrolysis are in the same ratio as their equivalent weights.

Faraday's laws may be understood if we consider the following electrode reactions in which metals are deposited electrochemically:

$$\tfrac{1}{2}\,Cu^{2+} + e^- \rightarrow \tfrac{1}{2}\,Cu$$

$$Ag^+ + e^- \rightarrow Ag$$

In both reactions the quantity of metal deposited depends upon the number of electrons passing through the circuit to the cathode, where the reduction occurs. One mole of electrons will deposit half a mole of copper, or one mole of silver; or 31.8 grams of copper and 108 grams of silver. If the number of electrons (quantity of electric charge) is doubled, the metal deposited will be doubled, in accord with the principles of stoichiometry. Stoichiometry also requires that certain weight ratios hold between reacting elements. It was these fixed ratios, which were measured experimentally, that led Faraday to the concept of equivalent weights.

Redox equivalents and equivalent weights will be discussed further in Section 17.9.

17.4 ELECTROCHEMICAL STOICHIOMETRY

Half-reaction equations, involving quantity of charge, may be used in stoichiometry problems in the same manner as other equations. The quantity of charge is amperes × seconds. This product may be converted to moles of electrons by using the relation 96,500 ampere-seconds = 1 mole of electrons. Consistent with the stoichiometry pattern (page 47), the given quantity is converted to moles of a given species in the half-reaction equation; from that you proceed to moles of the wanted species; and finally you convert to the desired units for that species. The following example illustrates the method:

Example 17.1. How many grams of copper metal will be deposited from a copper(II) plating solution in 45.0 minutes at a current of 8.40 amperes?

With any stoichiometry problem there must be an equation. In this problem, copper(II) ion is reduced to metallic copper. Write the equation.

17.1a. $Cu^{2+} + 2e^- \rightarrow Cu$.

The "given quantity" in this problem is quantity of charge, the product of amperes × seconds. The problem gives amperes and minutes. Begin by converting ampere-minutes to ampere-seconds—setup only.

17.1b. $8.40 \text{ amps} \times 45.0 \cancel{\text{ minutes}} \times \frac{60 \text{ seconds}}{\cancel{\text{minute}}}$

Next comes the conversion from ampere seconds to moles of electrons, based on Equation 17.1. This, or its converse, appears in every electrochemical stoichiometry problem. Extend the setup to that point.

17.1c.

$$8.40 \cancel{\text{ amps}} \times 45.0 \cancel{\text{ minutes}} \times \frac{60 \cancel{\text{ seconds}}}{\cancel{\text{minute}}} \times \frac{1 \text{ mole electrons}}{96{,}500 \cancel{\text{ amp seconds}}}$$

You now have the moles of electrons. Completion of the problem is routine by the stoichiometric pattern. Go all the way.

17.1d. 7.46 grams of copper.

$$8.40 \cancel{\text{ amps}} \times 45.0 \cancel{\text{ min}} \times \frac{60 \cancel{\text{ sec}}}{1 \cancel{\text{ min}}} \times \frac{1 \cancel{\text{ mole e}^-}}{96{,}500 \cancel{\text{ amp sec}}} \times \frac{1 \cancel{\text{ mole Cu}}}{2 \cancel{\text{ moles e}^-}} \times \frac{63.5 \text{ g Cu}}{\cancel{\text{mole Cu}}} = 7.46 \text{ g Cu}$$

Gases may be "electrodeposited" as well as metals. . . .

Example 17.2. In the electrolytic decomposition of water, calculate the STP volume of hydrogen released in one hour by a current of 7.10 amperes. The reduction equation is $2\ H^+ + 2\ e^- \rightarrow H_2$.

Except for the last step, the solution of this problem is identical to the solution of Example 17.1—and the last step became familiar back in Chapter 4 (page 51). Complete the problem.

17.2a. 2.97 liters.

$$7.10 \text{ amps} \times 1 \text{ hr} \times \frac{3600 \text{ sec}}{1 \text{ hr}} \times \frac{1 \text{ mole e}^-}{96{,}500 \text{ amp-sec}} \times \frac{1 \text{ mole H}_2}{2 \text{ mole e}^-} \times \frac{22.4\ l \text{ H}_2}{1 \text{ mole H}_2} = 2.97\ l \text{ H}_2$$

The electroplating engineer is more apt to be interested in specifying time or amperes required for a given plating purpose.

Example 17.3. What current is required to deposit 2000 grams of lead per hour from a Pb^{2+} plating solution?

First, the equation. . . .

17.3a. $Pb^{2+} + 2\ e^- \rightarrow Pb$.

The given quantity in this example is in two units, both mass and time, combined as a rate of deposition in grams of lead per hour. From that starting point the grams can be converted to moles of lead; then to moles of electrons; and to ampere-seconds by the methods of the previous examples. At that point your units will be ampere-seconds/hour. We'll leave it to you to figure out the final time conversion that will yield the desired answer in amperes.

17.3b. 518 amperes.

$$\frac{2000 \text{ g Pb}}{\text{hr}} \times \frac{1 \text{ mole Pb}}{207 \text{ g Pb}} \times \frac{2 \text{ moles e}^-}{1 \text{ mole Pb}} \times \frac{96{,}500 \text{ amp-sec}}{1 \text{ mole e}^-} \times \frac{1 \text{ hr}}{3600 \text{ sec}} = 518 \text{ amperes}$$

Perhaps Example 17.3 may be matched to the stoichiometric pattern more closely if we think of the grams of lead as the given quantity, which is first converted to moles of lead, then to moles of electrons, and finally to *quantity of charge* in ampere-seconds. In other words, the desired "quantity" is an amount of charge, rather than grams or liters of a chemical species. But quantity of charge in ampere-seconds has two components, or two factors in a mathematical sense. If you divide ampere-seconds by one of its factors, seconds, the result is in amperes, as in Example 17.3. If you divide ampere-seconds by the other factor, amperes, the result must be time in seconds. We'll solve for this "other half" of the ampere-second duo in Example 17.4:

Example 17.4. How many hours must an electrolytic system be operated at a current of 1500 amperes to produce 40 kilograms of metallic sodium from molten sodium chloride?

Again, begin with the reduction half-reaction equation.

17.4a. $Na^+ + e^- = Na$.

Now start with the given quantity in kilograms of sodium—and it is kilograms, not grams—and take it through the three steps of the stoichiometric pattern to quantity of charge in ampere-seconds. Setup only, please. . . .

17.4b.

$$40 \text{ kg Na} \times \frac{10^3 \text{ g}}{\text{kg}} \times \frac{1 \text{ mole Na}}{23.0 \text{ g Na}} \times \frac{1 \text{ mole } e^-}{1 \text{ mole Na}} \times \frac{96{,}500 \text{ amp-sec}}{1 \text{ mole } e^-}$$

At this point you have quantity of charge as the product of two factors. The problem statement gives you one of the factors. How do you find the other factor? Complete the setup all the way to the required time units and solve for the answer.

17.4c. 31.1 hours.

$$40 \text{ kg Na} \times \frac{10^3 \text{ g}}{\text{kg}} \times \frac{1 \text{ mole Na}}{23.0 \text{ g Na}} \times \frac{1 \text{ mole } e^-}{1 \text{ mole Na}} \times \frac{96{,}500 \text{ amp-sec}}{1 \text{ mole } e^-}$$

$$\times \frac{1}{1500 \text{ amps}} \times \frac{1 \text{ hr}}{3600 \text{ sec}} = 31.1 \text{ hours}$$

At the other electrode there should be quite a bit of chlorine that also has commercial value. . . .

Example 17.5. How many liters of chlorine, measured at 25°C and 20 atmospheres, are produced in the above reaction when the cell is operated for 31.1 hours at 1500 amperes?

There are two differences between this problem and Example 17.2. The first appears in the half-reaction equation for the conversion of chloride ion to chlorine gas. . . .

17.5a. $2 Cl^- \rightarrow Cl_2 + 2 e^-$

We note that this time the deposition of chlorine is an *oxidation* reaction, occurring at the anode, rather than a reduction reaction at the cathode, as all other electrodepositions have been. This is not uncommon for gases; the "other" gas released in the electrodecomposition of water (Example 17.2) is oxygen, and it is released by anodic oxidation. Faraday's laws and their application to electrochemical stoichiometry are the same for oxidation and reduction half-reactions.

The other variation in this problem is that it asks for volume at other than STP conditions. You may wish to refer to Example 8.9, page 106, if you need a refresher in this area. Complete the problem.

17.5b. $V = 1.06 \times 10^3$ liters of chlorine.

$$31.1 \text{ hr} \times 1500 \text{ amps} \times \frac{3600 \text{ sec}}{1 \text{ hr}} \times \frac{1 \text{ mole } e^-}{96{,}500 \text{ amp-sec}} \times \frac{1 \text{ mole } Cl_2}{2 \text{ moles } e^-} \times$$

$$\frac{0.0821 \, l\text{-atm}}{°\text{K-mole}} \times \frac{298 \, °\text{K}}{20 \text{ atm}} = 1.06 \times 10^3 \text{ liters } Cl_2$$

Alternate setup:

$$31.1 \text{ hr} \times 1500 \text{ amps} \times \frac{3600 \text{ sec}}{1 \text{ hr}} \times \frac{1 \text{ mole } e^-}{96{,}500 \text{ amp-sec}} \times \frac{1 \text{ mole } Cl_2}{2 \text{ moles } e^-} \times$$

$$\frac{22.4 \, l \; Cl_2}{1 \text{ mole } Cl_2} \times \frac{(273+25)°\text{K}}{273°\text{K}} \times \frac{1 \text{ atm}}{20 \text{ atm}} = 1.06 \times 10^3 \, l \; Cl_2$$

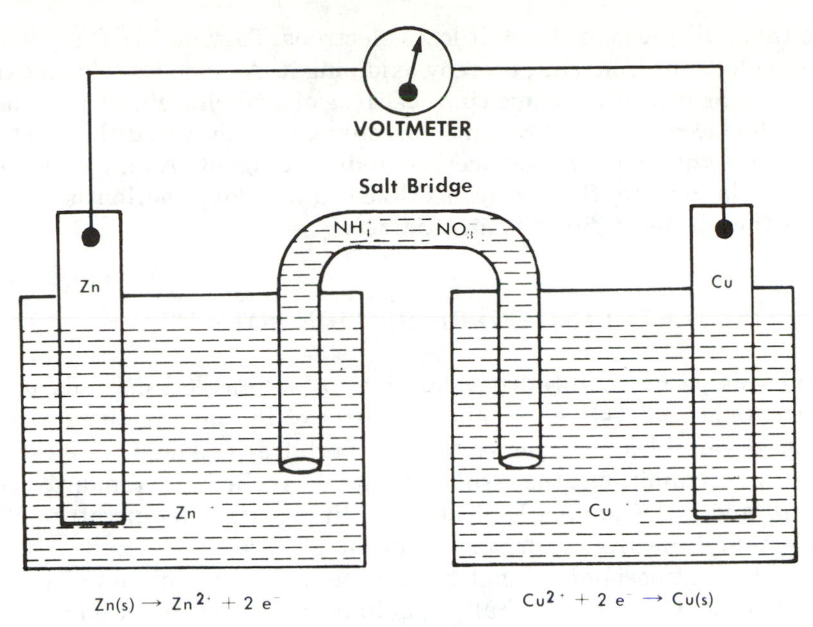

Figure 17.2.

17.5 VOLTAIC CELLS

Figure 17.2 illustrates a typical voltaic cell. In this example a piece of zinc is immersed in a solution of Zn^{2+} ions in one container, and a piece of copper is immersed in a solution of Cu^{2+} ions in another. The solutions are connected by a "salt bridge," an ionic solution that permits migration of charged ions without allowing free mixing of the Zn^{2+} and Cu^{2+} solutions. A solution of NH_4NO_3 makes a suitable salt bridge. When the zinc and copper are connected externally with a metallic conductor there is a spontaneous flow of electrons from the zinc to the copper. Insertion of a voltmeter in this external circuit measures the "electromotive force" that drives the electrons through the wire.

Let's look, for a moment, at these related terms, voltage or volt, and potential or potential difference. Electrical energy, or electrical work, is expended or absorbed as charged particles move through a circuit. **If 1 joule of energy is required or released in moving 1 coulomb of charge from one point to another in a circuit, the potential difference between those points is 1 volt.** This definition yields joules per coulomb as the "units" of voltage, and it is in these units that it must be considered in calculations. Voltage is sometimes referred to as electromotive force, or emf.

As in the electrolytic cell, oxidation and reduction reactions occur at the electrodes of a voltaic cell. The zinc anode in Figure 17.2 is oxidized: $Zn \rightarrow Zn^{2+} + 2\,e^-$. Copper(II) ions are reduced at the cathode: $Cu^{2+} + 2\,e^- \rightarrow Cu$. The two half-reaction equations may be added to produce the redox equation:

$$\begin{array}{r} Zn \rightarrow Zn^{2+} + 2e^- \\ Cu^{2+} + 2e^- \rightarrow Cu \qquad \\ \hline Zn + Cu^{2+} \rightarrow Zn^{2+} + Cu \end{array}$$

In this cell zinc is oxidized; it loses electrons. To what? To Cu^{2+}. The Cu^{2+} takes the electrons from zinc, thereby oxidizing it. As such Cu^{2+} is an **oxidizer,** or **oxidizing agent.** At the same time, *because* of oxidizing zinc, Cu^{2+} was itself reduced. Reduced by what? By zinc, or course; zinc delivered the electrons to Cu^{2+}. Zinc is therefore the **reducer,** or **reducing agent.** As a consequence of this interrelationship, **the species oxidized in a redox reaction is a reducer, whereas the species reduced is an oxidizer.**

17.6 STANDARD REDUCTION POTENTIALS

The voltage or potential developed by a voltaic cell depends upon the temperature of the system and the activities of the ions in the electrolytes. When these variables are adjusted to a set of arbitrary standards, referred to as **standard state conditions,** the resulting voltage is called the **standard electrode potential,** and is designated E°. These conditions are a temperature of 25°C; pure elemental electrodes, or, in the event of gaseous electrodes, a partial pressure of one atmosphere of that gas; and solutions at "unit activity" in their respective ions. As before, we shall think in terms of concentration rather than activity. Standard state conditions, then, call for ionic concentrations of 1.00 molar.

The voltmeter in Figure 17.2 measures the potential difference between zinc and copper electrodes. Standard state potentials are tabulated relative to an arbitrarily selected zero voltage, which is assigned to the reduction of hydrogen ion, $2\ H^+(aq) + 2\ e^- \rightarrow H_2(g)$. The "standard hydrogen electrode" used for such comparisons consists of a piece of inert platinum over which hydrogen gas is bubbled at a partial pressure of 1.00 atmosphere in a solution that is 1.00 molar in H^+. E° between the standard hydrogen electrode and some other electrode is called the **standard reduction potential** of the other electrode. It is positive if the other electrode functions as the cathode (reduction half-reaction), and negative if the other electrode behaves as the anode (oxidation half-reaction).

The standard reduction potentials of a selected number of electrodes are tabulated in the Appendix on page 294. These are listed in order of increasing E°. Both the order and use of this table may be illustrated by the Cu-Zn voltaic cell previously described. Isolated from the table, these two half-reactions and their corresponding E° values are

$$Zn^{2+} + 2\,e^- \rightarrow Zn \qquad E° = -0.76$$

$$Cu^{2+} + 2\,e^- \rightarrow Cu \qquad E° = +0.34$$

In the cell, Cu^{2+} gains electrons and Zn loses them in the spontaneous reaction. Cu^{2+} has a stronger affinity for electrons than Zn^{2+}. It is said, therefore, to be the stronger of the two oxidizers. As this example illustrates, the table is arranged with oxidizers on the left, with the weakest oxidizer at the top.

Of the reducers, it is zinc that releases its electrons most easily, not copper. Zinc, then, is said to be the stronger reducer of the two. The table lists the reducers on the right side in order of their decreasing strength, with the weakest at the bottom.

The reactions in the table are all reduction reactions. In the cell, zinc is oxidized. To express what is happening, we must *reverse* the $Zn^{2+} \rightarrow Zn$ reduction reaction, which also changes the sign of E°. Writing both oxidation and reduction equations with their E° values, they can be summed up for the redox reaction:

$$
\begin{array}{ll}
Zn \rightarrow Zn^{2+} + 2e^- & E° = +0.76 \\
Cu^{2+} + 2e^- \rightarrow Cu & E° = +0.34 \\
\hline
Cu^{2+} + Zn \rightarrow Cu + Zn^{2+} & E° = +1.10
\end{array}
$$

If the half reactions and E° values are known, the equation for any redox reaction may be written and the E° found. If the E° is positive, the reaction under standard conditions occurs spontaneously in the direction in which it is written; if negative, it proceeds spontaneously in the reverse direction. The sign of E°, then, becomes a criterion by which the spontaneous direction of the reaction at standard conditions may be determined.

On the basis of this information. . . .

Example 17.6. Write the equation for the possible redox reaction between bromine and the chloride ion, calculate its E°, and indicate whether the forward or reverse reaction will occur spontaneously at standard conditions.

First, select from the table the two half-reactions. Decide which will be the reduction reaction and write it as it appears in the table; but reverse the other and write it as an oxidation equation beneath the first equation. Include the appropriate E° values. Add the equations and voltages, and predict the direction in which the reaction will be spontaneous.

17.6a. E° = −0.29 volt. The reverse reaction is spontaneous.

$$
\begin{array}{ll}
Br_2 + 2\,e^- \rightarrow 2\,Br^- & E° = +1.07 \\
2\,Cl^- \rightarrow 2\,e^- + Cl_2 & E° = -1.36 \\
\hline
2Cl^- + Br_2 \rightarrow Cl_2 + 2Br^- & E° = -0.29
\end{array}
$$

According to the problem, bromine is reduced to bromide ion, so its equation may be taken directly from the table. Chloride ion, the other reactant, is oxidized to chlorine. The reduction equation must be reversed, which also changes the sign of E°. The total E° being negative, the reverse reaction will occur spontaneously.

Example 17.7. Write the equation for the possible redox reaction between Fe^{2+} and acidified $Cr_2O_7^{2-}$, calculate E°, and determine the direction in which the reaction will occur spontaneously at standard conditions.

The Fe^{2+} ion appears twice in the table, once as an oxidizer at E° = −0.44, and again as a reducer at E° = +0.77. Which, if either, is a possible redox reactant with $Cr_2O_7^{2-}$?

17.7a. $Cr_2O_7^{2-}$ is an oxidizer. Fe^{2+} must, therefore, act as a reducer. Now write the oxidation and reduction half-reaction equations.

17.7b.

$$Fe^{2+} \rightarrow Fe^{3+} + e^- \qquad E^\circ = -0.77$$

$$Cr_2O_7^{2-} + 14\,H^+ + 6\,e^- \rightarrow 2\,Cr^{3+} + 7\,H_2O \qquad E^\circ = +1.33$$

At this point something new arises. Addition of the two equations above will result in a net of five electrons on the left side of the equation. But in redox reactions there are no excesses or deficiencies in electrons. This condition is corrected by multiplying either or both the reduction and oxidation equations by whatever factors that will make the electron coefficients the same. They then cancel each other, as in Example 17.6. Select such a factor or factors, perform the necessary multiplication or multiplications, and rewrite the oxidation and reduction equations so their electron coefficients are the same. Add the resulting equations, find E°, and predict the spontaneous direction.

17.7c.

$$6\,Fe^{2+} \rightarrow 6\,Fe^{3+} + 6\,e^- \qquad E^\circ = -0.77$$

$$Cr_2O_7^{2-} + 14\,H^+ + 6\,e^- \rightarrow 2\,Cr^{3+} + 7\,H_2O \qquad E^\circ = +1.33$$

$$6\,Fe^{2+} + Cr_2O_7^{2-} + 14\,H^+ \rightarrow 6\,Fe^{3+} + 2\,Cr^{3+} + 7H_2O \qquad E^\circ = +0.56$$

The reaction is spontaneous to the right, as indicated by the positive value of E°.

You probably came up with the correct equation, but you may have handled the voltage calculation incorrectly. This would occur if you multiplied E° for the oxidation reaction by 6, as you did the oxidation equation. Voltages, unlike reaction energies, are not proportional to the quantity of chemicals involved. Fundamentally voltages are measurements of energy *per electron* transferred in a chemical reaction. While reaction energy depends upon quantity, energy *per electron* does not. An analogy might be that two gallons of water weigh twice as much as one, but the weight *per gallon* is the same regardless of quantity. Therefore do not multiply voltages when multiplying half-reaction equations.

Example 17.8. Write the equation for the possible redox reaction between the Sn(II) ion and elemental aluminum. Calculate E° and predict the direction of the spontaneous reaction.

You have the background. Work out the entire problem.

17.8a.

$$2\,Al \rightarrow 2\,Al^{3+} + 6\,e^- \qquad E^\circ = +1.66$$

$$3\,Sn^{2+} + 6\,e^- \rightarrow 3\,Sn \qquad E^\circ = -0.14$$

$$3\,Sn^{2+} + 2\,Al \rightarrow 3\,Sn + 2\,Al^{3+} \qquad E^\circ = +1.52$$

The reaction is spontaneous to the right.

In this problem there was again a choice between two places in the table where Sn^{2+} appeared. It had to function as an oxidizer for aluminum, and was chosen accordingly. Also in this problem both equations had to be multiplied. The electron charge for one equation was two, and for the other, three. Multiplications by three and two respectively brought both to six.

17.7 FREE ENERGY AND CELL POTENTIAL

Throughout the foregoing section we have been using the sign of cell potential as a criterion for predicting whether or not a redox reaction could be spontaneous as written. In Chapter 12 we found that the thermodynamic criterion for spontaneity is a negative value for ΔG, the change in free energy. If both criteria are valid, there should be a relationship between them. And indeed there is—a very important one.

The electrical work performed by a galvanic cell is the product of the quantity of charge (coulombs) transferred times voltage. We saw on page 257 that voltage is a measure of joules/coulomb. Therefore,

$$w_{elec} = \text{coulombs} \times \text{volts} = \text{coulombs} \times \frac{\text{joules}}{\text{coulomb}} = \text{joules}$$

In an electrochemical cell coulombs of charge can be found by multiplying the moles of electrons transferred, n, by the faraday constant, 96,500 coulombs per mole of electrons (see Equation 17.1, page 253). Accordingly,

$$w_{elec} = n\mathcal{F}E \qquad (17.2)$$

where n is the moles of electrons, $\mathcal{F}$ is the faraday constant, and E is the cell potential.

Unfortunately not all of the energy available can be captured as useful work in the practical operation of an electrochemical cell. Some is dissipated as heat. More is lost overcoming conditions that arise in the vicinity of the electrodes in any cell through which current passes, the effect of which is to reduce cell potential. However, by means of a device known as a potentiometer, it is possible to measure the voltage of a cell when *no current is flowing*, at which time there are neither heat losses nor electrode interferences caused by current. Accordingly, Equation 17.2 measures the *maximum* work a cell can deliver, and is generally written

$$w_{max} = n\mathcal{F}E \qquad (17.3)$$

Add to this the thermodynamic fact that maximum work is the negative of the free energy change—i.e., $w_{max} = -\Delta G$—and it follows that

$$\Delta G = -n\mathcal{F}E \qquad (17.4)$$

Equation 17.4 is a most important relationship. With it we can make a simple laboratory determination of the free energy change of a reaction. If the measurement is conducted at 25°C, with all solutions at one molar concentration (unit activity) and gases at one atm, the voltage measured is the standard cell potential and Equation 17.4 becomes

$$\Delta G^\circ = -n\mathcal{F}E^\circ \qquad (17.5)$$

You must be careful of units in using Equation 17.5. Straight substitution of moles of electrons, the faraday constant and voltage will yield ΔG° in joules. Conversion to calories or kilocalories may be made by the factors:*

$$1 \text{ cal} = 4.18 \text{ joules}; \qquad 1 \text{ kcal} = 4{,}180 \text{ joules} \qquad (17.6)$$

*Equation 17.5 may be written $\Delta G^\circ = -23.06nE^\circ$, in which the constant converts joules to kilocalories, as the unit of ΔG°.

Example 17.9. Calculate $\Delta G°$ in kilocalories for

(a) $2\ Cl^- + Br_2 \rightarrow Cl_2 + 2\ Br^-$ (Example 17.6);

(b) $6\ Fe^{2+} + Cr_2O_7^{2-} + 14\ H^+ \rightarrow 6\ Fe^{3+} + 2\ Cr^{3+} + 7\ H_2O$ (Example 17.7).

On the basis of your answers, predict whether the reaction will be spontaneous in the forward or reverse direction at standard state. Correlate your predictions with those made for Examples 17.6 and 17.7.

The method is straight substitution into Equation 17.5, plus use of Equation 17.6. Proceed. . . .

17.9a. (a) $\Delta G° = +14$ kcal; (b) $\Delta G° = -78$ kcal.

Reaction (a) spontaneous in reverse direction based on negative E° and positive $\Delta G°$; reaction (b) spontaneous in forward direction based on positive E° and negative $\Delta G°$.

(a) $\Delta G° = -n\mathcal{F}E° = (-2)(96{,}500)(-0.30) = 57{,}900$ joules

$$57{,}900\ \cancel{\text{joules}} \times \frac{1\ \text{kcal}}{4180\ \cancel{\text{joules}}} = 14\ \text{kcal}$$

(b) $\Delta G° = -n\mathcal{F}E° = (-6)(96{,}500)(0.56) = -324{,}000$ joules

$$-324{,}000\ \cancel{\text{joules}} \times \frac{1\ \text{kcal}}{4180\ \cancel{\text{joules}}} = -78\ \text{kcal}$$

Example 17.10. Find the standard free energy change for the reaction $NO_3^- + 4\ H^+ + 3\ Ag \rightarrow NO + 2\ H_2O + 3\ Ag^+$.

To begin, we must find E° and n. Without the half-reactions, neither is apparent. But examine the net ionic equation carefully and see if you cannot separate it into two half-reaction equations which, when added, will yield the equation given. If you cannot, get help from Table IX, page 294. (You might even be able to find n from an examination of the given equation, before you identify the half-reactions. . . .)

17.10a. n = 3.

$$NO_3^- + 4\ H^+ + 3\ e^- \rightarrow NO + 2\ H_2O$$

$$3\ Ag \rightarrow 3\ Ag^+ + 3\ e^-$$

Your best chance of spotting n and one of the half-reaction equations is the oxidation of silver to silver ion. It's a simple reaction, with only one species on both sides of the equation. The change in oxidation number is +1, and the coefficients of 3 indicate a transfer of three electrons. What's left of the given equation after removing the silver oxidation is the reduction half-reaction.

Now that you have the half-reaction equations and *n*, find E° and calculate $\Delta G°$.

17.10b. $\Delta G° = -46{,}000$ joules $= -11$ kcal.

$$NO_3^- + 4\,H^+ + 3\,e^- \rightarrow NO + 2\,H_2O \qquad E^\circ = +0.96$$

$$3\,Ag \rightarrow 3\,Ag^+ + 3\,e^- \qquad E^\circ = -0.80$$

$$NO_3^- + 4\,H^+ + 3\,Ag \rightarrow NO + 2\,H_2O + 3\,Ag^+ \qquad E^\circ = +0.16$$

$$\Delta G^\circ = -n\mathcal{F}E^\circ = -3 \times 96{,}500 \times 0.16 = -46{,}000 \text{ joules} = -11 \text{ kcal}$$

17.8 THE NERNST EQUATION

In Chapter 13, Equation 13.4 (page 191) expresses ΔG in terms of K_p and Q, the latter being a ratio having the form of K. Its value depends upon the nonequilibrium partial pressures of the gases in the reaction. A parallel equation may be derived for solution equilibria, involving concentrations in moles/liter:

$$\Delta G = -RT \ln K + RT \ln Q. \qquad (17.7)$$

If R is expressed in joules/mole−°K, ΔG is in joules, as it is in Equation 17.4 where it is equated to $-n\mathcal{F}E$. Equating these things, both equal to ΔG, to each other gives

$$-n\mathcal{F}E = -RT \ln K + RT \ln Q.$$

Dividing both sides by $-n\mathcal{F}$ yields a form of the Nernst Equation:

$$E = \frac{RT}{n\mathcal{F}} \ln K - \frac{RT}{n\mathcal{F}} \ln Q \qquad (17.8)$$

A more useful form of this equation at 298°K is found by evaluating $\frac{RT}{\mathcal{F}}$, converting joules to kilocalories, and natural logarithms to base 10 logarithms. The result is

$$E = \frac{0.0592}{n} \log K - \frac{0.0592}{n} \log Q. \qquad (17.9)$$

At standard state, with all concentrations equal to 1 mole/liter, gaseous partial pressures at 1 atmosphere, and temperature at 25°C, the cell voltage is the standard potential, E°, and Q has a value of 1. Because log 1 = 0, the second term becomes 0 and the equation becomes

$$E^\circ = \frac{0.0592}{n} \log K. \qquad (17.10)$$

At this point we see that easy-to-measure E° provides a direct path to another quantity, the equilibrium constant which is difficult to find by any other method.

Example 17.11. Calculate K for $Zn + Cu^{2+} \rightarrow Zn^{2+} + Cu$.

The values of E° and n must be found in solving this problem. E° was calculated earlier to be 1.10. What is n?

17.11a. n = 2.

This can be seen at a glance for this reaction. It takes 2 e^- to convert Zn to Zn^{2+} or Cu^{2+} to Cu. Further, there are 2 electrons in each half-reaction equation.

Insert the proper values into Equation 17.10 and solve for log K.

17.11b. log K = 37.2.

$$\log K = \frac{2 \times 1.10}{0.0592} = 37.2.$$

Thinking of log K as 0.2 + 37 and taking antilogs of both sides yields K in exponential notation.

17.11c. $K = 2 \times 10^{37}$.

$$K = 10^{37.2} = 10^{0.2} \times 10^{37} = 1.6 \times 10^{37} = 2 \times 10^{37}$$

Example 17.12. Calculate K for $2\ Cl^- + Br_2 = 2\ Br^- + Cl_2$.

In Example 17.6, E° was found to be −0.29 volt. The value of n should be readily seen. Solve for log K, using Equation 17.10.

17.12a. log K = −9.8.

From

$$E° = \frac{0.0592}{n} \log K;$$

$$-0.29 = \frac{0.0592}{2} \log K$$

and

$$\log K = \frac{2(-0.29)}{0.0592} = -9.8$$

This situation is slightly different from Example 17.11 in that log K is negative. All pH values are negative logarithms, and you convert freely from pH to $[H^+]$. The method is identical: find the value of K.

17.12b. $K = 2 \times 10^{-10}$.

Note that the K values found in Examples 17.11 and 17.12 confirm the predictions of reaction spontaneity for the two systems. For $Zn + Cu^{2+} \rightarrow Zn^{2+} + Cu$, E° = 1.10 and $K = 2 \times 10^{37}$. E° predicts a spontaneous forward reaction, and the very large value of K indicates that Cu^{2+} as a reactant will be almost totally consumed. For $2\ Cl^- + Br_2 \rightarrow Cl_2 + 2\ Br^-$, however, E° is −0.29. The negative value predicts spontaneity to the left. With a K of about 10^{-10}, the reverse reaction is clearly favored.

It is interesting to note that the use of standard reduction potentials to calculate equilibrium constants is not limited to redox reactions. Solubility product constants, for example, are readily found from E° values, whereas their determination by concentration measurements is virtually impossible because of the extremely dilute solutions normally encountered. Devices known as reference electrodes are used for this purpose. The reduction reaction associated with a silver-silver chloride reference electrode is

$$AgCl + e^- = Ag + Cl^- \qquad E° = +0.22 \text{ volt}$$

If the reduction equation for silver is reversed so it becomes an oxidation equation we have

$$Ag = Ag^+ + e^- \qquad E° = -0.80 \text{ volt}$$

Combining these equations and their standard electrode potentials yields

$$\begin{array}{rcll} AgCl + e^- & = & Ag + Cl^- & E° = +0.22 \text{ volt} \\ Ag & = & Ag^+ + e^- & E° = -0.80 \text{ volt} \\ \hline AgCl & = & Ag^+ + Cl^- & E° = -0.58 \text{ volt} \end{array}$$

The added equations represent the solubility equilibrium for silver chloride. From the E° value associated with it, K_{sp} may be calculated.

Example 17.13. Calculate K_{sp} for AgCl if E° = −0.58 volt for $AgCl = Ag^+ + Cl^-$.

The procedure is as it was in the two previous examples. Proceed.

17.13a. $K_{sp} = 1.6 \times 10^{-10}$

$$-0.58 = \frac{0.0592}{1} \log K; \qquad \log K = -9.80$$

$$K = 10^{-9.80} = 10^{0.20} \times 10^{-10} = 1.6 \times 10^{-10}$$

17.9 EFFECT OF CONCENTRATION ON CELL VOLTAGE

If $\frac{0.0592}{n} \log K$ in equation 17.9 is replaced by its equal, E°, the equation becomes

$$E = E° - \frac{0.0592}{n} \log Q. \qquad (17.11)$$

Equation 17.11 permits the calculation of cell potential at 25°C if solution concentrations are known.

Example 17.14. Calculate the potential of a silver-copper cell in which $[Ag^+] = 0.50$ and $[Cu^{2+}] = 1.0$.

It always helps to have an equation—and you'll need the half-reaction equations to find E°. Go that far.

17.14a. E° = 0.46 volt.

$$\begin{array}{lr} 2\ Ag^+ + 2\ e^- \rightarrow 2\ Ag & E° = +0.80 \\ Cu \rightarrow Cu^{2+} + 2\ e^- & E° = -0.34 \\ \hline 2\ Ag^+ + Cu \rightarrow 2\ Ag + Cu^{2+} & E° = +0.46 \end{array}$$

Recalling that Q has the form of K, but the value determined by actual concentrations, calculate Q from the information given. Careful. . . .

17.14b. Q = 4.0.

$$Q = \frac{[Cu^{2+}]}{[Ag^+]^2} = \frac{1.0}{0.50^2} = 4.0$$

Now you are ready to substitute into Equation 17.11 and solve for the cell voltage.

17.14c. E = 0.44 volt.

$$E = E° - \frac{0.0592}{2} \log 4.0 = 0.46 - 0.02 = 0.44 \text{ volt}$$

Example 17.15. Find the theoretical cell potential of an aluminum-nickel cell in which $[Ni^{2+}] = 0.80$ and $[Al^{3+}] = 0.020$.

First find n and E°.

17.15a. n = 6; E° = 1.41 volts

$$\begin{array}{lr} 2\ Al(s) \rightarrow 2\ Al^{3+} + 6\ e^- & E° = +1.66 \\ 3\ Ni^{2+} + 6\ e^- \rightarrow 3\ Ni(s) & E° = -0.25 \\ \hline 2\ Al(s) + 3\ Ni^{2+} \rightarrow 2\ Al^{3+} + 3\ Ni(s) & E° = +1.41 \end{array}$$

Now calculate Q.

17.15b. Q = 7.8×10^{-4}.

$$Q = \frac{[Al^{3+}]^2}{[Ni^{2+}]^3} = \frac{(0.020)^2}{(0.80)^3} = \frac{4.0 \times 10^{-4}}{0.512} = 7.8 \times 10^{-4}$$

Now substitute appropriate values into Equation 17.11 and solve.

17.15c. E = 1.45 volts.

$$E = E° - \frac{0.0592}{n} \log Q = 1.41 - \frac{0.0592}{6} \log(7.8 \times 10^{-4})$$

$$= 1.41 - \frac{0.0592}{6} (\log 7.8 + \log 10^{-4})$$

$$= 1.41 - \frac{0.0592}{6} (-3.11)$$

$$= 1.41 + 0.0307 = 1.44$$

Notice that this time Q is less than 1, and therefore log Q is negative. This results in a value for E that is greater than E°.

17.10 REDOX TITRATIONS

The concepts of equivalent and equivalent weight were introduced in the context of acid-base reactions in Chapter 9. In redox reactions, **one equivalent is defined as that quantity of a reagent that is involved in the transfer of one mole of electrons in a redox reaction.** Note that, as before, the equivalent is related to a specific chemical reaction. The number of equivalents in one mole of a reactant is the number of moles of electrons transferred by one mole of reactant. More simply, the number of *equivalents per mole can be found by calculating the total change in oxidation number per formula unit.* The following examples illustrate the concept:

Reaction	Total oxidation number change	Equivalents per mole
$MnO_4^- \rightarrow MnO_2$	$+7 \rightarrow +4$ or -3	3 eq/mole MnO_4^- 3 eq/mole MnO_2
$MnO_4^- \rightarrow Mn^{2+}$	$+7 \rightarrow +2$ or -5	5 eq/mole MnO_4^- 5 eq/mole Mn^{2+}
$Cr_2O_7^{2-} \rightarrow 2\ Cr^{3+}$	$+6 \rightarrow +3$ or -3	6 eq/mole $Cr_2O_7^{2-}$ 6 eq/2 moles Cr^{3+} or 3 eq/mole Cr^{3+}

The equivalent weight of a species in a redox reaction is, as in acid-base reactions, the number of grams per equivalent. It may be found by dividing the molar weight by equivalents per mole:

$$\frac{\text{grams}}{\text{mole}} \div \frac{\text{equivalents}}{\text{mole}} = \frac{\text{grams}}{\text{mole}} \times \frac{\text{moles}}{\text{equivalent}} = \frac{\text{grams}}{\text{equivalent}} \quad (17.12)$$

Example 17.16. What are the equivalent weights of the species listed below?

a. SO_2 in $SO_2 \rightarrow SO_4^{2-}$
b. $SnCl_4$ in $SnCl_4 \rightarrow Sn^{2+}$
c. HNO_3 as a source of NO_3^- in $NO_3^- \rightarrow NO_2$
d. KI as a source of I^- in $I_2 \rightarrow 2\ I^-$

e. I_2 in $I_2 \rightarrow 2\ I^-$
f. $K_2Cr_2O_7$ as a source of $Cr_2O_7^{2-}$ in $Cr_2O_7^{2-} \rightarrow 2\ Cr^{3+}$

Set up and solve in accord with Equation 17.12.

17.16a.

a. $$\frac{64.1\text{ g } SO_2}{\text{mole}} \times \frac{1\text{ mole}}{2\text{ equivalents}} = 32.0\ \frac{\text{grams } SO_2}{\text{equivalent}}$$

b. $$\frac{261\text{ g } SnCl_4}{1\text{ mole}} \times \frac{1\text{ mole}}{2\text{ equivalents}} = 130\text{ g } SnCl_4/\text{equivalent}$$

c. $$\frac{63\text{ g } HNO_3}{1\text{ mole}} \times \frac{1\text{ mole}}{1\text{ equivalent}} = 63\text{ g } HNO_3/\text{equivalent}$$

d. $$\frac{167\text{ g KI}}{1\text{ mole}} \times \frac{1\text{ mole}}{1\text{ equivalent}} = 167\text{ g KI/equivalent}$$

e. $$\frac{254\text{ g } I_2}{1\text{ mole}} \times \frac{1\text{ mole}}{2\text{ equivalents}} = 127\text{ g } I_2/\text{equivalent}$$

f. $$\frac{294\text{ g } K_2Cr_2O_7}{1\text{ mole}} \times \frac{1\text{ mole}}{6\text{ equivalents}} = 49.0\text{ g } K_2Cr_2O_7/\text{equivalent}$$

Inasmuch as the number of electrons lost by one species in a redox reaction is equal to the number gained by another, the number of equivalents of both species is the same. In fact, the previous generalization that the number of equivalents of all species in a reaction must be equal remains valid. Another valid statement is that the number of equivalents is equal to volume × normality, where normality is equivalents per liter for a specific reaction. Consequently, $V_1N_1 = V_2N_2$ holds for redox reactions.

Example 17.17. What is the normality of a Fe^{2+} solution if a 25.0-ml sample requires 29.3 ml of 0.442 N $KMnO_4$ in this titration reaction: $5\ Fe^{2+} + MnO_4^- + 8\ H^+ \rightarrow 5\ Fe^{3+} + Mn^{2+} + 4\ H_2O$?

Don't look for something difficult. Set up and solve.

17.17a. N = 0.518 N Fe^{2+}.

$$V_1N_1 = V_2N_2; \quad (25.0\text{ ml}) \times N = (29.3\text{ ml}) \times 0.442; \quad N = 0.518$$

As long as the normalities are stated *for the reaction used,* the $V_1N_1 = V_2N_2$ relation holds.

How would you go about preparing such a solution of $FeSO_4$?

Example 17.18. How many grams of $FeSO_4 \cdot (NH_4)_2SO_4 \cdot 6\ H_2O$ are required to prepare 500 ml of 0.500 N Fe^{2+} solution for the reaction in Example 17.17? (Note: $FeSO_4$ as such is not used for the preparation of such solutions; instead the compound given above is used.)

First, how many equivalents per mole of $FeSO_4 \cdot (NH_4)_2SO_4 \cdot 6\ H_2O$?

17.18a. One equivalent per mole. One mole of $FeSO_4 \cdot (NH_4)_2SO_4 \cdot 6\ H_2O$ yields one mole of Fe^{2+}, which then undergoes an oxidation number change of 1 according to the equation. Therefore there is one equivalent per mole.

Now complete the problem in the manner developed in Chapter 9 (see page 134).

17.18b.

$$0.500\ \cancel{\ell} \times \frac{0.500\ \cancel{eq}\ FeSO_4 \cdot (NH_4)_2SO_4 \cdot 6\ H_2O}{\cancel{\ell}} \times \frac{1\ \cancel{mole}}{1\ \cancel{eq}} \times \frac{392\ g}{1\ \cancel{mole}}$$

$$= 98.0\ g\ FeSO_4 \cdot (NH_4)_2SO_4 \cdot 6\ H_2O$$

PROBLEMS

Section 17.4

17.1. In a laboratory experiment a student deposits nickel from a Ni^{2+} solution for 35.0 minutes at 0.65 amperes. Calculate the theoretical yield of nickel. •

17.2. Copper sulfate plating solutions are used for a variety of industrial purposes, including the making of electrotypes (printing plates) and stampers from which phonograph records are pressed. How many pounds of copper anode will be dissolved into the Cu^{2+} solution in an electrotyping tank over an eight hour working day if the average current through the tank is 3000 amperes?

17.3. Elemental aluminum is obtained electrolytically from alumina, Al_2O_3, which is dissolved in a "solvent" of molten cryolite, $AlF_3 \cdot 3\ NaF$. How many pounds of aluminum can be recovered in a seven hour day in a system in which current is held constant at 50,000 amperes?

17.4. Chromium plating, the kind done on automobile bumpers and other hardware, is performed with a solution in which chromium is present in the +6 oxidation state. The method has a poor "cathode efficiency," meaning that only about 15% of the current at the cathode yields chromium, and the other 85% is wasted in other reduction processes. If a

17.38. Electrodeposited cadmium is widely used to protect iron and steel products from corrosion. While the actual process involves a complex ion of cadmium, the reduction is adequately represented by $Cd^{2+} + 2\ e^- \rightarrow Cd$. Calculate the grams of cadmium deposited by a current of 75 amperes in an 8.0 minute plating cycle.

17.39. "Galvanized" iron is iron that has been coated with zinc to protect it from electrolytic corrosion. Assuming the zinc is reduced to the metallic state from an oxidation state of +2, what is the rate at which zinc is plated in kilograms/hour in a tank through which 2600 amperes pass?

17.40. Gold plating is performed for functional purposes, as in electronics applications, and for its decorative value, as on jewelry. Deposition is from a bath in which the gold is at a +1 oxidation state in a complex ion. How many grams of gold will be deposited in a three hour production run at 18 amperes? What will be its value at \$174 per troy ounce (1 troy ounce = 31.1 grams)?

17.41. "Electroforming" is the name given to the process which creates a metallic object of intricate shape by plating a heavy metal coating onto a plastic mold made electroconductive by a layer of about two millionths of an inch of silver, and then removing the mold. Paint spray masks, used for production painting of different colors on selected areas of plastic toys

bumper plating line is powered by 18,000 amperes, how many kilograms of chromium will be deposited in an eight hour shift?

17.5 Parts about to be electroplated are frequently cleaned by electrolysis in an alkaline solution. The process is essentially the decomposition of water. The bubbles of hydrogen or oxygen released at the surface of the part—it may be either anode or cathode or both in a given electrocleaning cycle—assist in dislodging foreign objects that would interfere with good adherence of the electrodeposit. How many liters of oxygen, measured at 40°C and 733 torr, are liberated in a two minute cycle at 350 amperes? The oxidation half-reaction may be represented by $4\ OH^- \rightarrow O_2 + 2\ H_2O + 4\ e^-$.

17.6. A Down's cell has been built for the purpose of making sodium by the electrolysis of molten NaCl. Production is figured at 20 pounds of sodium per hour. What ampere rating must be specified for the rectifier that will power this installation?

17.7. Hydrogen and oxygen are produced commercially by the electrolytic decomposition of water in a 15% NaOH solution. Specify the current rating of a rectifier that will yield 1500 liters of hydrogen, measured at 20.0 atmospheres and 25°C, per 8 hour shift. The reduction equation in an alkaline solution is $2\ H_2O + 2\ e^- \rightarrow H_2 + 2\ OH^-$.

17.8. In electroforming a paint spray mask from a Ni^{2+} solution, it is determined that 1000 grams of nickel must be deposited to produce adequate thickness at its weakest point. The mask is to be deposited at 50.0 amperes. How many hours will be required?

17.9. A chemistry teacher is planning to demonstrate the electrolysis of water in a Hoffman apparatus that is capable of collecting a maxi-

and models, are one example. A Ni^{2+} solution from which such masks are electroformed operates at 95% cathode efficiency, in which 95% of the current deposits nickel and 5% releases hydrogen gas from the acidic solution. Calculate the grams of nickel deposited and liters of hydrogen released at STP by 1500 amperes in 21 hours.

17.42. Magnesium is extracted from sea water by precipitating it as $Mg(OH)_2$, neutralizing it with HCl, and evaporating the resulting $MgCl_2$ solution to dryness. The solid $MgCl_2$ is melted and electrolyzed, yielding magnesium pigs 99.8% pure. The chlorine oxidized at the anode is recycled and converted to HCl for the neutralization step. How many liters of chlorine measured at 0.920 atm and 140°C, are released in 7.50 hours from a 10,000 ampere installation?

17.43. When copper is plated on iron or steel, an alkaline bath, in which copper is present in a +1 oxidation state as part of a complex ion, is used. What current rating is required on a rectifier that is to deposit 70.0 pounds of copper from such a solution in a 7.50 hour day?

17.44. Caustic soda (NaOH) and chlorine are manufactured simultaneously by the electrolysis of aqueous NaCl. The overall equation for the reaction is

$$2\ NaCl(aq) + 2\ H_2O(l) \rightarrow 2\ NaOH(aq) + Cl_2(g) + H_2(g).$$

What ampere capacity is required in the generator that will produce 1000 kilograms of NaOH per 24 hour day? (Hint: By stoichiometry, determine the amount of chlorine that corresponds with the NaOH.)

17.45. The ends of heavy copper bus bars are sometimes silver plated where they are to be bolted together to assure better electrical contact and minimum heat losses. On one such occasion it was determined that 12.5 grams of silver were required, and that it could be deposited from a Ag^+ solution at 7.00 amperes. How many minutes should the bars be left in the tank?

17.46. A complete new charge of copper anodes in a Cu^{2+} plating solution weighs 900 pounds. How many days will they last at 10.0

mum of 100 ml of either gas produced, hydrogen or oxygen. He will work with a current of 0.50 amperes. He also plans to set a timer to sound when 85 milliliters of the larger volume gas has collected, this volume to be measured at a temperature of 25°C and a partial pressure of 720 torr. For how many minutes should the timer be set?

production hours per day if we assume a steady current of 2700 amperes?

Section 17.6. Problems 17.10–17.14 and 17.47–17.51: Using the standard Reduction Potential Table on page 294 as a source of half-reaction equations and E° values, (a) write the half-reaction equations for each of the redox reactions indicated; (b) modify them, if necessary, in such a way as to yield the balanced net ionic equation when they are added; (c) compute E° for the reaction as written; and (d) predict the favored direction of the reaction.

17.10. $Fe + Sn^{2+} \rightarrow Fe^{2+} + Sn$

17.11. $Fe^{2+} + Sn^{2+} \rightarrow Fe + Sn^{4+}$

17.12. $Al + Na^{+} \rightarrow Al^{3+} + Na$

17.13. $Pt + Cl^{-} + NO_3^{-} + H^{+} \rightarrow PtCl_6^{2-} + NO + H_2O$

17.14. $Cr_2O_7^{2-} + H^{+} + Mn^{2+} \rightarrow Cr^{3+} + MnO_2 + H_2O$

17.47. $Fe^{2+} + Sn^{2+} \rightarrow Fe^{3+} + Sn$

17.48. $Fe^{3+} + Sn^{2+} \rightarrow Fe^{2+} + Sn^{4+}$

17.49. $Mg + Al^{3+} \rightarrow Mg^{2+} + Al$

17.50. $Au + Cl^{-} + NO_3^{-} + H^{+} \rightarrow AuCl_4^{-} + NO_2 + H_2O$

17.51. $MnO_4^{-} + H^{+} + Hg \rightarrow Mn^{2+} + H_2O + Hg_2^{2+}$

Section 17.7. Problems 17.15–17.19 and 17.52–17.56: Using the results of Problems 17.10–17.14 and 17.47–17.51, calculate ΔG° for the reaction of the problem number given, and predict the favored direction. Compare your prediction with that based on E°.

17.15. Problem 17.10.

17.16. Problem 17.11.

17.17. Problem 17.12.

17.18. Problem 17.13.

17.19. Problem 17.14.

17.52. Problem 17.47.

17.53. Problem 17.48.

17.54. Problem 17.49.

17.55. Problem 17.50.

17.56. Problem 17.51.

Section 17.8. Problems 17.20–17.24 and 17.57–17.61: Using the results of Problems 17.10–17.14 and 17.47–17.51, calculate K for the reaction of the problem number given and predict the favored direction. Compare your prediction with those based on E° and ΔG°.

17.20. Problem 17.10.

17.21. Problem 17.11.

17.22. Problem 17.12.

17.23. Problem 17.13.

17.24. Problem 17.14.

17.57. Problem 17.47.

17.58. Problem 17.48.

17.59. Problem 17.49.

17.60. Problem 17.50.

17.61. Problem 17.51.

17.25. If E° = +0.355 volt for the half-reaction $AgIO_3 + e^- \rightarrow Ag + IO_3^-$, determine the solubility product constant for silver iodate. (The reduction potential for Ag^+ may be found in Table IX, page 294.)

17.26. If E° = +0.01 volt for the reduction half-reaction $Ag(S_2O_3)_2^{3-} + e^- \rightarrow Ag + 2\ S_2O_3^{2-}$, calculate K_d for $Ag(S_2O_3)_2^{3-} = Ag^+ + 2\ S_2O_3^{2-}$.

17.62. The standard state reduction potential for the calomel electrode, one of the electrodes commonly found in pH meters, is +0.27 volt. The reduction half-reaction is $Hg_2Cl_2 + 2\ e^- \rightarrow 2\ Hg + 2\ Cl^-$. Using the E° of the mercury(I) ion, Hg_2^{2+}, from Table IX, page 294, find K_{sp} for mercury(I) chloride: $Hg_2Cl_2 = Hg_2^{2+} + 2\ Cl^-$.

17.63. Determine K_d for the dissociation of the complex ion $AuCl_4^-$: $AuCl_4^- = Au^{3+} + 4\ Cl^-$. The necessary data may be found in Table IX (page 294).

Section 17.9 Problems 17.27, 17.28, 17.64 and 17.65: Calculate the theoretical voltages developed by the cells described, in which pure electrodes are immersed in solutions of their own ions having the concentrations given. Also predict the direction of the spontaneous reaction.

17.27. Cell equation: $Fe + Co^{2+} \rightarrow Fe^{2+} + Co$. $[Fe^{2+}] = 0.010$; $[Co^{2+}] = 0.50$.

17.28. Cell equation: $2\ Al + 3\ Sn^{2+} \rightarrow 2\ Al^{3+} + 3\ Sn$. $[Al^{3+}] = 0.020$; $[Sn^{2+}] = 0.10$.

17.29. The calomel reference electrode in pH meters involves the reduction half-reaction $Hg_2Cl_2 + 2\ e^- \rightarrow 2\ Hg + 2\ Cl^-$. If the solution is 1 molar in Cl^-, the reduction potential against a standard hydrogen electrode is +0.280 volt. But the electrode most commonly used in pH meters is saturated in KCl, which fixes $[Cl^-]$ at about 4.6 moles per liter. Calculate the emf of the saturated calomel electrode against the standard hydrogen electrode.

17.64. Cell equation: $Ni + 2\ Ag^+ \rightarrow Ni^{2+} + 2\ Ag$. $[Ni^{2+}] = 0.80$; $[Ag^+] = 0.10$.

17.65. Cell equation: $Ni + Sn^{2+} \rightarrow Ni^{2+} + Sn$. $[Ni^{2+}] = 1.30$; $[Sn^{2+}] = 1.0 \times 10^{-5}$.

17.66. A concentration cell is one in which both reduction and oxidation half-cells have the same metal immersed in a solution of its own ions. The solution concentrations are different, however, and this alone is responsible for the emf developed by the cell. The net ionic equation of such a cell with copper electrodes in copper(II) solutions might be represented as Cu^{2+} (aq, 1 M) $\rightarrow$ Cu^{2+} (aq, 0.001 M). Calculate the potential of the cell and predict whether the forward or reverse reaction is favored relative to the equation as written.

Section 17.10 Problems 17.30–17.32 and 17.67–17.69: Compute (a) the equivalents per mole and (b) the equivalent weight of the substance shown for the redox change given.

17.30. $KMnO_4$ in $MnO_4^- \rightarrow MnO_2$.

17.31. Au in $Au(s) \rightarrow AuCl_4^-(aq)$.

17.32. $H_2C_2O_4 \cdot 2\ H_2O$ in $C_2O_4^{2-} \rightarrow CO_2$.

17.67. KIO_3 in $IO_3^- \rightarrow I_2$.

17.68. H_3AsO_3 in $H_3AsO_3 \rightarrow H_3AsO_4$.

17.69. Hg_2Cl_2 in $Hg_2Cl_2 \rightarrow 2\ Hg$.

Questions 17.33–17.37 represent a sequence of operations in an analytical procedure commonly used in college laboratories. The numerical results from one step do not extend to the next step. Problems 17.70–17.74 are, for the most part, unrelated to each other, nor do they match the corresponding problem in the left column.

17.33. What quantity of $KMnO_4$ is required to prepare 500 ml 0.10 N $KMnO_4$ for the reaction $MnO_4^- \rightarrow Mn^{2+}$?

17.34. The $KMnO_4$ solution of Problem 17.33 is to be standardized against a solution of $Na_2C_2O_4$, which can be prepared much more accurately than the $KMnO_4$ solution. How many grams of $Na_2C_2O_4$ must be dissolved in a 250 ml solution to make a solution 0.10 N in $C_2O_4^{2-}$? The equation for the standardization reaction is $5\ C_2O_4^{2-} + 2\ MnO_4^- + 16\ H^+ \rightarrow 10\ CO_2 + 2\ Mn^{2+} + 8\ H_2O$.

17.35. In preparing a solution as in Problem 17.34, the quantity of $Na_2C_2O_4$ actually diluted to 250.0 ml was 1.734 grams. Calculate the normality of this solution for the standardization of the $KMnO_4$.

17.36. The normality of a $Na_2C_2O_4$ solution prepared as in Problem 17.35 was 0.1035. A 50.00 ml sample of this solution required 48.20 ml of $KMnO_4$ solution for standardization by the equation in Problem 17.34. What was the normality of the $KMnO_4$?

17.37. A $KMnO_4$ solution standardized as in Problem 17.36 has a normality of 0.1074. It took 37.8 ml of this solution to oxidize 283 milligrams of an unidentified reducing agent in a reaction in which it was known that the MnO_4^- was converted to Mn^{2+}, as in the standardization reaction. What is the equivalent weight of the reducing agent?

17.70. Calculate the grams of $K_2Cr_2O_7$ required to prepare 250 ml 0.200 N $K_2Cr_2O_7$ for the reaction $Cr_2O_7^{2-} + 14\ H^+ + 6\ e^- \rightarrow 2\ Cr^{3+} + 7\ H_2O$.

17.71. In a redox titration between Ce^{4+} and H_3AsO_3, 12.5 ml of the H_3AsO_3 solution react with 10.0 ml 0.770 *N* Ce^{4+} by the equation $2\ Ce^{4+} + H_3AsO_3 + H_2O \rightarrow 2\ Ce^{3+} + H_3AsO_4 + 2\ H^+$. Find the normality of the H_3AsO_3.

17.72. How many grams of H_3AsO_3 are in 100 milliliters of the solution analyzed in Problem 17.71?

17.73. Arsenic(III) oxide, As_2O_3, is a primary standard for iodine solutions. When the oxide is placed in water it reacts to form H_3AsO_3 by the equation $As_2O_3 + 3\ H_2O \rightarrow 2\ H_3AsO_3$. What is the normality of an iodine solution if 25.8 milliliters are required in a titration with a solution containing 0.178 gram of pure As_2O_3 if the standardization equation is $H_3AsO_3 + I_2 + H_2O \rightarrow H_3AsO_4 + 2\ I^- + 2\ H^+$?

17.74. A solution is 0.050 M in BrO_3^-. What is its normality when used as an oxidizer in a titration in which it undergoes the following half-reaction: $BrO_3^- + 6\ H^+ + 6\ e^- \rightarrow Br^- + 3\ H_2O$?

APPENDIX I

REVIEW OF MATHEMATICS*

Arithmetic and Algebra

1. Reciprocals. For any non-zero number, n, the reciprocal is $\frac{1}{n}$. For the fraction $\frac{a}{b}$, the reciprocal is $\frac{b}{a}$ $(a \neq 0 \neq b)$.

If any number is multiplied by its reciprocal, the product is equal to 1: $\frac{a}{b} \times \frac{b}{a} = \frac{ab}{ab} = 1$. In this sense the reciprocal of a number is referred to as its multiplicative inverse.

2. Multiplication. Multiplication is commutative. This means that the product is the same regardless of the order in which factors are used: $a \times b = b \times a$. Multiplication is also associative. When three or more factors are to be multiplied, they may be grouped in any possible way as well as taken in any order $a \times b \times c = (a \times b) \times c = a \times (b \times c) = (a \times c) \times b$.

In multiplying two or more fractions, the numerator of the product is the product of the numerators; and the denominator of the product is the product of the denominators: $\frac{a}{b} \times \frac{c}{d} = \frac{ac}{bd}$.

3. Division. A fraction is an indicated division: $\frac{a}{b} = a \div b$.

The process of dividing by a non-zero number, n, is essentially the same as multiplying by the reciprocal of the number: $a \div b = a \times \frac{1}{b}$. To divide by a fraction, you multiply by its inverse—"invert and multiply":

$$\frac{a}{b/c} = a \div \frac{b}{c} = a \times \frac{c}{b} = \frac{ac}{b}.$$

Notice that the denominator of the denominator fraction, c, appears in the numerator of the result.

---

*This section of the Appendix serves as a brief reminder of mathematical facts and concepts that arise in general chemistry problems. For a more detailed review of pertinent mathematics you are referred to Masterton and Slowinski, *Elementary Mathematical Preparation for General Chemistry* (W. B. Saunders Company, 1974).

Any non-zero number divided by itself yields 1 as the quotient: $\frac{n}{n} = 1$. It follows that any fraction in which the numerator equals the denominator has the value 1: if $x + y = z$, then $\frac{x + y}{z} = 1 = \frac{z}{x + y}$; if $m \times n = w$, then $\frac{m \times n}{w} = 1 = \frac{w}{m \times n}$.

4. Multiplication by One. If a quantity is multiplied by 1, the product is equal to the original quantity: $n \times 1 = n$. The number 1 may assume any form, which may change the form of the product. In its most frequent chemical application, conversion of units, 1 takes the form of a fraction in which the numerator is equal to the denominator. Thus if $m = bn$, the quantity am may be expressed in terms of n by multiplying by 1 in the form $\frac{bn}{m}$: $am \times \frac{bn}{m} = abn$.

5. "Cancellation." Factors common to the numerator and denominator of a fraction may be cancelled: $\frac{\not{a}z}{\not{a}y} = \frac{z}{y}$. If two or more fractions are to be multiplied, and the numerator of one contains a factor that appears in the denominator of another, those factors may be cancelled: $\frac{a\not{n}}{b} \times \frac{c}{\not{n}} = \frac{ac}{b}$. In both cases, cancellation is essentially multiplication by 1:

$$\frac{az}{ay} = \frac{a}{a} \times \frac{z}{y} = 1 \times \frac{z}{y} = \frac{z}{y}$$

$$\frac{an}{b} \times \frac{c}{n} = \frac{ac}{b} \times \frac{n}{n} = \frac{ac}{b} \times 1 = \frac{ac}{b}$$

6. Exponentials. A number raised to a power is an exponential. The power, or exponent, tells how many times the number is to be used as a factor. Thus $a^2 = a \times a$; $b^5 = b \times b \times b \times b \times b$; and $c^n = c \times c \times c \times \ldots$ until c is used as a factor n times.

An exponential having an exponent of zero equals 1: $n^0 = 1$ ($n \neq 0$).

In multiplying exponentials of the same base, exponents are added: $a^m \times a^n = a^{m+n}$. In dividing exponentials of the same base, the denominator exponent is subtracted from the numerator exponent: $\frac{a^m}{a^n} = a^{m-n}$. This operation is related to the fact that an exponential may be moved from the denominator to the numerator of a fraction, or vice versa, by changing its sign:

$$\frac{1}{M^n} = M^{-n} \quad \text{and} \quad N^k = \frac{1}{N^{-k}}$$

Solving an Equation for an Unknown

Solving an equation for an unknown involves manipulating the equation in such a manner that the unknown is left as the only item on one side of the equation, all known items being on the other side. "Manipulating" an equation takes on several forms, but the essential feature is that *whatever operations are performed on one side of the equation must be performed on the other.* The objective in each of the following examples is to solve the equation for the unknown x.

Example		Method
$x + a = b$ $x + a - a = b - a$ $x = b - a$	$x - a = b$ $x - a + a = b + a$ $x = b + a$	Add or subtract a and simplify
$b = a - x$ $b + x - b = a - x + x - b$ $x = a - b$		Add $x - b$ and simplify
$ax = b$ $\frac{\cancel{a}x}{\cancel{a}} = \frac{b}{a}$ $x = \frac{b}{a}$	$\frac{x}{a} = b$ $\frac{x}{\cancel{a}} \cdot \cancel{a} = b \cdot a$ $x = ba$	Multiply or divide by a
$\frac{b}{c} = \frac{a}{x}$ $\frac{\cancel{b}}{\cancel{c}} \cdot \frac{\cancel{c}x}{\cancel{b}} = \frac{a}{\cancel{x}} \cdot \frac{c\cancel{x}}{b}$ $x = \frac{ac}{b}$.		Multiply by $\frac{cx}{b}$

Another common approach to the last example is to "cross multiply." Thus if $\frac{b}{c} = \frac{a}{x}$, then $bx = ac$. Dividing both sides by b yields $x = \frac{ac}{b}$.

A quadratic equation may be solved for a variable by means of the quadratic formula. If the equation is in the form $ax^2 + bx + c = 0$, then

$$x = \frac{-b \pm \sqrt{b^2 - 4ac}}{2a}.$$

Proportionalities

If one quantity is directly proportional to another, a change in either yields a proportional change in the other. For example, the circumference of a circle is proportional to the diameter. Halve one, and the other is halved; triple one, and the other is tripled. A proportional relationship may be written in the form $c \propto d$. Proportionalities may be converted to equations by introducing an appropriate proportionality constant. For the circumference-diameter proportionality, the proportionality constant is pi, π. Thus $c \propto d$ becomes $c = \pi d$.

If one quantity is proportional to the reciprocal of another they are said to be inversely proportional to each other. If a is inversely proportional to b, $a \propto \frac{1}{b}$.

Exponential Notation

Calculations with the very large and very small numbers encountered in chemistry are simplified through exponential notation, in which the numbers are expressed as the product of a decimal factor and an exponential factor. The decimal factor usually is selected so its value is between 1 and 10; the exponential is ten raised to some integral power. One way of converting a number in its usual form to exponential notation is to multiply the number by 1 in the form $10^n \times 10^{-n}$, in which n is the number of places the decimal must be moved to fix the decimal factor between 1 and 10. Proper grouping of factors and simplification yields the desired expression. For example:

180,000,000

To set the decimal factor between 1 and 10 the decimal must be moved 8 places left. Therefore,

$180{,}000{,}000 \times 10^{-8} \times 10^{8}$

$(180{,}000{,}000 \times 10^{-8}) \times 10^{8}$

1.8×10^{8}

0.000 000 000 025

To set the decimal factor between 1 and 10 the decimal must be moved 11 places right. Therefore,

$0.000\,000\,000\,025 \times 10^{11} \times 10^{-11}$

$(0.000\,000\,000\,025 \times 10^{11}) \times 10^{-11}$

2.5×10^{-11}

In multiplying and dividing numbers, some or all of which are expressed in exponential notation, decimal factors and exponential factors are treated in groups, with a final adjustment made to express the result in the desired form:

$$\frac{(3.56 \times 10^{-6})(4.82)(193)}{(0.362)(2.09 \times 10^{5})(9.32 \times 10^{-17})} = \frac{(3.56)(4.82)(193)}{(0.362)(2.09)(9.32)} \times \frac{10^{-6}}{(10^{5})(10^{-17})}$$

$$= 470 \times (10^{-6} \times 10^{-5} \times 10^{17})$$

$$= (470 \times 10^{-2})(10^{2}) \times 10^{6}$$

$$= 4.70 \times (10^{2})(10^{6})$$

$$= 4.70 \times 10^{8}$$

As in arithmetic, addition and subtraction of exponentials requires that digit values (hundreds, units, tenths, etc.) be aligned vertically. This may be achieved by adjusting decimals and exponents so the exponents of all numbers are the same, and then proceeding as in ordinary arithmetic. Shown below is the addition of $6.44 \times 10^{-7} + 1.3900 \times 10^{-5}$ in three ways: with exponents of 10^{-5}, with exponents of 10^{-7}, and as decimal numerals.

1.3900×10^{-5}	=	1.3900×10^{-5}	=	139.00×10^{-7}	=	0.000013900
6.44×10^{-7}	=	0.0644×10^{-5}	=	6.44×10^{-7}	=	0.000000644
		1.4544×10^{-5}	=	145.44×10^{-7}	=	0.000014544

APPENDIX II

LOGARITHMS*

The Nature of Logarithms

A logarithm of a number is the exponent to which a certain base must be raised to express the number. In our number system the most convenient base is 10. Therefore,

$$\text{if } n = 10^x, \text{ then } \log n = \log 10^x = x \qquad (II.1)$$

In chemistry, log n is the same as $\log_{10}n$, the base 10 being understood even if not identified by a subscript.

The quantity 100 can be expressed as 10^2. Log 100 is therefore log 10^2, or 2. Similarly, log 1000 = log 10^3 = 3. Logarithms can also be negative. 0.1 can be written 10^{-1}. It follows that log 0.1 = log 10^{-1} = −1. The logarithm of 1 is zero. This is apparent when we recognize that any base raised to the zeroth power is 1. Thus $1 = 10^0$, and $\log 1 = \log 10^0 = 0$.

The above examples suggest correctly that the logarithm of any multiple or submultiple of 10 is an integer. The logarithms of other numbers generally involve decimal fractions: log 2 = 0.3010 to four decimal places. This means that $10^{0.3010} = 2$. Similarly, log 3 = 0.4771, meaning $10^{0.4771} = 3$. Four place logarithms (logarithms to four decimal places) of numbers from 1 to 10 are given in Table II at the end of this section. Three place logarithms may be determined from the L and C or D scales of most 10 inch slide rules.

Multiplication and Division by Logarithms

Being exponents, logarithms are governed by the laws of exponents. In the multiplication of exponentials, exponents are added: $a^m \times a^n = a^{m+n}$. Similarly, 2 × 3 becomes

$$10^{0.3010} \times 10^{0.4771} = 10^{0.3010+0.4771} = 10^{0.7781}.$$

This should lead us to expect that log 6 = 0.7781, an expectation that is confirmed in Table II, allowing for a 0.0001 round off. This example illustrates a fundamental fact regarding logarithms:

$$\log xy = \log x + \log y. \qquad (II.2)$$

*This section of the Appendix presents a brief review of the logarithmic operations used in this book. For a more detailed treatment of the application of logarithms in chemistry the reader is referred to Masterton and Slowinski, *Elementary Mathematical Preparation for General Chemistry* (W. B. Saunders Company, 1974).

In the division of exponentials the denominator exponent is subtracted from the numerator exponent. The problem $\frac{3}{2} = 1.5$ can be set up in exponential form as

$$\frac{3}{2} = \frac{10^{0.4771}}{10^{0.3010}} = 10^{0.4771-0.3010} = 10^{0.1761}.$$

0.1761 should therefore be the logarithm of 1.5—and checking Table II we find this to be the case. This problem is an example of the fact that

$$\log \frac{x}{y} = \log x - \log y. \quad \text{(II.3)}$$

Finding the Logarithm of a Number

Exponential notation is a helpful tool in finding the logarithm of a number that is not in the range from 1 to 10. Take 17, for example. Expressed in exponential notation, log 17 is the logarithm of 1.7×10^1, or $\log (1.7 \times 10^1)$. According to Equation II.2,

$$\log 17 = \log (1.7 \times 10^1) = \log 1.7 + \log 10^1.$$

The log of 1.7 from the table is 0.2304; the log of 10^1 is 1. Thus $\log 17 = \log. 1.7 + \log 10^1 = 0.2304 + 1 = 1.2304$.

An important feature about logarithms becomes apparent if we extend our example with 17 to finding the logarithms of 170 and 1700:

$$\log 170 = \log(1.7 \times 10^2) = \log 1.7 + \log 10^2 = 0.2304 + 2 = 2.2304$$

$$\log 1700 = \log(1.7 \times 10^3) = \log 1.7 + \log 10^3 = 0.2304 + 3 = 3.2304$$

Including the log of 1.7 in a summary of results to this point we see:

$$\log 1.7 = \log(1.7 \times 10^0) = 0.2304$$

$$\log 17 = \log(1.7 \times 10^1) = 1.2304$$

$$\log 170 = \log(1.7 \times 10^2) = 2.2304$$

$$\log 1700 = \log(1.7 \times 10^3) = 3.2304$$

Notice that, following the decimal point, all logarithms have the digits 2304, matching the logarithm of 1.7, or 0.2304. This number—the number after the decimal point in a logarithm—is called the **mantissa.** The integer before the decimal point, called the **characteristic,** is simply the exponent of 10. It locates the decimal in the number whose logarithm is shown. Finding the logarithm of a number expressed in exponential notation is summarized in the equation:

$$\log (A \times 10^n) = \log A + n. \quad \text{(II.4)}$$

Equation II.4 is useful for numbers between 0 and 1 also. For example,

$$\log 0.000017 = \log (1.7 \times 10^{-5}) = 0.2304 + (-5) = -4.7696.$$

The mantissa is not 0.2304 this time, but 1.0000 − 0.2304, as it will be for the logarithm of all numbers 1.7×10^n where n is negative.

Finding the Antilogarithm of a Logarithm

The number that corresponds with a given logarithm is called the **antilogarithm,** or **antilog,** of a given logarithm. Reversing Equation II.1, if $x = \log n$, then the antilog of $x = 10^x = n$. Thus the antilog of 0.3010 is $10^{0.3010}$, or 2, meaning that 2 is the number whose logarithm is 0.3010. Similarly, $10^{0.4771}$, or 3, is the antilog of 0.4771; $10^{0.2304}$, or 1.7, is the antilog of 0.2304; and $10^{-4.7696}$, or 0.000017, is the antilog of −4.7696. If the logarithm of a number is known as a decimal fraction, it may be located in Table II and its corresponding antilog identified.

To find the exponential form of a number, n, whose logarithm is expressed as a decimal number, the following steps may be followed:

STEPS	Example: Find n, the antilog of 3.2304
1. Write the number as an exponential of 10, in which the logarithm is the exponent.	$n = 10^{3.2304}$
2. Rewrite the logarithm (exponent) as the sum of a **positive decimal fraction** and the **largest integer less than** the logarithm.	Positive decimal fraction ↓ Integer ↓ $n = 10^{0.2304\ +\ 3}$
3. Rewrite the exponential as the product of two exponentials ($a^{m+n} = a^m \times a^n$).	$n = 10^{0.2304} \times 10^3$
4. Replace the factor with the fractional exponent with its antilog from Table II.	$n = 1.7 \times 10^3$

Selection of the decimal fraction and integer is a bit trickier if the logarithm is negative. For example, consider the antilog of −6.62. The antilog may be expressed as $10^{-6.62}$ (Step 1). Now we rewrite the exponent as the sum of a *positive* decimal fraction and the *closest integer smaller than* (more negative than) *the logarithm.* That integer is −7, the next integer less than −6.62. The positive decimal fraction is found by subtracting the integer from the logarithm: $-6.62 - (-7) = 0.38$. This give us $10^{0.38\ +(-7)} = 10^{0.38-7}$ (Step 2). The antilog is expressed as the product of exponentials in $10^{0.38} \times 10^{-7}$ (Step 3), and the antilog of the fractional exponent is substituted for the first factor, yielding 2.4×10^{-7} (Step 4).

Raising a Number to the n-th Power by Logarithms

To square an exponential, you double the exponent:

$$(m^a)^2 = m^a \times m^a = m^{a+a} = m^{2a}.$$

A logarithm is an exponent. Therefore to square a number by logarithms you find its logarithm (an exponent) and double it, yielding the logarithm of the square of the number. The antilog is therefore the quantity sought. Using 3^2 as an example, $\log 3 = 0.4771$. Doubling, $2 \log 3 = 2 \times 0.4771 = 0.9542$. From Table II, antilog $0.9542 = 9$.

To cube an exponential you triple the exponent: $(a^m)^3 = a^{3m}$. The log of the cube of a number is 3 times the log of the number. In general, to raise a number to the n-th power—$(a^m)^n = a^{nm}$. By logarithms, the logarithm of the n-th power of a number is n times the logarithm of the number:

$$\log a^n = n \log a. \qquad \text{(II.5)}$$

For example, to find 3^4 by logarithms,

$$\log 3^4 = 4 \log 3 = 4 \times 0.4771 = 1.9084$$

$$\text{Antilog } 1.9084 = 10^{0.9804+1} = 8.1 \times 10^1 = 81$$

$$\text{Confirming, } 3^4 = 3 \times 3 \times 3 \times 3 = 81$$

To find the n-th root of a number is the same as raising the number to the $\frac{1}{n}$ power: $\sqrt{x} = x^{1/2}$; $\sqrt[3]{y} = y^{1/3}$; $\sqrt[n]{z} = z^{1/n}$. Equation II.5 can be used in the same manner as before, except that n is a fraction. As an example, take the seventh root of 128, which may be expressed as $\sqrt[7]{128}$ or $128^{1/7}$. Substituting into Equation II.5,

$$\log 128^{1/7} = \frac{1}{7} \log 128 = \frac{1}{7} \log (1.28 \times 10^2) = \frac{1}{7} (2.1072) = 0.3010.$$

Antilog 0.3010 = 2.

Confirming, $2^7 = 2 \times 2 \times 2 \times 2 \times 2 \times 2 \times 2 = 128$.

Logarithms and Significant Figures

According to Table II, the logarithm of 4.58 grams, a measured quantity stated in three significant figures, is 0.6609—a four significant figure number. To be consistent in significant figures it is necessary to round off the logarithm to 0.661. If the same measurement is identified as 458 centigrams, or 4.58×10^2 cg, which is still a three significant figure number, Table II again identifies 0.6609 as the mantissa—but the characteristic is 2. Thus the log of 458 is 2.6609. Rounded off to three significant figures, this logarithm is 2.661. The logarithm of the same three significant figure measurement expressed as 4.58×10^{12} picograms is 12.661 to three significant figures. These examples illustrate the fact that *in logarithms only the mantissa is significant*; it should contain the same number of digits as there are significant figures in the measured quantity. The characteristic, as usual, serves only to locate the decimal or to identify the exponent of 10 when the quantity is expressed in exponential notation; it is not significant.

Within this book, all applications of logarithms are related to equilibrium constants. Usually these are expressed in two significant figures, occasionally one or three, and sometimes only to the closest order of magnitude. If the K values associated with problems involving logarithms are in two significant figures, two place logarithm tables are presumably adequate. Slide rules yield three place logarithms, and Table II gives them to four places.

The logarithm of a two or three significant figure number may be read directly from a slide rule or Table II; the logarithm of 4.52, for example, is 0.655 as read from a slide rule, or 0.6551 from the table. In the reverse direction, we shall state the antilog of 0.7689 to two significant figures. The logarithm must be rounded off to 0.769 for the slide rule. Opposite 0.769 on the L scale we read 5.87 on the C or D scale, which must now be rounded off to 5.9 in two significant figures. Using the table we find the antilog of 0.7689 lies between 5.87 (log 5.87 = 0.7686) and 5.88 (log 5.88 = 0.7694). In either case the two significant figure round off will be 5.9. If an antilog with three significant figures is required, we need only note that 0.7689 is closer to the log of 5.87 than 5.88, and therefore select 5.87. From these examples we conclude that the availability of three or four place logarithms virtually eliminates all need for interpolation when used for two and three significant figure quantities.

Rounding off logarithms *within* a calculation can be deceiving. Cumulative roundoffs that happen to be in the same direction can produce an answer quite different from that produced by calculating with three or four place logarithms and rounding off the end result to the proper number of significant figures. It is recommended that you carry no fewer than three places through logarithmic calculations, and four if you use the table. Only the final answer should be rounded off.

Natural (Base e) Logarithms

The entire discussion to this point has involved common, or base 10, logarithms. Nature, however, yields many relationships involving natural logarithms, in which the base is the constant e, 2.718 . . . . The two bases are distinguished in either of two ways: the base e logarithm of X may be written as $\log_e X$, and the base 10 logarithm as $\log_{10} X$; or the natural logarithm is shown as ln X, and the common logarithm simply as log X. Because of our base 10 number system, base 10 logarithms are more convenient to use. It is therefore customary to convert equations involving natural logarithms to common logarithms by the relationship

$$\ln X = 2.303 \log X. \quad \text{(II.6)}$$

TABLE II LOGARITHMS

	0	1	2	3	4	5	6	7	8	9
1.0	.0000	.0043	.0086	.0128	.0170	.0212	.0253	.0294	.0334	.0374
1.1	.0414	.0453	.0492	.0531	.0569	.0607	.0645	.0682	.0719	.0755
1.2	.0792	.0828	.0864	.0899	.0934	.0969	.1004	.1038	.1072	.1106
1.3	.1139	.1173	.1206	.1239	.1271	.1303	.1335	.1367	.1399	.1430
1.4	.1461	.1492	.1523	.1553	.1584	.1614	.1644	.1673	.1703	.1732
1.5	.1761	.1790	.1818	.1847	.1875	.1903	.1931	.1959	.1987	.2014
1.6	.2041	.2068	.2095	.2122	.2148	.2175	.2201	.2227	.2253	.2279
1.7	.2304	.2330	.2355	.2380	.2405	.2430	.2455	.2480	.2504	.2529
1.8	.2553	.2577	.2601	.2625	.2648	.2672	.2695	.2718	.2742	.2765
1.9	.2788	.2810	.2833	.2856	.2878	.2900	.2923	.2945	.2967	.2989
2.0	.3010	.3032	.3054	.3075	.3096	.3118	.3139	.3160	.3181	.3201
2.1	.3222	.3243	.3263	.3284	.3304	.3324	.3345	.3365	.3385	.3404
2.2	.3424	.3444	.3464	.3483	.3502	.3522	.3541	.3560	.3579	.3598
2.3	.3617	.3636	.3655	.3674	.3692	.3711	.3729	.3747	.3766	.3784
2.4	.3802	.3820	.3838	.3856	.3874	.3892	.3909	.3927	.3945	.3962
2.5	.3979	.3997	.4014	.4031	.4048	.4065	.4082	.4099	.4116	.4133
2.6	.4150	.4166	.4183	.4200	.4216	.4232	.4249	.4265	.4281	.4298
2.7	.4314	.4330	.4346	.4362	.4378	.4393	.4409	.4425	.4440	.4456
2.8	.4472	.4487	.4502	.4518	.4533	.4548	.4564	.4579	.4594	.4609
2.9	.4624	.4639	.4654	.4669	.4683	.4698	.4713	.4728	.4742	.4757
3.0	.4771	.4786	.4800	.4814	.4829	.4843	.4857	.4871	.4886	.4900
3.1	.4914	.4928	.4942	.4955	.4969	.4983	.4997	.5011	.5024	.5038
3.2	.5051	.5065	.5079	.5092	.5105	.5119	.5132	.5145	.5159	.5172
3.3	.5185	.5198	.5211	.5224	.5237	.5250	.5263	.5276	.5289	.5302
3.4	.5315	.5328	.5340	.5353	.5366	.5378	.5391	.5403	.5416	.5428
3.5	.5441	.5453	.5465	.5478	.5490	.5502	.5514	.5527	.5539	.5551
3.6	.5563	.5575	.5587	.5599	.5611	.5623	.5635	.5647	.5658	.5670
3.7	.5682	.5694	.5705	.5717	.5729	.5740	.5752	.5763	.5775	.5786
3.8	.5798	.5809	.5821	.5832	.5843	.5855	.5866	.5877	.5888	.5899
3.9	.5911	.5922	.5933	.5944	.5955	.5966	.5977	.5988	.5999	.6010
4.0	.6021	.6031	.6042	.6053	.6064	.6075	.6085	.6096	.6107	.6117
4.1	.6128	.6138	.6149	.6160	.6170	.6180	.6191	.6201	.6212	.6222
4.2	.6232	.6243	.6253	.6263	.6274	.6284	.6294	.6304	.6314	.6325
4.3	.6335	.6345	.6355	.6365	.6375	.6385	.6395	.6405	.6415	.6425
4.4	.6435	.6444	.6454	.6464	.6474	.6484	.6493	.6503	.6513	.6522
4.5	.6532	.6542	.6551	.6561	.6571	.6580	.6590	.6599	.6609	.6618
4.6	.6628	.6637	.6646	.6656	.6665	.6675	.6684	.6693	.6702	.6712
4.7	.6721	.6730	.6739	.6749	.6758	.6767	.6776	.6785	.6794	.6803
4.8	.6812	.6821	.6830	.6839	.6848	.6857	.6866	.6875	.6884	.6893
4.9	.6902	.6911	.6920	.6928	.6937	.6946	.6955	.6964	.6972	.6981

	0	1	2	3	4	5	6	7	8	9
5.0	.6990	.6998	.7007	.7016	.7024	.7033	.7042	.7050	.7059	.7067
5.1	.7076	.7084	.7093	.7101	.7110	.7118	.7126	.7135	.7143	.7152
5.2	.7160	.7168	.7177	.7185	.7193	.7202	.7210	.7218	.7226	.7235
5.3	.7243	.7251	.7259	.7267	.7275	.7284	.7292	.7300	.7308	.7316
5.4	.7324	.7332	.7340	.7348	.7356	.7364	.7372	.7380	.7388	.7396
5.5	.7404	.7412	.7419	.7427	.7435	.7443	.7451	.7459	.7466	.7474
5.6	.7482	.7490	.7497	.7505	.7513	.7520	.7528	.7536	.7543	.7551
5.7	.7559	.7566	.7574	.7582	.7589	.7597	.7604	.7612	.7619	.7627
5.8	.7634	.7642	.7649	.7657	.7664	.7672	.7679	.7686	.7694	.7701
5.9	.7709	.7716	.7723	.7731	.7738	.7745	.7752	.7760	.7767	.7774
6.0	.7782	.7789	.7796	.7803	.7810	.7818	.7825	.7832	.7839	.7846
6.1	.7853	.7860	.7868	.7875	.7882	.7889	.7896	.7903	.7910	.7917
6.2	.7924	.7931	.7938	.7945	.7952	.7959	.7966	.7973	.7980	.7987
6.3	.7993	.8000	.8007	.8014	.8021	.8028	.8035	.8041	.8048	.8055
6.4	.8062	.8069	.8075	.8082	.8089	.8096	.8102	.8109	.8116	.8122
6.5	.8129	.8136	.8142	.8149	.8156	.8162	.8169	.8176	.8182	.8189
6.6	.8195	.8202	.8209	.8215	.8222	.8228	.8235	.8241	.8248	.8254
6.7	.8261	.8267	.8274	.8280	.8287	.8293	.8299	.8306	.8312	.8319
6.8	.8325	.8331	.8338	.8344	.8351	.8357	.8363	.8370	.8376	.8382
6.9	.8388	.8395	.8401	.8407	.8414	.8420	.8426	.8432	.8439	.8445
7.0	.8451	.8457	.8463	.8470	.8476	.8482	.8488	.8494	.8500	.8506
7.1	.8513	.8519	.8525	.8531	.8537	.8543	.8549	.8555	.8561	.8567
7.2	.8573	.8579	.8585	.8591	.8597	.8603	.8609	.8615	.8621	.8627
7.3	.8633	.8639	.8645	.8651	.8657	.8663	.8669	.8675	.8681	.8686
7.4	.8692	.8698	.8704	.8710	.8716	.8722	.8727	.8733	.8739	.8745
7.5	.8751	.8756	.8762	.8768	.8774	.8779	.8785	.8791	.8797	.8802
7.6	.8808	.8814	.8820	.8825	.8831	.8837	.8842	.8848	.8854	.8859
7.7	.8865	.8871	.8876	.8882	.8887	.8893	.8899	.8904	.8910	.8915
7.8	.8921	.8927	.8932	.8938	.8943	.8949	.8954	.8960	.8965	.8971
7.9	.8976	.8982	.8987	.8993	.8998	.9004	.9009	.9015	.9020	.9026
8.0	.9031	.9036	.9042	.9047	.9053	.9058	.9063	.9069	.9074	.9079
8.1	.9085	.9090	.9096	.9101	.9106	.9112	.9117	.9122	.9128	.9133
8.2	.9138	.9143	.9149	.9154	.9159	.9165	.9170	.9175	.9180	.9186
8.3	.9191	.9196	.9201	.9206	.9212	.9217	.9222	.9227	.9232	.9238
8.4	.9243	.9248	.9253	.9258	.9263	.9269	.9274	.9279	.9284	.9289
8.5	.9294	.9299	.9304	.9309	.9315	.9320	.9325	.9330	.9335	.9340
8.6	.9345	.9350	.9355	.9360	.9365	.9370	.9375	.9380	.9385	.9390
8.7	.9395	.9400	.9405	.9410	.9415	.9420	.9425	.9430	.9435	.9440
8.8	.9445	.9450	.9455	.9460	.9465	.9469	.9474	.9479	.9484	.9489
8.9	.9494	.9499	.9504	.9509	.9513	.9518	.9523	.9528	.9533	.9538
9.0	.9542	.9547	.9552	.9557	.9562	.9566	.9571	.9576	.9581	.9586
9.1	.9590	.9595	.9600	.9605	.9609	.9614	.9619	.9624	.9628	.9633
9.2	.9638	.9643	.9647	.9652	.9657	.9661	.9666	.9671	.9675	.9680
9.3	.9685	.9689	.9694	.9699	.9703	.9708	.9713	.9717	.9722	.9727
9.4	.9731	.9736	.9741	.9745	.9750	.9754	.9759	.9763	.9768	.9773
9.5	.9777	.9782	.9786	.9791	.9795	.9800	.9805	.9809	.9814	.9818
9.6	.9823	.9827	.9832	.9836	.9841	.9845	.9850	.9854	.9859	.9863
9.7	.9868	.9872	.9877	.9881	.9886	.9890	.9894	.9899	.9903	.9908
9.8	.9912	.9917	.9921	.9926	.9930	.9934	.9939	.9943	.9948	.9952
9.9	.9956	.9961	.9965	.9969	.9974	.9978	.9983	.9987	.9991	.9996

APPENDIX III

VALUES OF CHEMICAL THERMODYNAMIC PROPERTIES*

TABLE III VALUES OF THERMODYNAMIC PROPERTIES

Symbols:

ΔH_f° The standard heat of formation of a substance from its elements at 25°C in kcal/mole and kJ/mole.

G_f° The standard free energy of formation of a substance from its elements at 25°C in kcal/mole and kJ/mole.

S° The entropy of a substance at 25°C in cal/mole-degree and joules/mole-degree.

Substance	ΔH_f° kcal/mole	ΔG_f° kcal/mole	S° cal/mole-°K	ΔH_f° kJ/mole	ΔG_f° kJ/mole	S° J/mole-°K
Aluminum						
Al(s)	0.00	0.00	6.77	0.00	0.00	28.33
Al_2O_3(s)	−399.09	−376.77	12.19	−1669.	−1576.4	51.00
Ammonium						
NH_3(g)	− 11.04	− 3.98	46.01	− 46.19	− 16.6	192.5
NH_3(aq)	− 19.32	− 6.37	26.3	− 80.83	− 26.6	110
NH_4Cl(s)	− 75.38	− 48.73	22.6	− 315.4	− 203.9	94.6
Bromine						
Br_2(l)	0.00	0.00	36.4	0.00	0.00	152
Br_2(g)	7.24	0.75	58.64	30.3	3.1	245.3
HBr(g)	− 8.66	− 12.72	47.44	− 36.2	− 53.22	198.5

Table continued on following page

*Data in the first three column following identification of the substance are from the *Handbook of Chemistry and Physics,* (The Chemical Rubber Publishing Company, 49th Edition). Values in the last three columns have been calculated from values in the first three columns on the basis of 1 calorie = 4.184 joules.

Substance	ΔH_f° $\frac{kcal}{mole}$	ΔG_f° $\frac{kcal}{mole}$	S° $\frac{cal}{mole\text{-}^\circ K}$	ΔH_f° $\frac{kJ}{mole}$	ΔG_f° $\frac{kJ}{mole}$	S° $\frac{J}{mole\text{-}^\circ K}$
Calcium						
Ca(s)	0.00	0.00	9.95	0.00	0.00	41.6
$CaCl_2$(s)	−190.0	−179.3	27.2	− 795.0	− 750.2	114
$CaCO_3$(s)	−288.45	−269.78	22.2	−1206.9	−1128.8	92.9
CaO(s)	−151.9	−144.4	9.5	− 635.5	− 604.2	40
$Ca(OH)_2$(s)	−235.80	−214.33	18.2	− 986.59	− 896.76	76.1
$CaSO_4$(s)	−342.42	−315.56	25.5	−1432.7	−1320.3	107
Carbon (Note: Organic compounds are listed separately at the end of this table.)						
C(s-graphite)	0.00	0.00	1.36	0.00	0.00	5.69
C(S-diamond)	0.45	0.69	0.58	1.9	2.9	2.4
CO(g)	− 26.42	− 32.81	47.30	− 110.5	− 137.3	197.9
CO_2(g)	− 94.05	− 94.26	51.06	− 393.5	− 394.4	213.6
Chlorine						
Cl_2(g)	0.00	0.00	53.29	0.00	0.00	223.0
HCl(g)	− 22.06	− 22.77	44.62	− 92.30	− 95.27	186.7
HCl(aq)	− 40.02	− 31.35	13.17	− 167.44	− 131.17	55.10
Copper						
Cu(s)	0.00	0.00	7.96	0.00	0.00	33.3
CuO(s)	− 37.1	− 30.4	10.4	− 155	− 127	43.5
Cu_2O(s)	− 39.84	− 34.98	24.1	− 166.7	− 146.4	101
CuS(s)	− 11.6	− 11.7	15.9	− 48.5	− 49.0	66.5
Cu_2S(s)	− 19.0	− 20.6	28.9	− 79.5	− 86.2	121
$CuSO_4$(s)	−184.00	−158.2	27.1	− 769.86	− 661.9	113
Fluorine						
F_2(g)	0.00	0.00	48.6	0.00	0.00	203
HF(g)	− 64.2	− 64.7	41.47	− 269	− 271	173.5
Hydrogen						
H_2(g)	0.00	0.00	31.21	0.00	0.00	130.6
H_2O(g)	− 57.80	− 54.64	45.11	− 241.8	− 228.6	188.7
H_2O(l)	− 68.32	− 56.69	16.72	− 285.8	− 237.2	69.96
H_2O_2(l)	− 44.84	− 28.2	—	− 187.6	− 118	—
Iodine						
I_2(s)	0.00	0.00	27.9	0.00	0.00	117
I_2(g)	14.88	4.63	62.28	62.26	19.4	260.6
IBr(g)	9.75	0.91	61.80	40.8	3.8	258.6
ICl(g)	4.20	− 1.32	59.12	17.6	− 5.52	247.4
HI(g)	6.20	0.31	49.31	25.9	1.3	206.3
Iron						
Fe(s)	0.00	0.00	6.49	0.00	0.00	27.2
Fe_2O_3(s)	−196.5	−177.1	21.5	− 822.2	− 741.0	90.0
Fe_3O_4(s)	−267.0	−242.4	35.0	−1117	−1014	146
Lead						
Pb(s)	0.00	0.00	15.51	0.00	0.00	64.89
PbO(s-red)	− 52.40	− 45.25	16.2	− 219.2	− 189.3	67.8
PbO_2(s)	−66.12	− 52.34	18.3	− 276.6	− 219.0	76
Pb_3O_4(s)	−175.6	−147.6	50.5	− 734.7	− 617.6	211.3
Magnesium						
Mg(s)	0.00	0.00	7.77	0.00	0.00	32.5
$MgCl_2$(s)	−153.40	−141.57	21.4	− 641.83	− 592.33	89.5
MgO(s)	−143.84	−136.13	6.4	− 601.83	− 569.57	27
$Mg(OH)_2$(s)	−221.00	−199.27	15.09	− 924.66	− 833.75	63.14

Table continued on opposite page

Substance	ΔH_f° kcal/mole	ΔG_f° kcal/mole	S° cal/mole-°K	ΔH_f° kJ/mole	ΔG_f° kJ/mole	S° J/mole-°K
Nitrogen						
$N_2(g)$	0.00	0.00	45.77	0.00	0.00	191.5
$NO(g)$	21.60	20.72	50.34	90.37	86.69	210.6
$NO_2(g)$	8.09	12.39	57.47	33.8	51.84	240.4
$HNO_3(l)$	− 41.40	− 19.10	37.19	− 173.2	− 79.91	155.6
Oxygen						
$O_2(g)$	0.00	0.00	49.00	0.00	0.00	205.02
$O_3(g)$	34.0	39.06	56.8	142	163.4	238
Phosphorus						
P(s-white)	0.00	0.00	10.6	0.00	0.00	44.4
$PCl_3(g)$	− 73.22	− 68.42	74.49	− 306.4	− 286.3	311.7
$PCl(g)$	− 95.35	− 77.59	84.3	− 398.9	− 324.6	353
Potassium						
$KCl(s)$	−104.18	− 97.59	19.76	− 435.89	− 408.32	82.68
$KClO_3(s)$	− 93.50	− 69.29	34.17	− 391.2	− 289.9	143.0
Silver						
$AgBr(s)$	− 23.78	− 22.39	25.60	− 99.50	− 93.68	107.1
$AgCl(s)$	− 30.36	− 26.22	22.97	− 127.0	− 109.7	96.11
$AgI(s)$	− 14.91	− 15.85	27.3	− 62.38	− 66.32	114
$Ag_2O(s)$	− 7.31	− 2.59	29.09	− 30.6	− 10.8	121.7
Sodium						
$Na(s)$	0.00	0.00	12.2	0.00	0.00	51.0
$NaCl(s)$	− 98.23	− 91.79	17.30	− 411.0	− 384.0	72.38
$Na_2CO_3(s)$	−270.3	−250.4	32.5	−1131	−1048	136
$NaF(s)$	−136.0	−129.3	14.0	− 569.0	− 541.0	58.6
$NaNO_3(s)$	−101.54	− 87.45	27.8	− 424.84	− 365.9	116
$NaOH(s)$	−101.99	—	—	− 426.73	—	—
Sulfur						
S(s-rhombic)	0.00	0.00	7.62	0.00	0.00	31.9
$H_2S(g)$	− 4.82	− 7.89	49.15	− 20.2	− 33.0	205.6
$SO_2(g)$	− 70.96	− 71.79	59.40	− 296.9	− 300.4	248.5
$SO_3(g)$	− 94.45	− 88.52	61.24	− 395.2	− 370.4	256.2
$H_2SO_4(l)$	−193.91	—	—	− 811.32	—	—
Zinc						
$Zn(s)$	0.00	0.00	9.95	0.00	0.00	41.6
$ZnCO_3(s)$	−194.2	−174.8	19.7	− 812.5	− 731.4	82.4
$ZnO(s)$	− 83.17	− 76.05	10.5	− 348.0	− 318.2	43.9
$ZnS(s)$	− 48.5	− 47.4	13.8	− 203	− 198	57.7
Organic Compounds						
Normal Alkanes						
$CH_4(g)$	− 17.89	− 12.14	44.50	− 74.85	− 50.79	186.2
$C_2H_6(g)$	− 20.24	− 7.86	54.85	− 84.68	− 32.9	229.5
$C_3H_8(g)$	− 24.82	− 5.61	64.51	− 103.8	− 23.5	269.9
$C_4H_{10}(g)$	− 29.81	− 3.75	74.10	− 124.7	− 15.7	310.0
$C_5H_{12}(l)$	− 35.00	− 1.96	83.27	− 146.4	− 8.20	348.4
$C_6H_{14}(l)$	− 39.96	0.05	92.45	− 167.2	0.2	386.8
$C_7H_{16}(l)$	− 44.89	2.09	101.64	− 187.8	8.74	425.26
$C_8H_{18}(l)$	− 49.82	4.14	110.82	− 208.4	17.3	463.67

Table continued on following page

Substance	ΔH_f° kcal/mole	ΔG_f° kcal/mole	S° cal/mole−°K	ΔH_f° kJ/mole	ΔG_f° kJ/mole	S° J/mole−°K
Alkenes						
$C_2H_4(g)$	12.50	16.28	52.54	52.3	68.12	219.8
$C_3H_6(g)$	4.88	14.99	63.80	20.4	62.72	266.9
C_4H_8(*cis*)	− 1.36	16.05	71.90	− 5.69	67.15	300.8
C_4H_8(*trans*)	− 2.41	15.32	70.86	− 10.1	64.10	296.5
C_5H_{10}(1−)	− 5.00	18.79	83.08	− 20.9	78.62	347.6
C_6H_{12}(1−)	− 9.96	20.80	92.25	− 41.7	87.03	386.0
Alkynes						
$C_2H_2(g)$	− 54.19	50.00	48.00	− 226.7	− 209.2	200.8
Halogenated Alkanes						
$CH_3Br(g)$	− 8.5	− 6.2	58.74	− 36	− 26	245.8
$CHBr_3(g)$	6	3.8	79.18	25	16	331.3
$CH_3Cl(g)$	− 19.6	− 14.0	55.97	− 82.0	− 58.6	234.2
$CH_2Cl_2(g)$	− 21	− 14.0	64.68	− 88	− 58.6	270.6
$CHCl_3(g)$	− 24	− 16.0	70.86	− 100	− 66.9	296.5
$CHCl_3(l)$	− 31.5	− 17.1	48.5	− 132	− 71.5	203
$CCl_4(l)$	− 33.3	− 16.4	51.25	− 139	− 68.6	214.4
$C_2H_5Cl(g)$	− 25.1	− 12.7	65.90	− 105	− 53.1	275.7
$C_2H_4Cl_2(l)$	− 39.7	− 19.2	49.84	− 166	− 80.3	208.5
$C_2H_4Br_2(l)$	− 19.3	− 4.94	53.37	− 80.8	− 20.7	233.3
Alcohols						
$CH_3OH(l)$	− 57.02	− 39.73	30.3	− 238.6	− 166.2	127
$C_2H_5OH(l)$	− 66.36	− 41.77	38.4	− 277.7	− 174.8	161
Aldehydes						
$HCHO(g)$	− 27.7	− 26.3	52.26	− 116	− 110	218.7
$CH_3CHO(g)$	− 39.76	− 31.96	63.5	− 166.4	− 133.7	266
Acids						
$HCOOH(l)$	− 97.8	− 82.7	30.82	− 409	− 346	129.0
$CH_3COOH(l)$	−116.4	− 93.8	38.2	− 487.0	− 392	160
Ethers						
CH_3OCH_3	− 44.3	− 27.3	63.72	− 185	− 114	266.6

APPENDIX IV

THERMAL PROPERTIES

TABLE IV THERMAL PROPERTIES OF SELECTED SUBSTANCES

Substance	Specific Heat (cal/g–°C)	Latent heat (calories per gram)		Melting Point (°C)	Boiling Point (°C)	Specific Heat (joules/g–°C)	Latent heat	
		Fusion	*Vaporization*				Fusion joules/g	Vaporization kilojoules/g
$H_2O(s)$	0.49	80		0		2.05	335	
$H_2O(l)$	1.00		540		100	4.18		2.26
$H_2O(g)$	0.48					2.01		
Na	0.29	27	1020	98	892	1.2	113	4.27
NaCl	0.21	124		801	1413	0.88	519	
Cu(s)	0.092	49		1083		0.38	205	
Cu(*l*)	0.10		1150		2595	0.42		4.81
Zn	0.092	24	420	419	907	0.38	100	1.76
Bi	0.029	13		271	1560	0.12	54	
Pb	0.031	5.5		327	1744	0.13	23	
Ni	0.106	74		1453	2732	0.444	310	

APPENDIX V

WATER VAPOR PRESSURE

TABLE V WATER VAPOR PRESSURE IN TORR

Temperature (°C)	Vapor Pressure	Temperature (°C)	Vapor Pressure
0	4.6	28	28.3
5	6.5	29	30.0
10	9.2	30	31.8
15	12.8	31	33.7
16	13.6	32	35.7
17	14.5	33	37.7
18	15.5	34	40.0
19	16.5	35	42.2
20	17.5	40	55.3
21	18.6	45	71.9
22	19.8	50	92.5
23	21.1	60	149.4
24	22.4	70	233.7
25	23.8	80	355.1
26	25.2	90	525.8
27	26.7	100	760.0

APPENDIX VI

IONIZATION CONSTANTS OF WEAK ACIDS

TABLE VI IONIZATION CONSTANTS

Name	Equation	K_a
Sulfuric acid	$H_2SO_4 = H^+ + HSO_4^-$	Large
Oxalic acid	$H_2C_2O_4 = H^+ + HC_2O_4^-$	5.9×10^{-2}
Sulfurous acid	$H_2SO_3 = H^+ + HSO_3^-$	1.7×10^{-2}
Hydrogen sulfate ion	$HSO_4^- = H^+ + SO_4^{2-}$	1.2×10^{-2}
Phosphoric acid	$H_3PO_4 = H^+ + H_2PO_4^-$	7.5×10^{-3}
Hydrofluoric acid	$HF = H^+ + F^-$	7.0×10^{-4}
Nitrous acid	$HNO_2 = H^+ + NO_2^-$	4.5×10^{-4}
Formic acid	$HCH_2O = H^+ + CH_2O^-$	2.1×10^{-4}
Benzoic acid	$HC_7H_5O_2 = H^+ + C_7H_5O_2^-$	6.6×10^{-5}
Hydrogen oxalate ion	$HC_2O_4^- = H^+ + C_2O_4^{2-}$	6.4×10^{-5}
Acetic acid	$HC_2H_3O_2 = H^+ + C_2H_3O_2^-$	1.8×10^{-5}
Propionic acid	$HC_3H_5O_2 = H^+ + C_3H_5O_2^-$	1.4×10^{-5}
Carbonic acid	$H_2CO_3 = H^+ + HCO_3^-$	4.2×10^{-7}
Hydrosulfuric acid	$H_2S = H^+ + HS^-$	1.0×10^{-7}
Dihydrogen phosphate ion	$H_2PO_4^- = H^+ + HPO_4^{2-}$	6.2×10^{-8}
Hydrogen sulfite ion	$HSO_3^- = H^+ + SO_3^{2-}$	5.6×10^{-8}
Hypochlorous acid	$HClO = H^+ + ClO^-$	3.2×10^{-8}
Boric acid	$H_3BO_3 = H^+ + H_2BO_3^-$	5.8×10^{-10}
Ammonium ion	$NH_4^+ = H^+ + NH_3$	5.6×10^{-10}
Hydrocyannic acid	$HCN = H^+ + CN^-$	4.0×10^{-10}
Hydrogen carbonate ion	$HCO_3^- = H^+ + CO_3^{2-}$	4.8×10^{-11}
Monohydrogen phosphate ion	$HPO_4^{2-} = H^+ + PO_4^{3-}$	1.7×10^{-12}
Water	$HOH = H^+ + OH^-$	1.0×10^{-14}
Hydrogen sulfide ion	$HS^- = H^+ + S^{2-}$	1.0×10^{-15}

APPENDIX VII

DISSOCIATION CONSTANTS OF COMPLEX IONS

TABLE VII DISSOCIATION CONSTANTS

	Reaction	K_d
MA_2		
	$AgCl_2^- = Ag^+ + 2\ Cl^-$	1×10^{-6}
	$Ag(NH_3)_2^+ = Ag^+ + 2\ NH_3$	4×10^{-8}
	$Ag(SCN)_2^- = Ag^+ + 2\ SCN^-$	1×10^{-10}
	$Ag(S_2O_3)_2^{3-} = Ag^+ + 2\ S_2O_3^{2-}$	1×10^{-13}
	$Ag(CN)_2^- = Ag^+ + 2\ CN^-$	1×10^{-21}
	$CuCl_2^- = Cu^+ + 2\ Cl^-$	3×10^{-6}
	$Cu(NH_3)_2^+ = Cu^+ + 2\ NH_3$	1×10^{-7}
MA_4		
	$CdCl_4^{2-} = Cd^{2+} + 4\ Cl^-$	4×10^{-3}
	$Cd(NH_3)_4^{2+} = Cd^{2+} + 4\ NH_3$	1×10^{-7}
	$Cd(CN)_4^{2-} = Cd^{2+} + 4\ CN^-$	1×10^{-19}
	$Cu(NH_3)_4^{2+} = Cu^{2+} + 4\ NH_3$	2×10^{-13}
	$Cu(CN)_4^{2-} = Cu^{2+} + 4\ CN^-$	1×10^{-25}
	$Ni(CN)_4^{2-} = Ni^{2+} + 4\ CN^-$	1×10^{-14}
	$PdCl_4^{2-} = Pd^{2+} + 4\ Cl^-$	1×10^{-13}
	$Zn(NH_3)_4^{2+} = Zn^{2+} + 4\ NH_3$	3×10^{-10}
	$Zn(OH)_4^{2-} = Zn^{2+} + 4\ OH^-$	3×10^{-16}
	$Zn(CN)_4^{2-} = Zn^{2+} + 4\ CN^-$	1×10^{-17}
MA_6		
	$Cr(OH)_6^{3-} = Cr^{3+} + 6\ OH^-$	1×10^{-38}
	$Co(NH_3)_6^{3+} = Co^{3+} + 6\ NH_3$	1×10^{-35}
	$Co(CN)_6^{3-} = Co^{3+} + 6\ CN^-$	1×10^{-64}
	$Fe(CN)_6^{3-} = Fe^{3+} + 6\ CN^-$	1×10^{-31}
	$Ni(NH_3)_6^{2+} = Ni^{2+} + 6\ NH_3$	2×10^{-9}

APPENDIX VIII

SOLUBILITY PRODUCT CONSTANTS

TABLE VIII SOLUBILITY PRODUCT CONSTANTS

Bromides	
$AgBr$	5.0×10^{-13}
$CuBr$	4.9×10^{-9}
Carbonates	
Ag_2CO_3	8.2×10^{-12}
$BaCO_3$	8.1×10^{-9}
$CaCO_3$	8.7×10^{-9}
Li_2CO_3	1.7×10^{-5}
$NiCO_3$	6.6×10^{-9}
$ZnCO_3$	1.4×10^{-11}
Chlorides	
$AgCl$	1.7×10^{-10}
$PbCl_2$	1.6×10^{-5}
$TlCl$	1.9×10^{-4}
Chromates	
Ag_2CrO_4	2.2×10^{-12}
$BaCrO_4$	2.4×10^{-10}
Fluorides	
BaF_2	1.7×10^{-6}
CaF_2	4.0×10^{-11}
MgF_2	6.5×10^{-9}
PbF_2	2.7×10^{-8}
SrF_2	2.8×10^{-9}
Hydroxides	
$Al(OH)_3$	2.0×10^{-32}
$Cd(OH)_2$	5.9×10^{-15}
$Ca(OH)_2$	5.5×10^{-5}
$Co(OH)_2$	2.0×10^{-16}
$Mg(OH)_2$	1.2×10^{-11}
Iodates	
$AgIO_3$	3.0×10^{-8}
$Cu(IO_3)_2$	7.4×10^{-8}
Iodides	
AgI	8.3×10^{-17}
PbI_2	7.1×10^{-9}
Sulfates	
Ag_2SO_4	1.6×10^{-5}
$BaSO_4$	1.1×10^{-10}
$CaSO_4$	1.9×10^{-4}
$PbSO_4$	1.6×10^{-8}
Sulfides	
Ag_2S	2.0×10^{-49}
CdS	7.8×10^{-27}
CuS	9×10^{-36}
NiS	3×10^{-19}
ZnS	1×10^{-21}
Phosphates	
Ag_3PO_4	1.3×10^{-20}
$Ca_3(PO_4)_2$	2.0×10^{-29}
$Zn_3(PO_4)_2$	9.1×10^{-33}
Miscellaneous	
$Pb(C_2O_4)_2$	2.7×10^{-11}
$AgSCN$	1.2×10^{-12}

APPENDIX IX

STANDARD REDUCTION POTENTIALS AT 25°C

TABLE IX STANDARD REDUCTION POTENTIALS

REDUCTION HALF-REACTION		E° (VOLTS)
$Li^+(aq) + e^-$	$\rightarrow Li(s)$	−3.05
$K^+(aq) + e^-$	$\rightarrow K(s)$	−2.93
$Ba^{2+}(aq) + 2e^-$	$\rightarrow Ba(s)$	−2.90
$Ca^{2+}(aq) + 2e^-$	$\rightarrow Ca(s)$	−2.87
$Na^+(aq) + e^-$	$\rightarrow Na(s)$	−2.71
$Mg^{2+}(aq) + 2\ e^-$	$\rightarrow Mg(s)$	−2.37
$Al^{3+}(aq) + 3\ e^-$	$\rightarrow Al(s)$	−1.66
$Zn^{2+}(aq) + 2\ e^-$	$\rightarrow Zn(s)$	−0.76
$Cr^{3+}(aq) + 3\ e^-$	$\rightarrow Cr(s)$	−0.74
$Fe^{2+}(aq) + 2\ e^-$	$\rightarrow Fe(s)$	−0.44
$PbSO_4(s) + 2\ e^-$	$\rightarrow Pb(s) + SO_4^{2-}(aq)$	−0.36
$Co^{2+}(aq) + 2\ e^-$	$\rightarrow Co(s)$	−0.28
$Ni^{2+}(aq) + 2\ e^-$	$\rightarrow Ni(s)$	−0.25
$AgI(s) + e^-$	$\rightarrow Ag(s) + I^-(aq)$	−0.15
$Sn^{2+}(aq) + 2\ e^-$	$\rightarrow Sn(s)$	−0.14
$Pb^{2+}(aq) + 2\ e^-$	$\rightarrow Pb(s)$	−0.13
$2\ H^+(aq) + 2\ e^-$	$\rightarrow H_2(g)$	0.00
$AgBr(s) + e^-$	$\rightarrow Ag(s) + Br^-(aq)$	0.10
$Sn^{4+}(aq) + 2\ e^-$	$\rightarrow Sn^{2+}(aq)$	0.15
$SO_4^{2-}(aq) + 4\ H^+(aq) + 2\ e^-$	$\rightarrow SO_2\ (g) + 2\ H_2O\ (l)$	0.20
$Cu^{2+}(aq) + 2\ e^-$	$\rightarrow Cu(s)$	0.34
$I_2(s) + 2\ e^-$	$\rightarrow 2\ I^-(aq)$	0.53
$PtCl_6^{2-}(aq) + 4\ e^-$	$\rightarrow Pt(s) + 6\ Cl^-(aq)$	0.60
$Fe^{3+}(aq) + e^-$	$\rightarrow Fe^{2+}(aq)$	0.77
$Hg_2^{2+}(aq) + 2e^-$	$\rightarrow 2\ Hg(l)$	0.79
$Ag^+(aq) + e^-$	$\rightarrow Ag(s)$	0.80
$NO_3^-\ (aq) + 2\ H^+(aq) + e^-$	$\rightarrow NO_2\ (g) + H_2O(l)$	0.80
$2\ Hg^{2+}(aq) + e^-$	$\rightarrow Hg_2^{2+}\ (aq)$	0.92
$NO_3^-(aq) + 4H^+\ (aq) + 3\ e^-$	$\rightarrow NO\ (g) + 2\ H_2O\ (l)$	0.96
$AuCl_4^-\ (aq) + 3\ e^-$	$\rightarrow Au(s) + 4\ Cl^-(aq)$	1.00
$Br_2(l) + 2\ e^-$	$\rightarrow 2\ Br^-\ (aq)$	1.07
$O_2(g) + 4\ H^+(aq) + 4\ e^-$	$\rightarrow 2\ H_2O\ (l)$	1.23
$MnO_2(s) + 4\ H^+(aq) + 2\ e^-$	$\rightarrow Mn^{2+}\ (aq) + 2\ H_2O\ (l)$	1.23
$Cr_2O_7^{2-}\ (aq) + 14\ H^+(aq) + 6\ e^-$	$\rightarrow 2\ Cr^{3+}\ (aq) + 7\ H_2O\ (l)$	1.33
$Cl_2(g) + 2\ e^-$	$\rightarrow 2\ Cl^-\ (aq)$	1.36
$Au^{3+}(aq) + 3\ e^-$	$\rightarrow Au(s)$	1.50
$MnO_4^-(aq) + 8\ H^+(aq) + 5\ e^-$	$\rightarrow Mn^{2+}(aq) + 4\ H_2O$	1.52
$F_2(g) + 2\ e^-$	$\rightarrow 2\ F^-(aq)$	2.87

APPENDIX X

SOLUTIONS TO PROBLEMS

Problem solutions in Appendix X are arranged in a "line setup" in which a series of fractional factors are arranged with numerators and denominators above and below a common horizontal line. Individual factors are separated by vertical lines. For example, the conversion of 1.5 miles to inches would normally be represented as

$$1.5 \text{ miles} \times \frac{5280 \text{ feet}}{1 \text{ mile}} \times \frac{12 \text{ inches}}{1 \text{ foot}} = 95{,}040 \text{ inches.}$$

The line setup for the same problem is

$$\frac{1.5 \text{ miles}}{} \left| \frac{5280 \text{ feet}}{1 \text{ mile}} \right| \frac{12 \text{ inches}}{1 \text{ foot}} = 95{,}040 \text{ inches.}$$

The line setup has no inherent advantage over the conventional fractional setup, but many students, scientists, and engineers find it convenient for organizing their work and arranging it neatly on a page. We suggest you try it, use it if it appeals to you, and otherwise discard it. Another alternative you may wish to consider is to isolate individual factors by symbols of enclosure, viz., parentheses:

$$1.5 \text{ miles} \left(\frac{5280 \text{ feet}}{1 \text{ mile}}\right) \left(\frac{12 \text{ inches}}{1 \text{ foot}}\right) = 95{,}040 \text{ inches.}$$

Chapter 2

2.1) $\frac{9.46 \text{ cm}}{} \left| \frac{10 \text{ mm}}{1 \text{ cm}} \right. = 94.6 \text{ mm}$

2.2) $\frac{734 \text{ ml}}{} \left| \frac{1\, l}{1000 \text{ ml}} \right. = 0.734\, l$

2.3) $\frac{0.603 \text{ g}}{} \left| \frac{100 \text{ cg}}{1 \text{ g}} \right. = 60.3 \text{ cg}$

2.4) $\frac{81.9 \text{ dm}}{} \left| \frac{1 \text{ m}}{10 \text{ dm}} \right. = 8.19 \text{ m}$

2.5) $\frac{9.30 \times 10^4 \text{ cm}}{} \left| \frac{1 \text{ km}}{10^5 \text{ cm}} \right. = 0.930 \text{ km}$

2.6) $\frac{8.99 \text{ mg}}{} \left| \frac{10^9 \text{ mg}}{1 \text{ mg}} \right. = 8.99 \times 10^9 \text{ mg}$

2.7) $\frac{56.7 \text{ mm}^3}{} \left| \frac{1 \text{ cm}^3}{10^3 \text{ mm}^3} \right. = 0.0567 \text{ cm}^3$

2.8) $\frac{36 \text{ in}}{} \left| \frac{2.54 \text{ cm}}{1 \text{ in}} \right. = 91.4 \text{ cm}$

2.9) $\frac{100 \text{ m}}{} \left| \frac{1.09 \text{ yd}}{1 \text{ m}} \right| \frac{3 \text{ ft}}{1 \text{ yd}} = 327 \text{ ft}$

2.10) $\dfrac{850 \text{ miles}}{} \left| \dfrac{1.61 \text{ km}}{1 \text{ mile}} \right. = 1370 \text{ km}$

2.11) $\dfrac{9.48 \text{ ft}}{} \left| \dfrac{12 \text{ in}}{1 \text{ ft}} \right| \dfrac{2.54 \text{ cm}}{1 \text{ in}} = 289 \text{ cm}$

2.12) $\dfrac{12 \text{ in}}{} \left| \dfrac{5.5 \text{ in}}{} \right| \dfrac{3.5 \text{ in}}{} \left| \dfrac{2.54^3 \text{ cm}^3}{1 \text{ in}^3} \right. = 3790 \text{ cm}^3$

2.13) $\dfrac{0.425 \text{ km}^2}{} \left| \dfrac{1^2 \text{ mile}^2}{1.61^2 \text{km}^2} \right| \dfrac{5280^2 \text{ ft}^2}{1^2 \text{ mile}^2} \left| \dfrac{1 \text{ acre}}{43{,}560 \text{ ft}^2} \right. = 105 \text{ acres}$

2.14) $\dfrac{1120 \text{ kg}}{} \left| \dfrac{2.2 \text{ lbs}}{1 \text{ kg}} \right. = 2460 \text{ lbs}$

2.15) $\dfrac{12 \text{ oz}}{} \left| \dfrac{1 \text{ lb}}{16 \text{ oz}} \right| \dfrac{454 \text{ g}}{1 \text{ lb}} = 340 \text{ g}$

2.16) $\frac{5}{9}\ (88-32) = 31°\text{C}$

2.17) $\frac{5}{9}\ (525-32) = 274°\text{C}$

2.18) $\frac{9}{5}\ (2000) + 32 = 3632°\text{F}$

2.19) $296-273 = 23°\text{C}; \quad \frac{9}{5}\ (23) + 32 = 73°\text{F}$

2.20) 3 2.21) 4 2.22) 3 2.23) 4.22 kg 2.24) 279 cm^3

2.25) 8.00×10^3 g 2.26) $2.86 + 3.9 + 0.896 + 246 = 253.7$ g

2.27) Two because of two significant figures in mass of ammonium sulfate:
$\dfrac{3.9}{253.7} \times 100 = 1.5\%$

2.28) $1510 \text{ g}/865 \text{ cm}^3 = 1.75 \text{ g/cm}^3$

2.29) $\dfrac{4.92 \text{ kg}}{4.60 \text{ cm} \times 10.3 \text{ cm} \times 13.2 \text{ cm}} \left| \dfrac{1000 \text{ g}}{1 \text{ kg}} \right. = 7.87 \text{ g/cm}^3$

2.30) $\dfrac{35.3 \text{ ml}}{} \left| \dfrac{0.790 \text{ g}}{1 \text{ ml}} \right. = 27.9 \text{ g}$

2.31) $\dfrac{16.723 \text{ g} - 12.047 \text{ g}}{17.003 \text{ g} - 12.047 \text{ g}} = \dfrac{4.676 \text{ g}}{4.956 \text{ g}} = 0.9435$

2.32) $\dfrac{50.0\, l}{} \left| \dfrac{1000 \text{ ml}}{1 l} \right| \dfrac{0.925 \text{ g}}{1 \text{ ml}} \left| \dfrac{1 \text{ lb}}{454 \text{ g}} \right. = 102 \text{ lb}$

2.33) $\dfrac{1 \text{ qt}}{} \left| \dfrac{1 \text{ gal}}{4 \text{ qts}} \right| \dfrac{3.785 l}{1 \text{ gal}} \left| \dfrac{1000 \text{ ml}}{1 l} \right| \dfrac{13.6 \text{ g}}{1 \text{ ml}} \left| \dfrac{1 \text{ lb}}{454 \text{ g}} \right. = 28.3 \text{ lb}$

2.34) (a) $\dfrac{557 \text{ lbs}}{1 \text{ ft}^3} \left| \dfrac{1 \text{ ft}^3}{30.5^3 \text{ cm}^3} \right| \dfrac{454 \text{ g}}{1 \text{ lb}} = 8.91 \text{ g/cm}^3 \cdot \text{Sp. g} = 8.91$

(b) $12.0 \text{ cm} \times 4.00 \text{ cm} \times 0.50 \text{ cm} \times \dfrac{8.91 \text{ g}}{1 \text{ cm}^3} = 2.1 \times 10^2 \text{ g}$

2.35) $\dfrac{2.50 \text{ g}}{} \left| \dfrac{1 \text{ troy oz}}{31.1 \text{ g}} \right| \dfrac{174 \text{ dollars}}{1 \text{ troy oz}} = 1.40 \times 10^3 \text{ dollars}$

Chapter 3

3.1) 35.5 3.2) $2 \times 35.5 = 71.0$ 3.3) $6.9 + 19.0 = 25.9$

3.4) $2(39.1) + 32.1 + 4(16.0) = 174$ 3.5) $137 + 2(14.0) + 6(16.0) = 261$

3.6) $24.3 + 2(35.5) + 12(1.0) + 6(16.0) = 203$

3.7) $\frac{78.2}{174} \times 100 = 44.9\%$ K: $\frac{32.1}{174} \times 100 = 18.4\%$ S

$\frac{64.0}{174} \times 100 = 36.8\%$ O

3.8) $\frac{137}{261} \times 100 = 52.5\%$ Ba; $\frac{28.0}{261} \times 100 = 10.7\%$N;

$\frac{96.0}{261} \times 100 = 36.8\%$ O

3.9) $\frac{24.3}{203} \times 100 = 12.0\%$ Mg: $\frac{71.0}{203} \times 100 = 35.0\%$ Cl

$\frac{12.0}{203} \times 100 = 5.91\%$ H: $\frac{96.0}{203} \times 100 = 47.3\%$ O

3.10) $$\begin{array}{c|c} 65.4\text{ g LiF} & 6.94\text{ g Li} \\ \hline & 25.9\text{ g LiF} \end{array} = 17.5\text{ g Li}$$

3.11) $$\begin{array}{c|c} 16.3\text{ g } K_2SO_4 & 78.2\text{ g K} \\ \hline & 174\text{ g } K_2SO_4 \end{array} = 7.33\text{ g K}$$

3.12) $$\begin{array}{c|c} 65.2\text{ g } Cl_2 & 1\text{ mole} \\ \hline & 71.0\text{ g} \end{array} = 0.918\text{ mole } Cl_2$$

3.13) $$\begin{array}{c|c} 52.0\text{ g } MgCl_2 \cdot 6\,H_2O & 1\text{ mole} \\ \hline & 203\text{ g} \end{array} = 0.256\text{ mole } MgCl_2 \cdot 6\,H_2O$$

3.14) $$\begin{array}{c|c} 0.353\text{ mole LiF} & 25.9\text{ g} \\ \hline & 1\text{ mole} \end{array} = 9.14\text{ g LiF}$$

3.15) $$\begin{array}{c|c|c} 4.00 \times 10^{12}\text{ atoms Al} & 1\text{ mole} & 27.0\text{ g} \\ \hline & 6.02 \times 10^{23}\text{ atoms} & 1\text{ mole} \end{array} = 1.79 \times 10^{-10}\text{ g}$$

3.16) $$\begin{array}{c|c|c|c|c} 0.500\text{ carat} & 200\text{ mg} & 1\text{ g} & 1\text{ mole} & 6.02 \times 10^{23}\text{ atoms} \\ \hline & 1\text{ carat} & 1000\text{ mg} & 12.0\text{ g} & 1\text{ mole} \end{array} = 5.02 \times 10^{21}\text{ atoms}$$

3.17) $$\begin{array}{c|c|c} 500\text{ g } H_2O & 1\text{ mole} & 6.02 \times 10^{23}\text{ molecules} \\ \hline & 18.0\text{ g} & 1\text{ mole} \end{array} = 1.67 \times 10^{25}\text{ molecules}$$

3.18) $$\begin{array}{c|c|c} 2.95 \times 10^{22}\text{ molecules} & 1\text{ mole} & 29\text{ g air} \\ \hline & 6.02 \times 10^{23}\text{ molecules} & 1\text{ mole molecules} \end{array} = 1.4\text{ g air}$$

3.19)

	Grams	Moles	Mole Ratio	Formula Ratio	
C	52.2	4.35	2.00	2	
H	13.0	13.0	5.96	6	C_2H_6O
O	34.8	2.18	1.00	1	

3.20)

	Grams	Moles	Mole Ratio	Formula Ratio	
Fe	11.89	0.213	1.00	2	Fe_2O_3
O	16.99−11.89=5.1	0.319	1.50	3	

3.21) 2.500 g Li + Cl + O − 1.170 g Li + Cl = 1.330 g O

$$\begin{array}{c|c} 3.963\text{ g AgCl} & 35.5\text{ g Cl} \\ \hline & 143.5\text{ g AgCl} \end{array} = 0.980\text{ g Cl}$$

2.500 g Li + Cl + O − 1.33 g O − 0.980 g Cl = 0.190 g Li

	Grams	Moles	Mole Ratio	Formula Ratio	
Li	0.190	0.0274	1.00	1	
Cl	0.980	0.276	1.01	1	$LiClO_3$
O	1.330	0.0831	3.03	3	

Chapter 4

4.1) $$\frac{3.40 \text{ moles } C_4H_{10}}{} \left| \frac{13 \text{ moles } O_2}{2 \text{ moles } C_4H_{10}} \right. = 22.1 \text{ moles } O_2$$

4.2) $$\frac{4.68 \text{ moles } C_4H_{10}}{} \left| \frac{8 \text{ moles } CO_2}{2 \text{ moles } C_4H_{10}} \right. = 18.7 \text{ moles } CO_2$$

4.3) $$\frac{0.568 \text{ mole } CO_2}{} \left| \frac{10 \text{ moles } H_2O}{8 \text{ moles } CO_2} \right. = 0.710 \text{ mole } H_2O$$

4.4) $$\frac{1.42 \text{ moles } O_2}{} \left| \frac{2 \text{ moles } C_4H_{10}}{8 \text{ moles } O_2} \right| \frac{58.0 \text{ g } C_4H_{10}}{1 \text{ mole } C_4H_{10}} = 20.6 \text{ g } C_4H_{10}$$

4.5) $$\frac{9.43 \text{ g } O_2}{} \left| \frac{1 \text{ mole } O_2}{32.0 \text{ g } O_2} \right| \frac{10 \text{ moles } H_2O}{13 \text{ moles } O_2} = 0.227 \text{ mole } H_2O$$

4.6) $$\frac{78.4 \text{ g } C_4H_{10}}{} \left| \frac{1 \text{ mole } C_4H_{10}}{58.0 \text{ g } C_4H_{10}} \right| \frac{8 \text{ moles } CO_2}{2 \text{ moles } C_4H_{10}} \left| \frac{44.0 \text{ g } CO_2}{1 \text{ mole } CO_2} \right. = 238 \text{ g } CO_2$$

4.7) $$\frac{43.8 \text{ g } H_2O}{} \left| \frac{1 \text{ mole } H_2O}{18.0 \text{ g } H_2O} \right| \frac{13 \text{ moles } O_2}{10 \text{ moles } H_2O} \left| \frac{32.0 \text{ g } O_2}{1 \text{ mole } O_2} \right. = 101 \text{ g } O_2$$

4.8) $$\frac{47.1 \text{ g Al}}{} \left| \frac{1 \text{ mole Al}}{27.0 \text{ g Al}} \right| \frac{1 \text{ mole } Fe_2O_3}{2 \text{ moles Al}} \left| \frac{160 \text{ g } Fe_2O_3}{1 \text{ mole } Fe_2O_3} \right. = 140 \text{ g } Fe_2O_3$$

4.9) $$\frac{100 \text{ g } C_{600}H_{1000}O_{500}}{} \left| \frac{1 \text{ mole } C_{600}H_{1000}O_{500}}{1.62 \times 10^4 C_{600}H_{1000}O_{500}} \right|$$

$$\frac{50 \text{ moles } C_{12}H_{22}O_{11}}{1 \text{ mole } C_{600}H_{1000}O_{500}} \left| \frac{342 \text{ g } C_{12}H_{22}O_{11}}{1 \text{ mole } C_{12}H_{22}O_{11}} \right. = 106 \text{ g } C_{12}H_{22}O_{11}$$

4.10) $$\frac{600 \text{ g } NaHCO_3}{} \left| \frac{1 \text{ mole } NaHCO_3}{84.0 \text{ g } NaHCO_3} \right| \frac{1 \text{ mole } H_2SO_4}{2 \text{ moles } NaHCO_3} \left| \frac{98.1 \text{ g } H_2SO_4}{1 \text{ mole } H_2SO_4} \right.$$

$$= 350 \text{ g } H_2SO_4$$

4.11) $$\frac{5.00 \text{ kg } Na_2SO_4}{} \left| \frac{10^3 \text{ g}}{1 \text{ kg}} \right| \frac{1 \text{ mole } Na_2SO_4}{142 \text{ g } Na_2SO_4} \left| \frac{4 \text{ moles NaCl}}{2 \text{ moles } Na_2SO_4} \right|$$

$$\frac{58.5 \text{ g NaCl}}{1 \text{ mole NaCl}} = 4.12 \times 10^3 \text{ g NaCl}$$

4.12) $Zn + S \rightarrow ZnS$

$$\frac{45.0 \text{ g Zn}}{} \left| \frac{1 \text{ mole Zn}}{65.4 \text{ g Zn}} \right. = 0.688 \text{ mole Zn—limiting reagent}$$

$$\frac{28.0 \text{ g S}}{} \left| \frac{1 \text{ mole S}}{32.1 \text{ g S}} \right. = 0.872 \text{ mole S}$$

$$\frac{0.688 \text{ mole Zn}}{} \left| \frac{1 \text{ mole ZnS}}{1 \text{ mole Zn}} \right| \frac{97.5 \text{ g ZnS}}{1 \text{ mole ZnS}} = 67.1 \text{ g ZnS}$$

$$\frac{(0.872 - 0.688) \text{ mole S remain}}{} \left| \frac{32.1 \text{ g}}{1 \text{ mole}} \right. = 5.91 \text{ g S remain}$$

4.13) $BaCl_2 + Na_2CrO_4 \rightarrow BaCrO_4 + 2\ NaCl$

$$\frac{1.46 \text{ } BaCl_2}{} \left| \frac{1 \text{ mole } BaCl_2}{208 \text{ g } BaCl_2} \right. = 7.02 \times 10^{-3} \text{ mole } BaCl_2\text{—limiting reagent}$$

$$\frac{2.14 \text{ g } Na_2CrO_4}{} \left| \frac{1 \text{ mole } Na_2CrO_4}{162 \text{ g } Na_2CrO_4} \right. = 13.2 \times 10^{-3} \text{ mole } Na_2CrO_4$$

$$\begin{array}{l|l|l} 7.02\times10^{-3}\text{ mole BaCl}_2 & 1\text{ mole BaCrO}_4 & 253\text{ g BaCrO}_4 \\ \hline & 1\text{ mole BaCl}_2 & 1\text{ mole BaCrO}_4 \end{array} = 1.78\text{ g BaCrO}_4$$

$$\begin{array}{l|l} (13.2-7.02)\,10^{-3}\text{ mole Na}_2\text{CrO}_4 & 162\text{ g Na}_2\text{CrO}_4 \\ \hline & 1\text{ mole Na}_2\text{CrO}_4 \end{array} = 1.00\text{ g Na}_2\text{CrO}_4\text{ remains}$$

4.14) $$\begin{array}{l|l|l|l} 41.9\text{ g Cu}_2\text{S} & 1\text{ mole Cu}_2\text{S} & 2\text{ moles Cu} & 63.5\text{ g Cu} \\ \hline & 159\text{ g Cu}_2\text{S} & 1\text{ mole Cu}_2\text{S} & 1\text{ mole Cu} \end{array} = 33.5\text{ g Cu theoretical}$$

$$\frac{29.2\text{ g yield}}{33.5\text{ g theoretical}}\times 100 = 87.2\%$$

4.15) $$\begin{array}{l|l|l|} 1\text{ ton P yield} & 100\text{ tons theoretical} & 1\text{ ton mole P} \\ \hline & 63\text{ tons yield} & 31.0\text{ tons P} \end{array}$$

$$\begin{array}{l|l} 1\text{ ton mole Ca}_3(\text{PO}_4)_2 & 310\text{ tons Ca}_3(\text{PO}_4)_2 \\ \hline 2\text{ ton moles P} & 1\text{ ton mole Ca}_3(\text{PO}_4)_2 \end{array} = 7.94\text{ tons Ca}_3(\text{PO}_4)_2$$

4.16) $$\begin{array}{l|l|l|l} 28.9\text{ g HgO} & 1\text{ mole HgO} & 1\text{ mole O}_2 & 22.4\ l\ \text{O}_2 \\ \hline & 217\text{ g HgO} & 2\text{ moles HgO} & 1\text{ mole O}_2 \end{array} = 1.49\ l\ \text{O}_2$$

4.17) $C_7H_{16} + 11\,O_2 \rightarrow 7\,CO_2 + 8\,H_2O$

$$\begin{array}{l|l|l|l} 72.0\text{ g C}_7\text{H}_{16} & 1\text{ mole C}_7\text{H}_{16} & 11\text{ moles O}_2 & 22.4\ l\ \text{O}_2 \\ \hline & 100\text{ g C}_7\text{H}_{16} & 1\text{ mole C}_7\text{H}_{16} & 1\text{ mole O}_2 \end{array} = 177\ l\ \text{O}_2$$

$$\begin{array}{l|l} 177\ l\ \text{O}_2 & 100\ l\text{ air} \\ \hline & 21\ l\ \text{O}_2 \end{array} = 840\ l\text{ air}$$

4.18) $2\,NO + O_2 \rightarrow 2\,NO_2$

$$\begin{array}{l|l} 0.345\ l\text{ NO} & 2\ l\text{ NO}_2 \\ \hline & 2\ l\text{ NO} \end{array} = 0.345\ l\text{ NO}_2$$

Chapter 5

[Answers in units of calories or kilocalories are accompanied by the equivalent answer in joules or kilojoules enclosed in brackets.]

5.1) $$Q = \begin{array}{l|l|l} 204\text{ g} & 0.031\text{ cal} & (64.9-22.8)^\circ\text{C} \\ \hline & \text{g–}^\circ\text{C} & \end{array} = 266\text{ cal }[1.11\text{ kJ}]$$

5.2) $$\Delta T = \begin{array}{l|l} 750\text{ cal} & \text{g–}^\circ\text{C} \\ \hline 454\text{ g} & 0.106\text{ cal} \end{array} = 15.6^\circ\text{C} \qquad 25.0 + 15.6 = 40.6^\circ\text{C}$$

5.3) $$Q = \begin{array}{l|l|l|l} 1000\text{ g CO}_2 & 1\text{ mole CO}_2 & 8.87\text{ cal} & (19-252)^\circ\text{C} \\ \hline & 44.0\text{ g CO}_2 & \text{mole–}^\circ\text{C} & \end{array}$$

$$\frac{1\text{ kcal}}{1000\text{ cal}} = -47.0\text{ kcal} \qquad [-197\text{ kJ}]$$

5.4) $$Q = \begin{array}{l|l} 17\text{ g} & 124\text{ cal} \\ \hline & 1\text{ g} \end{array} = 2100\text{ cal} = 2.1\text{ kcal }[8.8\text{ kJ}]$$

5.5) $$Q = \begin{array}{l|l|l} 250\text{ g NH}_3 & 1\text{ mole NH}_3 & 5.51\text{ kcal} \\ \hline & 17.0\text{ g NH}_3 & 1\text{ mole NH}_3 \end{array} = 81.0\text{ kcal }[339\text{ kJ}]$$

5.6) $$Q_{17^\circ\rightarrow 0^\circ} = \begin{array}{l|l|l} 265\text{ g} & 1.00\text{ cal} & (0.0-17.0)^\circ\text{C} \\ \hline & \text{g–}^\circ\text{C} & \end{array} = -4{,}500\text{ cal}$$

$$Q_{l\rightarrow s} = \begin{array}{l|l} 265\text{ g} & -80\text{ cal} \\ \hline & \text{g} \end{array} = -21{,}200\text{ cal}$$

$$Q_{0°\to-8°} = \frac{265\text{ g} \mid 0.49\text{ cal} \mid (-8-0)°\text{C}}{\mid \text{g–°C} \mid} = \frac{-1{,}040\text{ cal}}{-26{,}740\text{ cal}}\ [-111{,}900\text{ J}]$$

To two significant figures justified,

$-4.5 \times 10^3 - 21 \times 10^3 - 1 \times 10^3 = -27 \times 10^3 = -2.7 \times 10^4$ cal $[-1.1 \times 10^5$ J]

5.7)
$$\frac{120\text{ g }H_2O \mid 1.00\text{ cal} \mid (29.1-20.4)°\text{C}}{\mid \text{g–°C} \mid} +$$

$$\frac{110\text{ g }H_2O \mid 1.00\text{ cal} \mid (29.1-39.3)°\text{C}}{\mid \text{g–°C} \mid} + K_c\,(29.1-20.4)°\text{C} = 0$$

$$K_c = \frac{1122-1044}{29.1-20.4} = 9.0\text{ cal/°C [38 J/K]}$$

5.8)
$$\frac{72.0\text{ g }H_2O \mid 1.00\text{ cal} \mid (25.5-19.2)°\text{C}}{\mid \text{g–°C} \mid} +$$

$$\frac{5.2\text{ cal} \mid (25.5-19.2)°\text{C}}{°\text{C} \mid} + \frac{140\text{ g} \mid y\text{ cal} \mid (25.5-89.0)°\text{C}}{\mid \text{g–°C} \mid} = 0$$

$$453.6 + 32.8 - 8890\,y = 0$$

$$y = 0.055\text{ cal/g – °C}$$

5.9) 101 g $HC_2H_3O_2$ + 103 g NaOH = 204 g solution

$$\frac{204\text{ g} \mid 0.960\text{ cal} \mid 5.7°\text{C}}{\mid \text{g–°C} \mid} + \frac{7.4\text{ cal} \mid 5.7°\text{C}}{°\text{C} \mid} + \Delta H = 0$$

$\Delta H = -1116 - 42 = -1158$ cal [-4845 J]

$$\frac{100\text{ ml }HC_2H_3O_2\text{ soln} \mid 1.01\text{ g} \mid 6.00\text{ g }HC_2H_3O_2 \mid 1\text{ mole}}{\mid 1\text{ ml} \mid 100\text{ g }HC_2H_3O_2\text{ soln} \mid 60.0\text{ g}}$$

$$= 0.101\text{ mole }HC_2H_3O_2$$

$$\frac{100\text{ ml NaOH soln} \mid 1.03\text{ g} \mid 3.85\text{ g NaOH} \mid 1\text{ mole}}{\mid 1\text{ ml} \mid 100\text{ g NaOH soln} \mid 40.0\text{ g}} = 0.0991\text{ mole NaOH}$$

0.0991 mole H_2O formed.

$$\Delta H = \frac{-1158\text{ cal}}{0.0991\text{ mole}} = -11700\text{ cal/mole} = -11.7\text{ kcal/mole } [-49.0\text{ kJ/mole}]$$

To two significant figures, $\Delta H = -12$ kcal/mole [-50 kJ/mole]

Chapter 6

6.1)
$$\frac{85.0\text{ kcal} \mid 2\text{ moles }H_2O \mid 18.0\text{ ml}}{\mid 136.6\text{ kcal} \mid 1\text{ mole}} = 22.4\text{ ml}$$

6.2)
$$\frac{454\text{ g }C_6H_{12}O_6 \mid 1\text{ mole }C_6H_{12}O_6 \mid 673\text{ kcal}}{\mid 180\text{ g }C_6H_{12}O_6 \mid 1\text{ mole }C_6H_{12}O_6}$$

$$= 1.70 \times 10^3\text{ kcal } [7.10 \times 10^3\text{ kJ}]$$

6.3)
$$\frac{1500\text{ g }C_4H_{10} \mid 1\text{ mole} \mid 690\text{ kcal}}{\mid 58.0\text{ g} \mid 1\text{ mole }C_4H_{10}} = 1.78 \times 10^5\text{ kcal } [7.45 \times 10^5\text{ kJ}]$$

6.4) $\dfrac{454\text{ g Al}}{} \left| \dfrac{1\text{ mole}}{27.0\text{ g}} \right| \dfrac{235\text{ kcal}}{2\text{ moles Al}} \left| \dfrac{1\text{ kw–hr}}{860\text{ kcal}} \right. = 2.30\text{ kw–hr}$

6.5) $\Delta H = 8.09 - 21.60 = -13.51$ kcal [-56.5 kJ]

6.6) $\Delta H = -3(-196.5) = 589.5$ kcal [2.466 kJ]

6.7) $\Delta H = -7.31 + 2(-8.66) - 2(-23.78) - (-57.80) = +80.73$ kcal [+338 kJ]

6.8) $CH_4(g) + 2\ O_2(g) \rightarrow CO_2(g) + 2\ H_2O(l)$

$\Delta H_c = -94.05 + 2(-68.32) - (-17.89) = -212.8$ kcal [-890 kJ]

6.9) $C_2H_6(g) + \frac{7}{2}O_2(g) \rightarrow 2\ CO_2(g) + 3\ H_2O(l)$

$\Delta H_c = 2(-94.05) + 3(-68.32) - (-20.24) = -372.82$ kcal [-1560 kJ]

6.10) $SO_2(g) + \frac{1}{2}O_2(g) \rightarrow SO_3(g)$

$\Delta H = -94.45 - (-70.96) = -23.49$ kcal [-98.28 kJ]

$\dfrac{28.6\text{ g }SO_2}{} \left| \dfrac{1\text{ mole}}{64.1\text{ g}} \right| \dfrac{-23.49\text{ kcal}}{1\text{ mole }SO_2} = -10.5\text{ kcal } [-43.9\text{ kJ}]$

6.11)

	ΔH
$2\ C(s) + 2\ O_2(g) \rightarrow 2\ CO_2(g)$	$2(-94.05)$
$3\ H_2(g) + \frac{3}{2}\ O_2(g) \rightarrow 3\ H_2O(l)$	$3(-68.32)$
$2\ CO_2(g) + 3\ H_2O(l) \rightarrow C_2H_4(OH)_2(l) + \frac{5}{2}\ O_2(g)$	$+281.9$
$2\ C(s) + O_2(g) + 3\ H_2(g) \rightarrow C_2H_4(OH)_2(l)$	-111.2 kcal [-465.3 kJ]

6.12)

	ΔH
$HC_2H_3O_2(aq) + OH^-(aq) \rightarrow HOH(l) + C_2H_3O_2^-(aq)$	-11.7 kcal
$HOH(l) \rightarrow H^+(aq) + OH^-(aq)$	$+12.9$ kcal
$HC_2H_3O_2(aq) \rightarrow H^+(aq) + C_2H_3O_2^-(aq)$	$+\ 1.2$ kcal [5.0 kJ]

Chapter 7

7.1) $752 + 284 = 1036$ torr

7.2) $764 + \dfrac{35.0\text{ ft }H_2O}{} \left| \dfrac{305\text{ mm}}{\text{ft}} \right| \dfrac{1.02\text{ torr}}{13.6\text{ mm }H_2O} = 1565\text{ torr}$

7.3) $\dfrac{1.33\text{ atm}}{} \left| \dfrac{682\text{ ml}}{419\text{ ml}} \right. = 2.16\text{ atm}$

7.4) $\dfrac{14.2\text{ m}^3}{} \left| \dfrac{297°K}{335°K} \right. = 12.6\text{ m}^3$

7.5) $\dfrac{355\text{ psi}}{} \left| \dfrac{296°K}{255°K} \right. = 412\text{ psi}$

7.6) $\dfrac{8.42\ l}{} \left| \dfrac{725\text{ torr}}{760\text{ torr}} \right| \dfrac{273°K}{308°K} = 7.12\ l$

7.7) $\dfrac{0.140\text{ m}^3}{} \left| \dfrac{125\text{ psi}}{751\text{ torr}} \right| \dfrac{760\text{ torr}}{14.7\text{ psi}} \left| \dfrac{286°K}{306°K} \right. = 1.13\text{ m}^3$

7.8) $\dfrac{1.00\ l}{} \left| \dfrac{287°K}{408°K} \right| \dfrac{844\text{ torr}}{748\text{ torr}} = 0.794\ l$

7.9) $$\frac{644^\circ\text{K}}{} \Big| \frac{131 \text{ torr}}{51 \text{ atm}} \Big| \frac{1 \text{ atm}}{760 \text{ torr}} \Big| \frac{153\, l}{1\, l} = 333^\circ\text{K} = 60^\circ\text{C}$$

7.10) $$P = \frac{nRT}{V} = \frac{23.5 \text{ moles}}{9.81\, l} \Big| \frac{0.0821\, l\text{-atm}}{\text{mole-}^\circ\text{K}} \Big| \frac{296^\circ\text{K}}{} = 58.2 \text{ atm}$$

7.11) $$V = \frac{gRT}{P(MW)} = \frac{28.6 \text{ g } SO_2}{850 \text{ torr}} \Big| \frac{62.4\, l\text{-torr}}{\text{mole-}^\circ\text{K}} \Big| \frac{313^\circ\text{K mole}}{64.1 \text{ g}} = 10.3\, l$$

7.12) $$n = \frac{PV}{RT} = \frac{1.62 \text{ atm}}{290^\circ\text{K}} \Big| \frac{5.24\, l}{} \Big| \frac{\text{mole-}^\circ\text{K}}{0.0821\, l\text{-atm}} = 0.357 \text{ mole}$$

7.13) $$T = \frac{PV(MW)}{gR} = \frac{6.43 \text{ atm}}{} \Big| \frac{l}{10.3 \text{ g}} \Big| \frac{39.9 \text{ g}}{\text{mole}} \Big| \frac{\text{mole-}^\circ\text{K}}{0.0821\, l\text{-atm}} = \begin{array}{r} 303^\circ\text{K} \\ -\underline{273} \\ 30^\circ\text{C} \end{array}$$

7.14) $$g = \frac{PV(MW)}{TR} = \frac{4.76 \text{ atm}}{298^\circ\text{K}} \Big| \frac{6.64\, l}{} \Big| \frac{17.0 \text{ g}}{\text{mole}} \Big| \frac{\text{mole-}^\circ\text{K}}{0.0821\, l\text{-atm}} = 22.0 \text{ g}$$

7.15) $$MW = \frac{gRT}{PV} = \frac{1.31 \text{ g}}{l} \Big| \frac{62.4\, l\text{-torr}}{\text{mole-}^\circ\text{K}} \Big| \frac{293^\circ\text{K}}{749 \text{ torr}} = 32.0 \text{ g/mole}$$

7.16) $$\frac{g}{V} = \frac{P(MW)}{RT} = \frac{1 \text{ atm}}{298^\circ\text{K}} \Big| \frac{\text{mole-}^\circ\text{K}}{0.0821\, l\text{-atm}} \Big| \frac{29\text{g}}{1 \text{ mole}} = 1.2 \text{ g}/l$$

Chapter 8

8.1) $$\frac{28.4 \text{ g } C_3H_8}{} \Big| \frac{\text{mole}}{44.0 \text{ g}} \Big| \frac{22.4\, l}{\text{mole}} = 14.5\, l$$

8.2) $$\frac{1.50\, l\ N_2}{} \Big| \frac{\text{mole}}{22.4\, l} \Big| \frac{28.0 \text{ g}}{\text{mole}} = 1.88 \text{ g}$$

8.3) $$\frac{20.2 \text{ g Ne}}{\text{mole}} \Big| \frac{\text{mole}}{22.4\, l} = 0.902 \text{ g}/l$$

8.4) $$\frac{1.63 \text{ g}}{l} \Big| \frac{22.4\, l}{\text{mole}} = 36.5 \text{ g/mole}$$

8.5) $$\frac{2.63 \text{ g}}{2.10\, l} \Big| \frac{22.4\, l}{\text{mole}} = 28.1 \text{ g/mole}$$

8.6) $$MW = \frac{0.625 \text{ g}}{l} \Big| \frac{0.0821\, l\text{-atm}}{\text{mole-}^\circ\text{K}} \Big| \frac{1373^\circ\text{K}}{1.10 \text{ atm}} = 64.04 \text{ g/mole}$$

$$\frac{64.04 \text{ g}}{\text{mole S}} \Big| \frac{\text{mole atoms}}{32.1 \text{ g}} = 2 \text{ moles atoms/moles molecules}$$

Molecular formula: S_2

8.7)

	Grams	Moles	Mole Ratio
C	55.8	4.65	2
H	7.0	7.0	3
O	37.2	2.32	1

Empirical Formula: C_2H_3O; MW = 43 g/mole

$$MW = \frac{3.26 \text{ g}}{1.47\, l} \Big| \frac{0.0821\, l\text{-atm}}{\text{mole-}^\circ\text{K}} \Big| \frac{433^\circ\text{K}}{0.914 \text{ atm}} = 86.3 \text{ g/mole}$$

$\frac{86.3}{43} = 2$ Molecular Formula: $C_4H_6O_2$

8.8)

	Grams	**Moles**	**Mole Ratio**
C	92.31	7.69	1
H	7.69	7.69	1

Empirical Formula: CH
MW = 13

$$MW = \frac{4.35\ g}{4.16\ l} \Big| \frac{62.4\ l-torr}{mole-°K} \Big| \frac{295°K}{738\ torr} = 26.1\ g/mole$$

$$\frac{26.0}{13} = 2$$

Molecular Formula: C_2H_2

$$MW = \frac{1.88\ g}{l} \Big| \frac{62.4\ l-torr}{mole-°K} \Big| \frac{468°K}{702\ torr} = 78.2\ g/mole$$

$$\frac{78}{13} = 6$$

Molecular Formula: C_6H_6

8.9) $CH_4 + 2\ O_2 \rightarrow CO_2 + 2\ H_2O$

$$\frac{35.0\ l\ CH_4}{} \Big| \frac{749\ torr}{760\ torr} \Big| \frac{273°K}{295°K} \Big| \frac{mole}{22.4\ l} \Big| \frac{1\ mole\ CO_2}{1\ mole\ CH_4} \Big| \frac{44.0\ g}{1\ mole} = 62.7\ g\ CO_2$$

OR $n = \dfrac{PV}{RT}$

$$\frac{749\ torr}{} \Big| \frac{35.0\ l\ CO_2}{295°K} \Big| \frac{mole-°K}{62.4\ l-torr} \Big| \frac{1\ mole\ CO_2}{1\ mole\ CH_4} \Big| \frac{44.0\ g}{1\ mole} = 62.7\ g\ CO_2$$

8.10)

$$\frac{1000\ g\ CaCO_3 \cdot MgCO_3}{} \Big| \frac{mole}{184\ g} \Big| \frac{2\ moles\ CO_2}{1\ mole\ CaCO_3 \cdot MgCO_3} \Big| \frac{22.4\ l}{mole} \Big| \frac{760\ torr}{825\ torr} \Big| \frac{498°K}{273°K} = 409\ l\ CO_2$$

OR

$$\frac{1000\ g\ CaCO_3 \cdot MgCO_3}{} \Big| \frac{mole}{184\ g} \Big| \frac{2\ moles\ CO_2}{1\ mole\ CaCO_3 \cdot MgCO_3} \Big| \frac{62.4\ l-torr}{mole-°K} \Big| \frac{498°K}{825\ torr} = 409\ l\ CO_2$$

8.11)

$$\frac{155\ l\ O_2}{} \Big| \frac{2\ l\ NO_2}{1\ l\ O_2} = 310\ l\ NO_2$$

8.12)

$$\frac{500\ l\ CO}{} \Big| \frac{440\ torr}{160\ torr} \Big| \frac{295°K}{1973°K} \Big| \frac{0.5\ l\ O_2}{1\ l\ CO} = 103\ l\ O_2$$

8.13)

$$\frac{1000\ l\ air}{} \Big| \frac{747\ torr}{804\ torr} \Big| \frac{498°K}{292°K} \Big| \frac{6\ l\ PG}{5\ l\ air} = 1900\ l\ PG$$

8.14)

$$\frac{7.23\ l\ CH_4}{} \Big| \frac{1.35\ atm}{1.00\ atm} \Big| \frac{273°K}{280°K} \Big| \frac{mole}{22.4\ l} \Big| \frac{210.8\ kcal}{1\ mole\ CH_4} = 89.6\ kcal \quad [375\ kJ]$$

8.15)

	Grams	**Moles**	**Partial Pressure**
NH_3	24.0	1.41	(1.41/3.10)(642) = 292 torr
CO_2	21.0	0.48	(0.48/3.10)(642) = 99 torr
N_2	34.0	1.21	(1.21/3.10)(642) = 251 torr
		3.10	642 torr

8.16)

$$p_{Ne} = \frac{n_{Ne}RT}{V} = \frac{0.34\ mole}{4.60\ l} \Big| \frac{0.0821\ l-atm}{mole-°K} \Big| \frac{300°K}{} = 1.8\ atm$$

8.17) $p_{CO} = 0.2\%$ of 791 torr $= 0.002 \times 791 = 1.58$ torr

$$g = \frac{PV(MW)}{RT} = \frac{1.58\ torr}{393°K} \Big| \frac{1000\ l}{} \Big| \frac{28.0\ g}{mole} \Big| \frac{mole-°K}{62.4\ l-torr} = 1.80\ g\ CO$$

To one significant figure justified by 0.2%, 2 g CO.

8.18) $p_{O_2} = 786 - 19 = 767$ torr

8.19) $p_{O_2} = 755 - 36 = 719$ torr

$$\frac{80.4\text{ ml} \mid 719\text{ torr} \mid 273°\text{K}}{\mid 760\text{ torr} \mid 305°\text{K}} = 68.1\text{ ml}$$

8.20) $p_{O_2} = 769 - 30 = 739$ torr

$$\frac{2.63\,l\ O_2 \mid 739\text{ torr} \mid 273°\text{K} \mid \text{mole} \mid 2\text{ moles } KClO_3 \mid 122.6\text{ g}}{\mid 760\text{ torr} \mid 302°\text{K} \mid 22.4\,l \mid 3\text{ moles } O_2 \mid 1\text{ mole}} = 8.44\text{ g } KClO_3$$

OR

$$\frac{739\text{ torr} \mid 2.63\,l\ O_2 \mid \text{mole-}°\text{K} \mid 2\text{ moles } KClO_3 \mid 122.6\text{ g}}{\mid 302°\text{K} \mid 62.4\,l\text{-torr} \mid 3\text{ moles } O_2 \mid 1\text{ mole}} = 8.43\text{ g } KClO_3$$

8.21) $$\frac{\text{rate of } CH_4}{\text{rate of } C_2H_6} = \sqrt{\frac{30}{16}} = 1.37{:}1$$

8.22) $$\sqrt{\frac{28.0}{MW_{unk}}} = \frac{62}{51} = 1.22{:}\quad \frac{28.0}{MW_{unk}} = 1.48$$

$$MW_{unk} = \frac{28.0}{1.48} = 18.9\text{ g/mole}$$

8.23) $$\frac{\text{time } O_2}{45} = \sqrt{\frac{32}{2}} = 4$$

time $O_2 = 4 \times 45 = 180$ seconds

Chapter 9

9.1) $$\frac{18.5\text{ g solute}}{135\text{ g soln}} \times 100 = 13.7\%$$

9.2) $$\frac{65.0\text{ g solution} \mid 13.0\text{ g solute}}{\mid 100\text{ g soln}} = 8.45\text{ g solute}$$

9.3) $$\frac{20.0\text{ g } KNO_3 \mid 100\text{ g soln}}{\mid 12.0\text{ g } KNO_3} = 167\text{ g solution}$$

167 g soln − 20 g KNO_3 = 147 g H_2O

9.4) $$\frac{400\text{ ml soln} \mid 1.09\text{ g soln} \mid 16.0\text{ g solute}}{\mid 1.00\text{ ml soln} \mid 100\text{ g soln}} = 69.8\text{ g solute}$$

9.5) $$\frac{Z\text{ g } H_3BO_3}{100\text{ g } H_2O + Z\text{ g } H_3BO_3} \times 100 = 4.00 \qquad Z = 4.17\text{ g}$$

9.6)

	Grams	Moles	Mole Fraction
$HC_2H_3O_2$	12.0	0.200	0.200/8.53 = 0.023
H_2O	150	8.33	8.33/8.53 = 0.977
		8.53	

9.7) $$\frac{20.0\text{ g } C_{12}H_{22}O_{11} \mid 1\text{ mole } C_{12}H_{22}O_{11}}{0.100\text{ kg water} \mid 342\text{ g } C_{12}H_{22}O_{11}} = 0.585\text{ m}$$

9.8) $$\frac{0.080\text{ kg } H_2O \mid 4.00\text{ moles } CO(NH_2)_2 \mid 60.0\text{ g } CO(NH_2)_2}{\mid \text{kg } H_2O \mid 1\text{ mole } CO(NH_2)_2} = 19.2\text{ g } CO(NH_2)_2$$

9.9) $$\frac{90.9\text{ g } HC_2H_3O_2 \mid 1\text{ mole} \mid 1000\text{ ml } H_2O}{\mid 60.0\text{ g} \mid 1.40\text{ moles } HC_2H_3O_2} = 1080\text{ ml } H_2O$$

9.10) a) 50% each

b)

	Grams (assumed)	Moles	Mole Fraction
CH_3OH	100	3.13	3.13/5.30 = 0.591
C_2H_5OH	100	2.17	2.17/5.30 = 0.409
		5.30	

c) $\frac{3.13 \text{ moles } CH_3OH}{0.100 \text{ kg } C_2H_5OH} = 31.3 \text{ m}$

9.11) 1000 g H_2O = 55.5 moles H_2O
1.80 moles NH_4Cl
57.3 moles total

$X_{NH_4Cl} = \frac{1.80}{57.3} = 0.0314$

$\frac{1.80 \text{ moles } NH_4Cl \mid 53.5 \text{ g}}{\mid 1 \text{ mole}} = 96.3 \text{ g } NH_4Cl$

$\frac{96.3 \text{ g } NH_4Cl}{1000 \text{ g } H_2O + 96.3 \text{ g } NH_4Cl} \times 100 = 8.78\% \ NH_4Cl$

9.12)

	Grams	Moles	Mole Fraction
HCOOH	25.0	0.543	0.543/4.71 = 0.115
H_2O	75.0	4.17	4.17/4.71 = 0.885
		4.71	

$\frac{0.543 \text{ mole HCOOH}}{0.075 \text{ kg } H_2O} = 7.24 \text{ m}$

9.13) $\frac{23.5 \text{ g } Na_2SO_4 \mid 1 \text{ mole}}{0.600\, l \mid 142 \text{ g}} = 0.276 \text{ M}$

9.14) $\frac{120 \text{ g } Na_2SO_3 \cdot 5H_2O \mid 1 \text{ mole}}{1.25\, l \mid 248 \text{ g}} = 0.387 \text{ M}$

9.15) $\frac{0.400\, l \mid 0.800 \text{ mole } Na_2CO_3 \mid 106 \text{ g}}{\mid l \mid 1 \text{ mole}} = 33.9 \text{ g } Na_2CO_3$

9.16) $\frac{0.750\, l \mid 0.600 \text{ mole } HC_2H_3O_2 \mid 60.0 \text{ g}}{\mid l \mid 1 \text{ mole}} = 27.0 \text{ g } HC_2H_3O_2$

9.17) $\frac{0.0150 \text{ mole HCl} \mid 1000 \text{ ml}}{\mid 0.850 \text{ mole}} = 17.6 \text{ ml}$

9.18) $\frac{75.0 \text{ g } NH_3 \mid 1 \text{ mole} \mid 1000 \text{ ml}}{\mid 17.0 \text{ g} \mid 15 \text{ moles}} = 294 \text{ ml}$

9.19) $\frac{0.065\, l \mid 2.20 \text{ moles solute}}{\mid l} = 0.143 \text{ mole solute}$

9.20) $\frac{0.0293\, l \mid 0.482 \text{ mole } H_2SO_4}{\mid l} = 0.0141 \text{ mole } H_2SO_4$

9.21) $\frac{0.100\, l \text{ conc.} \mid 12 \text{ moles HCl}}{2.00\, l \text{ dil.} \mid 1\, l \text{ conc.}} = 0.600 \text{ M HCl}$

9.22) $\frac{0.500\, l \mid 6 \text{ moles } NH_3 \mid 1000 \text{ ml conc.}}{\mid l \mid 15 \text{ moles } NH_3} = 200 \text{ ml conc. } NH_3$

9.23) HF – 1 ; HSO_3^- – 1 ; $H_2C_2O_4$ – 2

9.24) $Ni(OH)_2$ – 2 ; LiOH – 1 ; $Zn(OH)_2$ – 2

9.25) $\frac{20.0 \text{ g HF}}{1 \text{ eq}}$; $\frac{81.1 \text{ g } HSO_3^-}{1 \text{ eq}}$; $\frac{90.0 \text{ g } H_2C_2O_4}{2 \text{ eq}} = 45.0 \text{ g/eq}$

9.26) $$\frac{92.7\text{ g }Ni(OH)_2}{2\text{ eq}} = 46.4\text{ g/eq} \;;\; \frac{23.9\text{ g LiOH}}{1\text{ eq}} \;;$$

$$\frac{99.4\text{ g }Zn(OH)_2}{2\text{ eq}} = 49.7\text{ g/eq}$$

9.27) $$\frac{0.196\text{ eq }HC_2H_3O_2}{} \Big| \frac{60\text{ g}}{1\text{ eq}} = 11.8\text{ g }HC_2H_3O_2$$

$$\frac{0.045\text{ eq }Ca(OH)_2}{} \Big| \frac{74.1\text{ g}}{2\text{ eq}} = 1.67\text{g }Ca(OH)_2$$

9.28) $$\frac{150\text{ g NaOH}}{} \Big| \frac{1\text{ eq}}{40.0\text{ g}} = 3.75\text{ eq NaOH}$$

$$\frac{42.0\text{ g HCOOH}}{} \Big| \frac{1\text{ eq}}{46.0\text{ g}} = 0.913\text{ eq HCOOH}$$

9.29) $$\frac{17.2\text{ g }HC_2H_3O_2}{0.300\ l} \Big| \frac{1\text{ eq}}{60.0\text{ g}} = 0.956\text{ N }HC_2H_3O_2$$

9.30) $$\frac{9.79\text{ g }NaHCO_3}{0.500\ l} \Big| \frac{1\text{ eq}}{84.0\text{ g}} = 0.233\text{ N }NaHCO_3$$

9.31) $$\frac{0.600\ l}{} \Big| \frac{2\text{ eq KOH}}{l} \Big| \frac{56.1\text{ g}}{1\text{ eq}} = 67.3\text{ g KOH}$$

9.32) $$\frac{0.250\ l}{} \Big| \frac{0.500\text{ eq }H_2C_2O_4}{l} \Big| \frac{126\text{ g }H_2C_2O_4 \cdot 2\,H_2O}{2\text{ eq }H_2C_2O_4} = 7.88\text{ g }H_2C_2O_4 \cdot 2\,H_2O$$

9.33) $$\frac{1\ l}{} \Big| \frac{0.50\text{ eq HCl}}{1\ l} \Big| \frac{1\text{ mole HCl}}{1\text{ eq HCl}} \Big| \frac{1000\text{ ml}}{12\text{ moles}} = 42\text{ ml }12\text{ M HCl}$$

9.34) $$\frac{0.620\text{ mole }HC_2H_3O_2}{l} \Big| \frac{1\text{ eq}}{1\text{ mole}} = 0.620\text{ N }HC_2H_3O_2$$

$$\frac{0.051\text{ mole }Ba(OH)_2}{l} \Big| \frac{2\text{ eq}}{1\text{ mole}} = 0.10\text{ N }Ba(OH)_2$$

9.35) $$\frac{2.15\text{ eq }HNO_3}{l} \Big| \frac{1\text{ mole}}{1\text{ eq}} = 2.15\text{ M }HNO_3$$

$$\frac{0.025\text{ eq }Sr(OH)_2}{l} \Big| \frac{1\text{ mole}}{2\text{ eq}} = 0.0125\text{ M }Sr(OH)_2 = 0.013\text{ M }Sr(OH)_2$$

Chapter 10

10.1) $AgNO_3 + Cl^- \rightarrow AgCl + NO_3^-$

$$\frac{0.050\ l}{} \Big| \frac{0.855\text{ mole }AgNO_3}{l} \Big| \frac{1\text{ mole AgCl}}{1\text{ mole }AgNO_3} \Big| \frac{144\text{ g}}{1\text{ mole}} = 6.16\text{ g AgCl}$$

10.2) $2\ NaF + Ba(NO_3)_2 \rightarrow BaF_2 + 2\ NaNO_3$

$$\frac{0.040\ l}{} \Big| \frac{0.436\text{ mole NaF}}{l} \Big| \frac{1\text{ mole }BaF_2}{2\text{ moles NaF}} \Big| \frac{175\text{ g}}{1\text{ mole}} = 1.53\text{ g }BaF_2$$

10.3) $2\ NaOH + CuSO_4 \rightarrow Cu(OH)_2 + Na_2SO_4$

$$\frac{0.025\ l}{} \Big| \frac{0.350\text{ mole NaOH}}{l} = 0.00875\text{ mole NaOH}$$ Limiting species

$$\frac{0.045\ l}{} \left| \frac{0.125 \text{ mole } CuSO_4}{l} \right. = 0.00563 \text{ mole } CuSO_4$$

$$\frac{0.00875 \text{ mole NaOH}}{} \left| \frac{1 \text{ mole } Cu(OH)_2}{2 \text{ moles NaOH}} \right| \frac{97.5 \text{ g}}{1 \text{ mole}} = 0.427 \text{ g } Cu(OH)_2$$

10.4) $2\ AgNO_3 + CaBr_2 \rightarrow 2\ AgBr + Ca(NO_3)_2$

$$\frac{0.035\ l}{} \left| \frac{0.128 \text{ mole } CaBr_2}{l} \right| \frac{2 \text{ moles } AgNO_3}{1 \text{ mole } CaBr_2} \left| \frac{1000 \text{ ml}}{0.415 \text{ mole } AgNO_3} \right. = 21.6 \text{ ml}$$

10.5) $HCl + NaOH \rightarrow NaCl + HOH$

$$\frac{0.020\ l}{} \left| \frac{0.809 \text{ mole NaOH}}{l} \right| \frac{1 \text{ mole HCl}}{1 \text{ mole NaOH}} \left| \frac{1000 \text{ ml}}{0.496 \text{ mole HCl}} \right. = 32.6 \text{ ml}$$

10.6) $2\ HCl + Na_2CO_3 \rightarrow 2\ NaCl + H_2O + CO_2$

$$\frac{1.24 \text{ g } Na_2CO_3}{} \left| \frac{\text{mole}}{106 \text{ g}} \right| \frac{2 \text{ moles HCl}}{1 \text{ mole } Na_2CO_3} \left| \frac{1000 \text{ ml}}{0.715 \text{ mole HCl}} \right. = 32.7 \text{ ml}$$

10.7) $2\ NaOH + H_2C_2O_4 \rightarrow Na_2C_2O_4 + 2\ HOH$

$$\frac{0.015\ l}{} \left| \frac{0.100 \text{ mole NaOH}}{l} \right| \frac{1 \text{ mole } H_2C_2O_4 \cdot 2\,H_2O}{2 \text{ moles NaOH}} \left| \frac{126 \text{ g}}{\text{mole}} \right. = 0.0945 \text{ g}$$

10.8)
$$\frac{0.050\ l}{} \left| \frac{1.20 \text{ moles HCl}}{l} \right| \frac{1 \text{ mole } Cl_2}{4 \text{ moles HCl}} \left| \frac{62.4\ l\text{-torr}}{\text{mole–°K}} \right| \frac{299\text{°K}}{740 \text{ torr}} = 0.378\ l\ Cl_2$$

OR

$$\frac{0.050\ l}{} \left| \frac{1.20 \text{ moles HCl}}{l} \right| \frac{1 \text{ mole } Cl_2}{4 \text{ moles HCl}} \left| \frac{22.4\ l}{\text{mole}} \right| \frac{760 \text{ torr}}{740 \text{ torr}} \left| \frac{299\text{°K}}{273\text{°K}} \right. = 0.378\ l\ Cl_2$$

10.9)
$$\frac{0.0168\ l\ AgNO_3}{0.0250\ l \text{ sample}} \left| \frac{0.629 \text{ mole } Ag^+}{l\ AgNO_3} \right| \frac{1 \text{ mole } Cl^-}{1 \text{ mole } Ag^+} = 0.423 \text{ M } Cl^-$$

10.10) $Pb^{2+} + 2I^- \rightarrow PbI_2$

$$\frac{0.060\ l}{} \left| \frac{0.322 \text{ mole } K^+ + I^-}{l} \right. = 0.0193 \text{ mole } K^+$$

$$= 0.0193 \text{ mole } I^- \text{ Limiting species}$$

$$\frac{0.020\ l}{} \left| \frac{0.530 \text{ mole } Pb(NO_3)_2}{l} \right| \frac{1 \text{ mole } Pb^{2+}}{1 \text{ mole } Pb(NO_3)_2} = 0.0106 \text{ mole } Pb^{2+}$$

$$\frac{0.0106 \text{ mole } Pb^{2+}}{} \left| \frac{2 \text{ moles } NO_3^-}{1 \text{ mole } Pb^{2+}} \right. = 0.0212 \text{ mole } NO_3^-$$

a) $$\frac{0.0193 \text{ mole } I^-}{} \left| \frac{1 \text{ mole } PbI_2}{2 \text{ moles } I^-} \right| \frac{461 \text{ g}}{\text{mole}} = 4.45 \text{ g } PbI_2$$

b) $$\frac{0.0193 \text{ mole } K^+}{0.080\ l} = 0.241 \text{ M } K^+$$

c) $$\frac{0.0212 \text{ mole } NO_3^-}{0.080\ l} = 0.265 \text{ M } NO_3^-$$

d) $$\frac{0.0193 \text{ mole } I^-}{} \left| \frac{1 \text{ mole } Pb^{2+}}{2 \text{ moles } I^-} \right. = 0.00965 \text{ mole } Pb^{2+} \text{ precipitated}$$

$$\frac{(0.0106 - 0.00965) \text{ mole } Pb^{2+} \text{ left}}{0.080\ l} = 0.119 \text{ M } Pb^{2+}$$

10.11) $$\frac{15.0 \mid 0.882}{12.8 \mid} = 1.03 \text{ N acid}$$

10.12) $$\frac{28.4 \mid 0.424}{25.0 \mid} = 0.482 \text{ N } NiCl_2$$

10.13) $$\frac{32.6 \mid 0.208}{20.0 \mid} = 0.339 \text{ N } H_2C_4H_4O_6$$

10.14) a) $$\frac{3.29 \text{ g } H_2C_2O_4 \cdot 2H_2O}{0.500\ l} \left| \frac{2 \text{ eq}}{126 \text{ g}} \right. = 0.104 \text{ N } H_2C_2O_4$$

b) $$\frac{25.0 \mid 0.104}{30.1 \mid} = 0.0864 \text{ N NaOH}$$

10.15) $$\frac{0.305 \text{ g } NH_2SO_3H}{0.0264\ l \text{ NaOH}} \left| \frac{1 \text{ eq}}{97.1 \text{ g } NH_2SO_3H} \right. = 0.119 \text{ N NaOH}$$

10.16) $$\frac{0.631 \text{ g acid}}{0.0156\ l \text{ NaOH}} \left| \frac{l \text{ NaOH}}{0.562 \text{ eq}} \right. = 72.0 \text{ g acid/eq}$$

Chapter 11

11.1) a)

	Grams	Moles	X
$C_6H_{12}O_6$	50.0	0.278	0.0477
H_2O	100	5.55	0.952
		5.83	

$P_{soln} = 0.952 \times 23.8 = 22.7$ torr

b) $$\Delta T_B = 0.52 \times \frac{0.278 \text{ mole } C_6H_{12}O_6}{0.100 \text{ kg } H_2O} = 1.45°C$$

$T_B = 100.00 + 1.45 = 101.45°C$

c) $$\Delta T_F = 1.86 \times \frac{0.278}{0.100} = 5.17°C \qquad T_F = 5.17°C$$

11.2) a)

	Grams	Moles	X
$CO(NH_2)_2$	10.0	0.167	0.0323
H_2O	90.0	5.00	0.967
		5.17	

$P_{soln} = 0.967 \times 17.5 = 16.9$ torr

b) $$\Delta T_B = 0.52 \times \frac{0.167 \text{ mole } CO(NH_2)_2}{0.0900 \text{ kg } H_2O} = 0.965°C$$

$T_B = 100.00 + 0.97 = 100.97°C$

c) $$\Delta T_F = 1.86 \times \frac{0.167}{0.0900} = 3.45°C \qquad T_F = -3.45°C$$

11.3) $$m = \frac{4.34 \text{ g } C_6H_4Cl_2}{0.0650 \text{ kg } C_{10}H_8} \left| \frac{\text{mole } C_6H_4Cl_2}{147 \text{ g } C_6H_4Cl_2} \right. = 0.454 \text{ m}$$

$\Delta T_F = 6.9 \times 0.454 = 3.1°C$

$T_F = 80.2 - 3.1 = 77.1°C$

11.4) $$m = \frac{40.0 \text{ g } HOOC(CH_2)_3COOH}{0.150 \text{ kg } H_2O} \left| \frac{1 \text{ mole } HOOC(CH_2)_3COOH}{132 \text{ g } HOOC(CH_2)_3COOH} \right. = 2.02 \text{ m}$$

$\Delta T_B = 0.52 \times 2.02 = 1.05 \qquad T_B = 101.05°C$

$\Delta T_F = 1.86 \times 2.02 = 3.76 \qquad T_F = -3.76°C$

11.5) $$m = \frac{1.40 \text{ g } CO(NH_2)_2}{0.0163 \text{ kg solvent}} \left| \frac{\text{mole } CO(NH_2)_2}{60.0 \text{ g } CO(NH_2)_2} \right. = 1.43$$

$$K_B = \frac{3.92}{1.43} = 2.74$$

11.6) $$m = \frac{32}{1.86} = 17.2 \text{ m}$$

$$\frac{8.50 \text{ kg } H_2O}{} \left| \frac{17.2 \text{ moles } C_3H_8O_3}{\text{kg } H_2O} \right| \frac{92.0 \text{ g}}{\text{mole}} \left| \frac{1 \text{ kg}}{10^3 \text{ g}} \right. = 13.5 \text{ kg}$$

11.7) $$m = \frac{0.84}{0.52} = 1.6 \text{ m}$$

11.8) $$m = \frac{1.18}{1.86} = 0.634 \text{ m}$$

$$\frac{26.0 \text{ g solute}/0.380 \text{ kg } H_2O}{0.634 \text{ mole solute/kg } H_2O} = 108 \text{ g/mole}$$

11.9) $\Delta T_F = 80.2 - 71.3 = 8.9°C$

$$m = \frac{8.9}{6.9} = 1.29 \text{ m}$$

$$\frac{12.0 \text{ g solute}/0.080 \text{ kg } C_{10}H_8}{1.29 \text{ moles solute/kg } C_{10}H_8} = 116 = 120 \text{ g/mole (2 sig. figs)}$$

11.10) $\Delta T_F = 2 \times 1.86 \times 0.10 = 0.37$ $\quad T_F = -0.37°C$

11.11) $$m = \frac{9.41 \text{ g } NaHSO_3}{1.00 \text{ kg } H_2O} \left| \frac{1 \text{ mole } NaHSO_3}{104 \text{ g } NaHSO_3} \right. = 0.0905 \text{ m}$$

$$i = \frac{0.33}{1.86} \left| \frac{}{0.0905} \right. = 1.96 \approx 2 \text{ moles ions per mole solute: } Na^+ \text{ and } HSO_3^-$$

11.12) $$\pi = \frac{4.00 \text{ g } C_{12}H_{22}O_{11}}{0.500\, l} \left| \frac{62.4\, l\text{–torr}}{\text{mole–°K}} \right| \frac{293°K}{} \left| \frac{\text{mole } C_{12}H_{22}O_{11}}{342 \text{ g } C_{12}H_{22}O_{11}} \right.$$

$$\pi = 428 \text{ torr}$$

11.13) $$\pi = \frac{0.089 \text{ mole}}{l} \left| \frac{62.4\, l\text{-torr}}{\text{mole–°K}} \right| \frac{298°K}{} = 1650 \text{ torr}$$

$$\frac{1650 \text{ mm Hg}}{} \left| \frac{13.6 \text{ mm soln}}{1.01 \text{ mm Hg}} \right| \frac{\text{m}}{10^3 \text{ mm}} = 22 \text{ meters} \qquad \text{(2 sig. figs.)}$$

11.14) a)

	Grams	Moles of Particles	X
NaCl	3.5	0.12	0.022
H_2O	96.5	5.36	0.98
		5.48	

$P_n = 0.98 \times 23.8 = 23.3$ torr

b) $$m = \frac{3.5 \text{ g NaCl}}{0.0965 \text{ kg } H_2O} \left| \frac{1 \text{ mole NaCl}}{58.5 \text{ g NaCl}} \right. = 0.62 \text{ m}$$

$\Delta T_F = 2 \times 1.86 \times 0.62 = 2.31°C$ $\quad T_F = -2.3°C$

c) $\Delta T_B = 2 \times 0.52 \times 0.62 = 0.64°C$ $\quad T_B = 100.64°C$

Chapter 12

12.1) $\Delta E = Q - W = +8.34 - (-0.71) = +9.05 \text{ kcal} \quad [37.9 \text{ kJ}]$

12.2)
$$W = P\Delta V = \frac{1.00 \text{ atm}}{} \left| \frac{5.20 \text{ moles}}{} \right| \frac{0.0821\ l\text{–atm}}{\text{mole–°K}} \left| \frac{240\text{°K}}{1.00 \text{ atm}} \right| \frac{0.0242 \text{ kcal}}{l\text{–atm}}$$

$$= 2.48 \text{ kcal } [10.4 \text{ kJ}]$$

12.3)
$$Q = \frac{84.8 \text{ g } H_2O}{} \left| \frac{1 \text{ mole}}{18.0 \text{ g}} \right| \frac{-9.72 \text{ kcal}}{1 \text{ mole}} = -45.8 \text{ kcal } [-192 \text{ kJ}]$$

$$W = -\frac{1.00 \text{ atm}}{} \left| \frac{84.8 \text{ g } H_2O}{} \right| \frac{1\text{mole}}{18.0 \text{ g}} \left| \frac{0.0821\ l\text{–atm}}{\text{mole–°K}} \right| \frac{373\text{°K}}{1.00 \text{ atm}} \left| \frac{0.0242 \text{ kcal}}{l\text{–atm}} \right.$$

$$= -3.49 \text{ kcal } [-14.6 \text{ kJ}]$$

$\Delta E = Q - W = -45.8 - (-3.49) = -42.3 \text{ kcal} \quad [-177 \text{ kJ}]$

12.4)
$$W = P\Delta V = \frac{1.00 \text{ atm}}{} \left| \frac{1 \text{ mole } CO_2}{} \right| \frac{0.0821\ l\text{–atm}}{\text{mole–°K}} \left| \frac{298\text{°K}}{1.00 \text{ atm}} \right| \frac{0.0242 \text{ kcal}}{l\text{–atm}}$$

$$= 0.592 \text{ kcal } [2.48 \text{ kJ}]$$

$\Delta E = Q - W = 42.5 - 0.592 = 41.9 \text{ kcal} \quad [175 \text{ kJ}]$

12.5)
$$W = RT\Delta n = \frac{0.00199 \text{ kcal}}{\text{mole–°K}} \left| \frac{1 \text{ mole}}{} \right| \frac{351\text{°K}}{} = 0.698 \text{ kcal } [2.92 \text{ kJ}]$$

OR
$$W = P\Delta V = \frac{1.00 \text{ mole}}{} \left| \frac{1.00 \text{ atm}}{} \right| \frac{0.0821\ l\text{–atm}}{\text{mole–°K}} \left| \frac{351\text{°K}}{1.00 \text{ atm}} \right| \frac{0.0242 \text{ kcal}}{l\text{–atm}}$$

$$= 0.697 \text{ kcal } [2.92 \text{ kJ}]$$

$\Delta E = Q - W = \Delta H - W = 9.40 - 0.698 = 8.70 \text{ kcal} \quad [36.4 \text{ kJ}]$

12.6)
$$\Delta H = Q = \frac{40.0 \text{ g Na}}{} \left| \frac{1 \text{ mole}}{23.0 \text{ g}} \right| \frac{630 \text{ cal}}{1 \text{ mole}} = 1100 \text{ cal} = 1.10 \text{ kcal } [4.60 \text{ kJ}]$$

$W = 0$ because $\Delta V \approx 0$. $\Delta E = Q - W = 1.10 - 0 = 1.10 \text{ kcal} \quad [4.60 \text{ kJ}]$

12.7) $CH_4(g) + 2\ O_2(g) \rightarrow CO_2(g) + 2\ H_2O(l)$

$\Delta H_c = -94.05 + 2(-68.32) - (-17.89) = -212.80 \text{ kcal } [-890 \text{ kJ}]$

12.8) $Q = \Delta H = -212.80 \text{ kcal } [-890 \text{ kJ}]$

$$W = RT\Delta n = \frac{0.00199 \text{ kcal}}{\text{mole–°K}} \left| \frac{298\text{°K}}{} \right| \frac{(1-3) \text{ moles}}{} = -1.19 \text{ kcal } [-4.98 \text{ kJ}]$$

$\Delta E = Q - W = -212.80 - (-1.19) = -211.61 \text{ kcal } [-885 \text{ kJ}]$

12.9) $\Delta H = -94.05 - 2(-143.84) = 193.63 \text{ kcal } [810 \text{ kJ}]$

$$\Delta E = \Delta H - RT\Delta n = 193.63 - \frac{0.00199 \text{ kcal}}{\text{mole–°K}} \left| \frac{298\text{°K}}{} \right| \frac{(1-0) \text{ mole}}{}$$

$$= 193.04 \text{ kcal} \quad [808 \text{ kJ}]$$

12.10)
$$\Delta H = \Delta E + RT\Delta n = 5.2 + \frac{0.00199 \text{ kcal}}{\text{mole–°K}} \left| \frac{298\text{°K}}{} \right| \frac{(1-0) \text{ mole}}{}$$

$$= 5.8 \text{ kcal } [24 \text{ kJ}]$$

12.11) $\Delta S° = 51.06 + 2(16.72) - 44.50 - 2(49.00) = -58.00 \text{ cal/°K } [-243 \text{ J/°K}]$

12.12) $\Delta S° = 2(16.72) - 2(31.2) - 49.00 = -77.96$ cal/°K [−326 J/°K]

12.13) $\Delta S = \frac{\Delta H}{T} = \frac{100\text{ g} \mid 1\text{ mole} \mid 1440\text{ cal} \mid}{\mid 18.0\text{ g} \mid 1\text{ mole} \mid 273°K} = 29.3$ cal/°K [123 J/°K]

12.14) $\Delta G° = -212.80 - 298(-0.05800) = -195.52$ kcal [−818 kJ]

12.15) $\Delta G° = -94.26 + 2(-56.69) - (-12.14) = -195.50$ kcal [−818 kJ]

12.16) $\Delta G° = \Delta H° = T\Delta S° = 5.8 - 298(0.05318) = -10.05$ kcal [−42.0 kJ]

12.17) $\Delta G° = \Delta H° - T\Delta S° = -21.36 - 298(0.11818) = -56.58$ kcal [−237 kJ]

The reaction is spontaneous at 298°K, and at all other temperatures with a negative ΔH and positive ΔS.

$\Delta G_{253°K} = -21.36 - 253(0.11818) = -51.26$ kcal [−214 kJ]

12.18) $Q = \Delta H = 2(34.0) = 68.0$ kcal [285 kJ]

$\Delta G° = 2(39.06) = 78.12$ kcal [327 kJ]

$\Delta S° = 2(56.8) - 3(49.00) = -33.4$ cal/°K [−140 J/°K]

$W = RT\Delta n = \frac{0.00199\text{ kcal} \mid 298°K \mid (2-3)\text{ moles}}{\text{mole–°K} \mid \mid} = -0.593$ kcal [−2.48 kJ]

$\Delta E = Q - W = 68.0 - (-0.6) = 68.6$ kcal [287 kJ]

12.19) $\Delta H = Q = 4(-94.05) + 2(-68.32) - 2(54.19) = -621.11$ kcal [−2600 kJ]

$\Delta G° = 4(-94.26) + 2(-56.69) - 2(50.00) = -590.42$ kcal [−2470 kJ]

$\Delta S° = 4(51.06) + 2(16.72) - 2(48.00) - 5(49.00) = -103.32$ cal/°K [−432 J/°K]

$W = RT\Delta n = \frac{0.00199\text{ kcal} \mid 298°K \mid (4-7)\text{ moles}}{\text{mole–°K} \mid \mid} = -1.78$ kcal [−7.45 kJ]

$\Delta E = Q - W = -621.22 - (-1.78) = -619.44$ kcal [−2592 kJ]

12.20) $\Delta G° = -71.79$ kcal ($\Delta G°$ for SO_2 from Table III) [−300.37 kJ]

12.21) $\Delta G° = 2(-88.52) - 2(-71.79) = -33.46$ kcal [−140.00 kJ]

12.22) $\Delta G° = 2(-88.52) = -177.04$ kcal [−740.74 kJ]

$2\ S(s) + 2\ O_2(g) \rightarrow 2\ SO_2(g)$ $\Delta G° = -143.58$ kcal

$2\ SO_2(g) + O_2(g) \rightarrow 2\ SO_3(g)$ $\Delta G° = -\ 33.46$ kcal

$2\ S(s) + 3\ O_2(g) \rightarrow 2\ SO_3(g)$ $\Delta G° = -177.04$ kcal [−740.74 kJ]

The result is the necessary consequence of Equation 12.16.

Chapter 13

13.1)

	$2\ SO_2$	+	O_2	=	$2\ SO_3$
I	6		4		0
R	−4		−2		+4
E	2		2		4

$$K = \frac{4^2}{2^2 \times 2} = 2$$

13.2)

	CO	+	Cl_2	=	$COCl_2$
I	3.00		5.00		0
R	−2.90		−2.90		+2.90
E	0.10		2.10		2.90

$$K = \frac{2.90}{0.10 \times 2.10} = 13.8$$

13.3)

	H_2	+	I_2	=	2 HI
I	4.6		7.4		0
R	−3.4		−3 4		+6.8
E	1.2		4.0		6.8

$$K = \frac{6.8^2}{1.2 \times 4.0} = 9.6$$

13.4)

	CO	+	H_2O	=	CO_2	+	H_2
I	2.4		2.4		0		0
R	−y		−y		+y		+y
E	2.4 − y		2.4 − y		y		y
	0.9		0.9		1.5		1.5

$$K = \frac{y^2}{(2.4 - y)^2} = 2.9$$

$$\frac{y}{2.4 - y} = 1.7$$

$$y = 1.5$$

13.5)

	PCl_5	=	PCl_3	+	Cl_2
I	12		0		0
R	−y		+y		+y
E	12 − y		y		y
	11.25		0.75		0.75

$$K = \frac{y^2}{12 - y} = 0.050$$

$$y = 0.75$$

13.6)

	N_2O_4	=	2 NO_2
I	4.00		0
R	−y		+2y
E	4.00 − y		2y
	3.92		0.164

$$K = \frac{(2y)^2}{4.00 - y} = 6.90 \times 10^{-3}$$

$$y = 0.082$$

13.7)

	H_2	+	I_2	=	2 HI
I	0		0		y + 3.6 = 8.0
R	+1.8		+1.8		−3.6
E	1.8		1.8		y

$$K = \frac{y^2}{1.8^2} = 6$$

$$y = 4.4$$

$$y + 3.6 = 8.0$$

13.8)

	CO	+	H_2O	=	CO_2	+	H_2
I	0.30		0.10 + y		0.20		0.60
R	−0.10		−0.10		+0.10		+0.10
E	0.20		y		0.30		0.70

$$K = \frac{0.20 \times 0.60}{0.30 \times 0.10} = 4.0$$

$$\frac{0.30 \times 0.70}{0.20\ y} = 4.0$$

$$y = 0.26$$

13.9)

	H_2	+	I_2	=	2 HI
I	0.20		0.20		1.20
R	+y		+y		−2y
E	0.20 + y		0.20 + y		1.20 − 2y

$$K = \frac{0.80^2}{0.20^2} = 16$$

$$\frac{(1.20 - 2y)^2}{(0.20 + y)^2} = 16$$

$$y = 0.0667$$

13.10) $\Delta G° = 2(-12.72) = -25.44$ kcal

$$\log K_p = \frac{2(-12.72)}{(-2.303)(0.00199)(298)} = 18.6$$

$$K_p = 10^{18.6} = 4 \times 10^{18};\ K_c = \frac{K_p}{(RT)^{\circ}} = \frac{K_p}{1} = 4 \times 10^{18}$$

The negative value of $\Delta G°$ and positive values of K_p and K_c indicate the equilibrium is favored in the forward direction

13.11) $\Delta G° = 2(-3.98)$

$$\log K_p = \frac{2(-3.98)}{(-2.303)(0.00199)(298)} = 5.83$$

$$K_p = 10^{5.83} = 6.8 \times 10^5;\quad K_c = \frac{6.8 \times 10^5}{(0.0821 \times 298)^{-2}} = 4.1 \times 10^8$$

13.12)

	N_2	+	$3\ H_2$	=	$2\ NH_3$
I	5.10		14.5		0
R	−4.40		−13.20		+8.80
E	0.70		1.3		8.80

$$K_p = \frac{8.80^2}{0.70 \times 1.3^3} = 50$$

13.13)

	H_2	+	I_2	=	$2\ HI$
I	0		0		6.40
R	+y		+y		−2y
E	y		y		6.40 − 2y
	0.729		0.729		4.94

$$K = \frac{(6.40 - 2y)^2}{y^2} = 46.0$$

$$\frac{6.40 - 2y}{y} = 6.78$$

$$y = 0.729$$

Chapter 14

14.1) a) $\frac{10^{-5}}{10^{-8}} = \frac{10^3}{1}$; b) $\frac{10^{-10}}{10^{-7}} = \frac{1}{10^3}$; c) $\frac{10^{-9}}{10^{-5}} = \frac{1}{10^4}$

14.2) $[H^+] = 10^{-2.3} = 5 \times 10^{-3}$; $[OH^-] = \frac{10 \times 10^{-15}}{5 \times 10^{-3}} = 2 \times 10^{-12}$

$pOH = -\log(2 \times 10^{-12}) = 11.7$

14.3) $pH = 14.0 - 12.1 = 1.9$; $[H^+] = 10^{-1.9} = 1.25 \times 10^{-2}$

$[OH^-] = \frac{10 \times 10^{-15}}{1.25 \times 10^{-2}} = 8.0 \times 10^{-13}$

14.4) $pH = -\log(4 \times 10^{-7}) = 6.4$; $pOH = 14.0 - 6.4 = 7.6$

$[OH^-] = 10^{-7.6} = 2.5 \times 10^{-8}$

14.5) $[H^+] = \frac{10 \times 10^{-15}}{5 \times 10^{-12}} = 2 \times 10^{-3}$; $pH = -\log(2 \times 10^{-3}) = 2.7$

$pOH = 14.0 - 2.7 = 11.3$

14.6) $[H^+] = 1 \times 10^{-5}$; $pH = 5$; $pOH = 14 - 5 = 9$; $[OH^-] = 1 \times 10^{-9}$

14.7) $\frac{0.0009\ g\ Mg(OH)_2}{0.100\ l} \Big| \frac{2\ moles\ OH^-}{58.3\ g\ Mg(OH)_2} = 3.09 \times 10^{-4}\ M\ OH^-\ (3 \times 10^{-4})$

$pOH = -\log(3.09 \times 10^{-4}) = 3.510\ (3.5)$; $pH = 10.490\ (10.5)$

$[H^+] = 10^{-10.490} = 3.24 \times 10^{-11}\ (3 \times 18^{11})$ (1 Sig. fig. shown in parentheses)

14.8) $[H^+] = \sqrt{9.55 \times 10^{-14}} = 3.09 \times 10^{-7}$

$pH = -\log(3.09 \times 10^{-7}) = 6.51$

14.9) 0.010 mole HCl in excess. pH = 2.0

14.10)

Solution	pH	pOH	$[OH^-]$	$[H^+]$
A	4.62	9.38	4.2×10^{-10}	2.4×10^{-5}
B	5.26	8.74	1.8×10^{-9}	5.6×10^{-6}
C	2.07	11.93	1.2×10^{-12}	8.6×10^{-3}
D	8.52	5.48	3.3×10^{-6}	3.0×10^{-9}

Chapter 15

15.1)

	$HCOOH =$	H^+	$+ HCOO^-$
I	0.50	0	0
R	−0.01	+0.01	+0.01
E	0.49	0.01	0.01

$$K = \frac{(0.01)^2}{0.49} = 2.0 \times 10^{-4}$$

15.2)

	$C_6H_5COOH =$	H^+	$+ C_6H_5COO^-$
I	0.20	0	0
R	−0.0036	+0.0036	+0.0036
E	0.20	0.0036	0.0036

$$[H^+] = 10^{-2.44} = 3.6 \times 10^{-3}$$

$$K = \frac{(3.6 \times 10^{-3})^2}{0.20} = 6.5 \times 10^{-5}$$

$$\frac{3.6 \times 10^{-3}}{0.20} \Big| \frac{10^2}{} = 1.8\% \text{ ionized}$$

15.3)

	$HL =$	H^+	$+ L^-$
I	1.0	0	0
R	−0.012	+0.012	+0.012
E	1.0	0.012	0.012

$$[H^+] = 10^{-1.93} = 1.17 \times 10^{-2}$$

$$K = \frac{0.012^2}{1.0} = 1.4 \times 10^{-4}$$

$$\frac{0.012}{1.0} \Big| \frac{10^2}{} = 1.2\% \text{ ionized}$$

15.4)

	$HNO_2 =$	H^+	$+ NO_2^-$
I	0.10	0	0
E	0.10 − y	y	y

$$K = \frac{y^2}{0.10 - y} = 4.5 \times 10^{-4}$$

$y = 6.5 \times 10^{-3}$:

$$pH = -\log(6.5 \times 10^{-3}) = 2.19$$

$$\frac{6.5 \times 10^{-3}}{0.10} \Big| \frac{10^{-2}}{} = 6.5\% \text{ ionized}$$

15.5)

	$HC_2H_3O_2 =$	H^+	$+ C_2H_3O_2^-$
I	0.30	0	0
E_a	0.30	y	y

$$K = \frac{y^2}{0.30} = 1.8 \times 10^{-5}$$

$y = 2.3 \times 10^{-3}$

$$pH = -\log(2.3 \times 10^{-3}) = 2.64$$

$$\frac{2.3 \times 10^{-3}}{0.30} \Big| \frac{10^2}{} = 0.77\% \text{ ionized}$$

15.6)

	$H_2CO_3 =$	H^+	$+ HCO_3^-$
I	0.033	0	0
E_a	0.033	y	y

$$K = \frac{y^2}{0.033} = 4.2 \times 10^{-7}$$

$y = 1.2 \times 10^{-4}$

$$pH = -\log(1.2 \times 10^{-4}) = 3.92$$

15.7)

	NO_2^-	$+ HOH =$	HNO_2	$+ OH^-$
I	0.50		0	0
E_a	0.50		y	y

$$K = \frac{y^2}{0.50} = \frac{10 \times 10^{-15}}{4.5 \times 10^{-4}}$$

$y = 3.3 \times 10^{-6} = [OH^-]$

$[H^+] = 3.0 \times 10^{-9}$

$$pH = -\log(3.0 \times 10^{-9}) = 8.52$$

15.8)

	NH_4^+ + HOH =	NH_3 +	H_3O^+
I	1.0	0	0
E_a	1.0	y	y

$K = y^2 = \frac{10 \times 10^{-15}}{1.8 \times 10^{-5}}$

$y = 2.34 \times 10^{-5}$

$pH = -\log(2.34 \times 10^{-5}) = 4.63$

15.9)

	$C_7H_5O_2^-$ + HOH =	$HC_7H_5O_2$ +	OH^-
I	0.25 0	0	0
E_a	0.25	y	y

$K = \frac{y^2}{0.25} = \frac{10 \times 10^{-15}}{6.6 \times 10^{-5}}$

$y = 6.15 \times 10^{-6}$

$[H^+] = 1.63 \times 10^{-8}$

$pH = -\log(1.63 \times 10^{-8}) = 7.79$

15.10)

	H_2SO_3 =	H^+ +	HSO_3^-
I	0.20	0	0
E	0.20 − y	y	y

$K = \frac{y^2}{0.20 - y} = 1.7 \times 10^{-2}$

$y = 5.0 \times 10^{-2}$

$pH = -\log(5.0 \times 10^{-2}) = 1.30$

	HSO_3^- =	H^+ +	SO_3^{2-}
I	0.05	0.05	0
E_a	0.05	0.05	z

$K = \frac{0.05z}{0.05} = z = 5.6 \times 10^{-8}$

$[H^+]$ is unchanged in second ionization

15.11)

	PO_4^{3-} + HOH =	HPO_4^{2-} +	OH^-
I	0.25	0	0
E	0.25−y	y	y

$K = \frac{y^2}{0.25 - y} = \frac{10 \times 10^{-15}}{1.7 \times 10^{-12}}$

$y = 3.5 \times 10^{-2} = [OH^-]$

$pOH = -\log(3.5 \times 10^{-2}) = 1.46$

$pH = 12.54$

	$H_2PO_4^-$ + HOH =	HPO_4^{2-} +	OH^-
I	0.035	0	0.035
E_a	0.035	z	0.035

$K = \frac{0.035z}{0.035} = \frac{10 \times 10^{-15}}{6.2 \times 10^{-8}}$

$z = 1.61 \times 10^{-7}$

No change in second ionization

15.12)

	HNO_2 =	H^+ +	NO_2^-
E	0.75	y	0.25

$K = \frac{0.25y}{0.75} = 4.5 \times 10^{-4}$

$y = 1.35 \times 10^{-3}$

$pH = -\log(1.35 \times 10^{-3}) = 2.87$

15.13) $\frac{28.0\ g\ NaC_2H_3O_2}{0.500\ l} \left| \frac{mole\ C_2H_3O_2^-}{82.0\ g\ NaC_2H_3O_2} \right. = 0.683\ mole\ C_2H_3O_2^-/liter$

	$HC_2H_3O_2$ =	H^+ +	$C_2H_3O_2^-$
E	0.12	y	0.683

$K = \frac{0.683y}{0.12} = 1.8 \times 10^{-5}$

$y = 3.16 \times 10^{-6}$

$pH = -\log(3.16 \times 10^{-6}) = 5.50$

15.14) $\frac{0.020\,l \;|\; 4.0 \text{ moles } H^+}{\;|\; l} = 0.080 \text{ mole } H^+$

$\frac{11.5 \text{ g } NaC_2H_3O_2 \;|\; 1 \text{ mole } C_2H_3O_2^-}{\;|\; 82.0 \text{ g } NaC_2H_3O_2} = 0.140 \text{ mole } C_2H_3O_2^-$

$0.080 \text{ mole } H^+ + 0.140 \text{ mole } C_2H_3O_2^- \rightarrow 0.080 \text{ mole } HC_2H_3O_2 + 0.060 \text{ mole } C_2H_3O_2^-$

$[H^+] = \frac{1.8 \times 10^{-5} \;|\; 0.080}{\;|\; 0.060} = 2.4 \times 10^{-5}$

$pH = -\log(2.4 \times 10^{-5}) = 4.62$ Volume has no bearing on the pH

15.15) $[H^+] = 10^{-4.80} = 1.59 \times 10^{-5}$

$\frac{[HC_7H_5O_2]}{[C_7H_5O_2^-]} = \frac{1.59 \times 10^{-5}}{6.6 \times 10^{-5}} = \frac{0.24 \text{ mole } HC_7H_5O_2}{1 \text{ mole } C_7H_5O_2^-}$

$\frac{16.0 \text{ g } HC_7H_5O_2 \;|\; 1 \text{ mole } HC_7H_5O_2 \;|\; 1 \text{ mole } C_7H_5O_2^- \;|}{\;|\; 122 \text{ g } HC_7H_5O_2 \;|\; 0.24 \text{ mole } HC_7H_5O_2 \;|}$

$\frac{144 \text{ g } NaC_7H_5O_2}{1 \text{ mole } C_7H_5O_2^-} = 79 \text{ g } NaC_7H_5O_2$

15.16) $[H^+] = 10^{-4.15} = 7.1 \times 10^{-5}$

$\frac{[C_2H_3O_2^-]}{[HC_2H_3O_2]} = \frac{1.8 \times 10^{-5}}{7.1 \times 10^{-5}} = \frac{0.253 \text{ mole } C_2H_3O_2^-}{1 \text{ mole } HC_2H_3O_2}$

$\frac{0.253 \text{ mole } C_2H_3O_2^- \;|\; 82 \text{ g } NaC_2H_3O_2}{\;|\; 1 \text{ mole } C_2H_3O_2^-} = 20.7 \text{ g } NaC_2H_3O_2$

$\frac{1 \text{ mole } HC_2H_3O_2 \;|\; 1000 \text{ ml}}{\;|\; 17 \text{ moles}} = 58.8 \text{ ml } 17 \text{ M } HC_2H_3O_2$

Any quantities such that $\frac{\text{g } NaC_2H_3O_2}{\text{ml } HC_2H_3O_2} = \frac{20.7}{58.8} = 0.35$

15.17)

	$Ag(S_2O_3)_2^{3-}$ =	Ag^+	+ $2\,S_2O_3^{2-}$
I	0	0.025	0.50
R	+0.025	−0.025	−0.050
E	0.025	y	0.45

$K = \frac{(0.45)^2 y}{0.025} = 1 \times 10^{-13}$

$y = 1.2 \times 10^{-14}$

$[Ag^+] = 1 \times 10^{-14}$

15.18) $\frac{4.0 \text{ g } CuSO_4 \;|\; 1 \text{ mole } Cu^{2+}}{l \;|\; 160 \text{ g } CuSO_4} = 0.025 \text{ M } Cu^{2+}$

$\frac{0.0400\,l \text{ conc} \;|\; 15 \text{ moles } NH_3}{l \;|\; l \text{ conc}} = 0.60 \text{ M } NH_3$

	$Cu(NH_3)_4^{2+}$ =	Cu^{2+}	+ $4\,NH_3$
I	0	0.025	0.60
R	+0.025	−0.025	−0.10
E	0.025	y	0.50

$K = \frac{(0.50)^4 y}{0.025} = 2 \times 10^{-13}$

$y = 8 \times 10^{-13} = [Cu^{2+}]$

15.19)

	$Co(NH_3)_6^{2+}$ =	Co^{2+}	+ $6\,NH_3$
I	0	0.050	0.300 + y
R	0.050	0.050	0.300
E	0.050	10^{-15}	y

$K = \frac{(10^{-15})(y^6)}{0.050} = 1 \times 10^{-35}$

$y = 2.8 \times 10^{-4}$

NH_3 required: $0.300 + 2.8 \times 10^{-4} = 0.300$ mole/l

$$\frac{0.250\,l}{} \left| \frac{0.300 \text{ mole } NH_3}{l} \right| \frac{1000 \text{ ml conc}}{15 \text{ moles } NH_3} = 5 \text{ ml conc } NH_3$$

2% volume increase is negligible

15.20)

	$Ni(NH_3)_6^{2+}$	= Ni^{2+}	+ 6 NH_3
I	0	0.24	y + 0.72
R	0.12	0.12	0.72
E	0.12	0.12	y

$$K = \frac{0.12\ y^6}{0.12} = 2 \times 10^{-9}$$

$$y = 3.55 \times 10^{-2} = 0.04$$

$$3.55 \times 10^{-2} = 0.04$$

NH_3 required: $0.72 + 0.04 = 0.76$ mole/l

$$\frac{0.050\,l}{} \left| \frac{0.76 \text{ mole } NH_3}{l} \right| \frac{1000 \text{ ml conc}}{15 \text{ moles } NH_3} = 2.5 \text{ ml}$$

Chapter 16

16.1) $K_{sp} = (8.8 \times 10^{-14})^2 = 7.7 \times 10^{-27}$

16.2) $$\frac{1.0 \times 10^{-3} \text{ g CuBr}}{0.100\,l} \left| \frac{\text{mole}}{144 \text{ g}} \right. = 6.9 \times 10^{-5} = [Cu^+] = [Br^-]$$

$K_{sp} = (6.9 \times 10^{-5})^2 = 4.8 \times 10^{-9}$

16.3) a) $CaCO_3$: $s^2 = 8.7 \times 10^{-9}$; $s = 9.3 \times 10^{-5}$

b) NiS: $s^2 = 3 \times 10^{-19}$; $s = 5.5 \times 10^{-10}$

c) $Cu(IO_3)_2$: $4s^3 = 7.4 \times 10^{-8}$; $s = 2.6 \times 10^{-3}$

16.4) $s = \sqrt{1.6 \times 10^{-8}} = 1.27 \times 10^{-4}$

$$\frac{0.400\,l}{} \left| \frac{1.27 \times 10^{-4} \text{ mole } PbSO_4}{l} \right| \frac{304 \text{ g}}{\text{mole}} = 1.5 \times 10^{-2} \text{ g } PbSO_4$$

16.5) $s = \sqrt{8.7 \times 10^{-9}} = 9.3 \times 10^{-5} = [Ca^{2+}]$

$$\frac{9.3 \times 10^{-5} \text{ mole } Ca^{2+}}{l} \left| \frac{10^3\,l}{1 \text{ million g } H_2O} \right| \frac{100 \text{ g } CaCO_3}{1 \text{ mole } Ca^{2+}} = 9.3 \text{ ppm}$$

16.6) $[Ag^+] = 0.75 \times 10^{-5}$: $[Br^-] = 2.4 \times 10^{-8}$

$IP = (0.75 \times 10^{-5})(2.4 \times 10^{-8}) = 1.8 \times 10^{-13} < 5.0 \times 10^{-13}$ No ppt.

16.7) $[Pb^{2+}] = \frac{150}{175}(8.2 \times 10^{-4}) = 7.0 \times 10^{-4}$

$[I^-] = \frac{25}{175}(4.2 \times 10^{-6}) = 0.60 \times 10^{-6}$

$IP = (7.0 \times 10^{-4})(0.60 \times 10^{-6})^2 = 2.52 \times 10^{-16} < 7.1 \times 10^{-9}$ No ppt.

16.8) $[Ca^{2+}] = \frac{65}{95}(0.28) = 0.19$

$[OH^-] = \frac{30}{95}(0.42) = 0.133$

$IP = (0.19)(1.33)^2 = 3.4 \times 10^{-3} > 5.5 \times 10^{-5}$ Ppt.

16.9) $s = [CO_3^{2-}] = \frac{8.1 \times 10^{-9}}{1.0 \times 10^{-2}} = 8.1 \times 10^{-7}$

16.10) $s = [Pb^{2+}] = \frac{1.6 \times 10^{-8}}{0.2 \times 10^{-1}} = 8.0 \times 10^{-7}$

$$\frac{0.400\,l \;\Big|\; 8.0 \times 10^{-7} \text{ mole } PbSO_4 \;\Big|\; 304 \text{ g}}{\;\Big|\; l \;\Big|\; \text{mole}} = 9.7 \times 10^{-5} \text{ g } PbSO_4$$

16.11) a) AgI because of smaller K_{sp}

b) $[Ag^+] = \frac{8.3 \times 10^{-17}}{1.0 \times 10^{-4}} = 8.3 \times 10^{-13}$

c) $[Ag^+] = \frac{1.7 \times 10^{-10}}{1.0 \times 10^{-4}} = 1.7 \times 10^{-6}$

d) $[I^-] = \frac{8.3 \times 10^{-17}}{1.7 \times 10^{-6}} = 4.9 \times 10^{-11}$

e) $\frac{4.9 \times 10^{-11} \;|\; 10^2}{1.0 \times 10^{-4} \;|\;} = 4.9 \times 10^{-5}\,\%$

16.12) a) $[Cl^-] = \frac{2.5 \times 10^{-3} \text{ mole}}{0.100\,l} = 2.5 \times 10^{-2}$

$[Ag^+] = \frac{1.7 \times 10^{-10}}{2.5 \times 10^{-2}} = 6.8 \times 10^{-9}$

$[Ag^+]$ per drop $= \frac{0.05}{100}(0.1) = 5 \times 10^{-5} > 6.8 \times 10^{-9}$

AgCl forms at first drop of 0.1 M $AgNO_3$

b) $[Ag^+]^2 = \frac{2.2 \times 10^{-12}}{2.6 \times 10^{-5}} = 8.46 \times 10^{-8}$; $[Ag^+] = 2.9 \times 10^{-4}$

c) $[Cl^-] = \frac{1.7 \times 10^{-10}}{2.9 \times 10^{-4}} = 5.9 \times 10^{-7}$

d) $\frac{5.9 \times 10^{-7} \;|\; 10^2}{2.5 \times 10^{-2} \;|\;} = 2.4 \times 10^{-3}\,\%$

16.13)

	AgSCN + 2 NH_3	=	$Ag(NH_3)_2^+$	+ SCN^-
I	0.50		0	0
E_a	0.50		y	y

$K = \frac{y^2}{0.50^2} = \frac{1.2 \times 10^{-12}}{4 \times 10^{-8}} = 3.0 \times 10^{-5}$

$y = 2.7 \times 10^{-3}$ mole/l

16.14) a) $s = \sqrt{6.6 \times 10^{-9}} = 8.1 \times 10^{-5}$

b)

	$NiCO_3$ + 6 NH_3	=	$Ni(NH_3)_6^{2+}$	+ CO_3^{2-}
I	1.0		0	0
E_a	1.0		y	y

$K = \frac{y^2}{1.0^6} = \frac{6.6 \times 10^{-9}}{2 \times 10^{-9}} = 3.3 \qquad y = 1.8$ moles/l

16.15)

	$Fe(OH)_3(s)$ +	$6\ CN^-$	$= Fe(CN)_6^{3-}$ +	$3\ OH^-$
I		$0.0036 + y$	0	—
R		0.0036	0.00060	—
E		y	0.00060	1.0×10^{-4}

$$K = \frac{(6.0 \times 10^{-4})(1.0 \times 10^{-4})^3}{y^6} = \frac{1.1 \times 10^{-36}}{1 \times 10^{-31}} = 1.1 \times 10^{-5}$$

$y = 1.95 \times 10^{-2}$; $y + 0.0036 = 2.31 \times 10^{-2}$

$$\frac{2.31 \times 10^{-2}\ \text{mole}\ CN^-}{l\ \text{soln}} \cdot \frac{1000\ \text{ml soln}}{0.25\ \text{mole}\ CN^-} = 93\ \text{ml}\ 0.25\ M\ KCN$$

In practical situations the nearly 10% dilution effect is compensated for by adding more than the calculated volume.

Chapter 17

17.1) $Ni^{2+} + 2\ e^- \rightarrow Ni$

$$\frac{35.0\ \text{min}}{} \cdot \frac{0.65\ \text{amp}}{} \cdot \frac{60\ \text{sec}}{1\ \text{min}} \cdot \frac{1\ \text{mole}\ e^-}{96{,}500\ \text{amp-sec}} \cdot \frac{1\ \text{mole Ni}}{2\ \text{moles}\ e^-} \cdot \frac{58.7\ \text{g}}{1\ \text{mole}} = 0.415\ \text{g Ni}$$

17.2) $Cu^{2+} + 2\ e^- \rightarrow Cu$

$$\frac{3000\ \text{amps}}{} \cdot \frac{8\ \text{hr}}{} \cdot \frac{3600\ \text{sec}}{1\ \text{hr}} \cdot \frac{1\ \text{mole}\ e^-}{96{,}500\ \text{amp-sec}} \cdot \frac{1\ \text{mole Cu}}{2\ \text{moles}\ e^-} \cdot \frac{63.5\ \text{g}}{1\ \text{mole}} \cdot \frac{1\ \text{lb}}{454\ \text{g}} = 62.6\ \text{lb Cu}$$

17.3) $Al^{3+} + 3\ e^- \rightarrow Al$

$$\frac{50{,}000\ \text{amps}}{} \cdot \frac{7\ \text{hrs}}{} \cdot \frac{3600\ \text{sec}}{1\ \text{hr}} \cdot \frac{1\ \text{mole}\ e^-}{96{,}500\ \text{amp-sec}} \cdot \frac{1\ \text{mole Al}}{3\ \text{moles}\ e^-} \cdot \frac{27.0\ \text{g}}{1\ \text{mole}} \cdot \frac{1\ \text{lb}}{454\ \text{g}} = 259\ \text{lb Al}$$

17.4) $Cr^{6+} + 6\ e^- \rightarrow Cr$

$$\frac{18{,}000\ \text{amps}}{} \cdot \frac{8\ \text{hr}}{} \cdot \frac{3600\ \text{sec}}{1\ \text{hr}} \cdot \frac{1\ \text{total mole}\ e^-}{96{,}500\ \text{amp-sec}} \cdot \frac{15\ \text{effective moles}\ e^-}{100\ \text{total moles}\ e^-}$$

$$\frac{1\ \text{mole Cr}}{6\ \text{moles}\ e^-} \cdot \frac{52.0\ \text{g}}{1\ \text{mole}} \cdot \frac{1\ \text{kg}}{1000\ \text{g}} = 6.98\ \text{kg Cr}$$

17.5)

$$\frac{350\ \text{amps}}{} \cdot \frac{2\ \text{min}}{} \cdot \frac{60\ \text{sec}}{1\ \text{min}} \cdot \frac{1\ \text{mole}\ e^-}{96{,}500\ \text{amp-sec}} \cdot \frac{1\ \text{mole}\ O_2}{4\ \text{moles}\ e^-}$$

$$\frac{22.4\ l}{1\ \text{mole}} \cdot \frac{760\ \text{torr}}{733\ \text{torr}} \cdot \frac{313°K}{273°K} = 2.90\ l\ O_2$$

17.6) $Na^+ + e^- \rightarrow Na$

$$\frac{20\text{ lb Na}}{1\text{ hr}}\left|\frac{454\text{ g}}{1\text{ lb}}\right|\frac{1\text{ mole}}{23.0\text{ g}}\left|\frac{1\text{ mole e}^-}{1\text{ mole Na}}\right|\frac{96{,}500\text{ amp-sec}}{1\text{ mole e}^-}\left|\frac{1\text{ hr}}{3600\text{ sec}}\right. = 10{,}600\text{ amps}$$

17.7)

$$\frac{1500\ l\ H_2}{7.50\text{ hrs}}\left|\frac{20.0\text{ atm}}{1\text{ atm}}\right|\frac{273°K}{298°K}\left|\frac{1\text{ mole }H_2}{22.4\ l\ H_2}\right|\frac{2\text{ moles e}^-}{1\text{ mole }H_2}\left|\frac{96{,}500\text{ amp-sec}}{1\text{ mole e}^-}\right|\frac{1\text{ hr}}{3600\text{ sec}} = 8770\text{ amps}$$

17.8) $Ni^{2+} + 2\,e^- \rightarrow Ni$

$$\frac{1000\text{ g Ni}}{50.0\text{ amps}}\left|\frac{1\text{ mole}}{58.7\text{ g}}\right|\frac{2\text{ moles e}^-}{1\text{ mole Ni}}\left|\frac{96{,}500\text{ amp-sec}}{1\text{ mole e}^-}\right|\frac{1\text{ hr}}{3600\text{ sec}} = 18.3\text{ hrs}$$

17.9) $2\,H^+ + 2\,e^- \rightarrow H_2$

$$\frac{0.085\ l\ H_2}{0.50\text{ amp}}\left|\frac{720\text{ torr}}{760\text{ torr}}\right|\frac{273°K}{298°K}\left|\frac{1\text{ mole }H_2}{22.4\ l\ H_2}\right|\frac{2\text{ moles e}^-}{1\text{ mole }H_2}\left|\frac{96{,}500\text{ amp-sec}}{1\text{ mole e}^-}\right|\frac{1\text{ min}}{60\text{ sec}} = 21.2\text{ min}$$

17.10)

Reaction	E°
$Sn^{2+} + 2\,e^- \rightarrow Sn$	E° = −0.14
$Fe \rightarrow Fe^{2+} + 2\,e^-$	E° = +0.44
$Sn^{2+} + Fe \rightarrow Sn + Fe^{2+}$	E° = +0.30

Forward direction favored.

17.11)

Reaction	E°
$Fe^{2+} + 2\,e^- \rightarrow Fe$	E° = −0.44
$Sn^{2+} \rightarrow Sn^{4+} + 2\,e^-$	E° = −0.15
$Fe^{2+} + Sn^{2+} \rightarrow Fe + Sn^{4+}$	E° = −0.59

Reverse direction favored.

17.12)

Reaction	E°
$3[Na^+ + e^- \rightarrow Na]$	E° = −2.71
$Al \rightarrow Al^{3+} + 3\,e^-$	E° = +1.66
$3\,Na^+ + Al \rightarrow 3\,Na + Al^{3+}$	E° = −1.05

Reverse direction favored.

17.13)

Reaction	E°
$4[NO_3^- + 4\,H^+ + 3\,e^- \rightarrow NO + 2\,H_2O]$	E° = +0.96
$3[Pt + 6\,Cl^- \rightarrow PtCl_6^{2-} + 4\,e^-]$	E° = −0.60
$4\,NO_3^- + 16\,H^+ + Pt + 6\,Cl^- \rightarrow 4\,NO + 8\,H_2O + PtCl_6^{2-}$	E° = +0.36

Forward direction favored.

17.14)

Reaction	E°
$Cr_2O_7^{2-} + 14\,H^+ + 6\,e^- \rightarrow 2\,Cr^{3+} + 7\,H_2O$	E° = +1.33
$3[Mn^{2+} + 2\,H_2O \rightarrow MnO_2 + 4\,H^+ + 2\,e^-]$	E° = −1.23
$Cr_2O_7^{2-} + 2\,H^+ + 3\,Mn^{2+} + 6\,e^- \rightarrow 2\,Cr^{3+} + H_2O + 3\,MnO_2$	E° = +0.10

Forward direction favored.

17.15) $\Delta G° = (-2)(96.5)(0.30) = 57.9\text{ kJ} = 13.8\text{ kcal}$

17.16) $\Delta G° = (-2)(96.5)(-0.59) = +114\text{ kJ} = +27.2\text{ kcal}$

17.17) $\Delta G° = (-3)(96.5)(-1.05) = +304 \text{ kJ} = +72.7 \text{ kcal}$

17.18) $\Delta G° = (-4)(96.5)(0.36) = -139 \text{ kJ} = -33.2 \text{ kcal}$

17.19) $\Delta G° = (-6)(96.5)(0.10) = -57.9 \text{ kJ} = -13.8 \text{ kcal}$

17.20) $\log K = \frac{2(0.30)}{0.0592} = 10.1; \quad K = 10^{10.1} = 1 \times 10^{10}$

17.21) $\log K = \frac{2(-0.59)}{0.0592} = -19.9; \quad K = 10^{-19.9} = 1 \times 10^{-20}$

17.22) $\log K = \frac{3(-1.05)}{0.0592} = -53.2; \quad K = 10^{-53.2} = 6 \times 10^{-54}$

17.23) $\log K = \frac{4(0.36)}{0.0592} = 24.3; \quad K = 10^{24.3} = 2 \times 10^{24}$

17.24) $\log K = \frac{6(0.10)}{0.0592} = 10.1; \quad K = 10^{10.1} = 1 \times 10^{10}$

17.25)

$$\begin{array}{lll} AgIO_3 + e^- \rightarrow Ag + IO_3^- & & E° = +0.355 \\ Ag \rightarrow Ag^+ + e^- & & E° = -0.80 \\ \hline AgIO_3 = Ag^+ + IO_3^- & & E° = -0.445 \end{array}$$

$\log K_{sp} = \frac{1(-0.445)}{0.0592} = -7.52; \quad K_{sp} = 10^{-7.52} = 3.0 \times 10^{-8}$

17.26)

$$\begin{array}{lll} Ag(S_2O_3)_2^{3-} + e^- \rightarrow Ag + 2\,S_2O_3^{2-} & & E° = +0.10 \\ Ag \rightarrow Ag^+ + e^- & & E° = -0.80 \\ \hline Ag(S_2O_3)_2^{3-} = Ag^+ + 2\,S_2O_3^{2-} & & E° = -0.70 \end{array}$$

$\log K_d = \frac{1(-0.70)}{0.0592} = -11.8; \quad K_d = 10^{-11.8} = 2 \times 10^{-12}$

17.27) $E = -0.28 + 0.44 - \frac{0.0592}{2} \log \frac{0.010}{0.50} = +0.21 \text{ volt}$

Forward reaction favored.

17.28) $E = -0.14 + 1.66 - \frac{0.0592}{6} \log \frac{0.020^2}{0.10^3} = +1.52 \text{ volts}$

Forward reaction favored.

17.29) $Hg_2Cl_2 + H_2 \rightarrow 2\,Hg + 2\,H^+ + 2\,Cl^-$ $\quad E = 0.280$ v

$E = 0.280 - \frac{0.0592}{2} \log 4.6^2 = 0.241 \text{ volt}$

17.30) $MnO_4^- \rightarrow MnO_2$ $\quad \Delta$ Ox. No. = 4 − 7 = −3; 3 eq/mole

$\frac{158 \text{ g } KMnO_4}{\text{mole}} \Big| \frac{\text{mole}}{3 \text{ eq}} = 52.7 \text{ g } KMnO_4/\text{eq}$

17.31) $Au \rightarrow AuCl_4^-$ $\quad \Delta$ Ox. No. = 3 − 0 = 3; 3 eq/mole

$\frac{197 \text{ g Au}}{\text{mole}} \Big| \frac{\text{mole}}{3 \text{ eq}} = 65.7 \text{ g/eq}$

17.32) $C_2O_4^{2-} \rightarrow CO_2$ $\quad \Delta$ Ox. No. = 4 − 3 = 1 per C atom: 2 eq $C_2O_4^{2-}$/mole

$\frac{126 \text{ g } H_2C_2O_4 \cdot 2\,H_2O}{\text{mole}} \Big| \frac{\text{mole}}{2 \text{ eq}} = 63.0 \text{ g/eq}$

17.33) $\dfrac{0.500\ l}{} \Big| \dfrac{0.10\ \text{eq KMnO}_4}{l} \Big| \dfrac{158\ \text{g KMnO}_4}{5\ \text{eq KMnO}_4} = 1.6\ \text{g KMnO}_4$

17.34) $\dfrac{0.250\ l}{} \Big| \dfrac{0.10\ \text{eq Na}_2\text{C}_2\text{O}_4}{l} \Big| \dfrac{134\ \text{g}}{2\ \text{eq}} = 1.7\ \text{g Na}_2\text{C}_2\text{O}_4$

17.35) $\dfrac{1.734\ \text{g Na}_2\text{C}_2\text{O}_4}{0.2500\ l} \Big| \dfrac{2\ \text{eq}}{134.0\ \text{g}} = 0.1035\ \text{N Na}_2\text{C}_2\text{O}_4$

17.36) $\dfrac{50.00 \times 0.1035}{48.20} = 0.1074\ \text{N KMnO}_4$

17.37) $\dfrac{0.283\ \text{g red agent}}{0.0378\ l} \Big| \dfrac{l}{0.1074\ \text{eq}} = 69.7\ \text{g/eq}$

INDEX

(t) indicates table